Innovative Economic Symposium (IES 2017)

SHS Web of Conferences Volume 39 (2017)

Ceske Budejovice, Czech Republic
19 October 2017

Editors:

Jan Vachal
Marek Vochozka
Jakub Horak

ISBN: 978-1-5108-5349-2

Printed from e-media with permission by:

Curran Associates, Inc.
57 Morehouse Lane
Red Hook, NY 12571

Some format issues inherent in the e-media version may also appear in this print version.

Printed by Curran Associates, Inc. (2018)

For additional information, please contact EDP Sciences – Web of Conferences at the address below.

EDP Sciences – Web of Conferences
17, Avenue du Hoggar
Parc d'Activité de Courtabœuf
BP 112
F-91944 Les Ulis Cedex A
France

Phone: +33 (0) 1 69 18 75 75
Fax: +33 (0) 1 69 28 84 91

contact-edps@webofconferences.org

Additional copies of this publication are available from:

Curran Associates, Inc.
57 Morehouse Lane
Red Hook, NY 12571 USA
Phone: 845-758-0400
Fax: 845-758-2633
Email: curran@proceedings.com
Web: www.proceedings.com

TABLE OF CONTENTS

Innovative Economic Symposium 2017

Strategic Partnership in International Trade

http://ies.vstecb.cz/

Editors:

- prof. Ing. Jan Váchal, CSc.
- doc. Ing. Marek Vochozka, MBA, Ph.D.
- Bc. Jakub Horák

Publication was created with the support of South Bohemian Region.

About the conference IES 2017

Date of the conference: 2017-10-19

The conference will be held under the auspices of President of the Czech Republic Ing. Miloš Zeman, CSc.

Motto

"The business enterprise has two–and only two–basic functions: marketing and innovation. Marketing and innovation produce results; all the rest are costs." Peter F. Drucker

Target

The conference brings together scientists in economics, international relations, finance and marketing both from China and Europe. The conference aims to exchange knowledge and discuss issues related to market mechanisms, international business and strategic decisions.

Primary Issues

Competitive strategies in international trade. Entrepreneurship, international trade and economic growth. Quantitative and qualitative methods in management decision making with a focus on international trade. Economic, social and environmental dimensions of strategic management in international business. Intellectual capital, knowledge and innovation in the development of international trade. Decision-making methods and optimization econometric modelling in corporate strategy. Consumer behavior, international marketing, marketing modelling.

Thematic Areas

Primary Issues

- Competitive strategies in international trade.

- Entrepreneurship, international trade and economic growth.
- Quantitative and qualitative methods in management decision making with a focus on international trade.
- Economic, social and environmental dimensions of strategic management in international business.
- Intellectual capital, knowledge and innovation in the development of international trade.
- Decision-making methods and optimization econometric modelling in corporate strategy.
- Consumer behavior, international marketing, marketing modelling.

Alternative Issues

- Risk management and portfolio optimization.
- Financial sustainability of SMEs.
- Sustainability in logistics / environmental management in the supply chain.
- Monetary policy / currency area theory.
- Foreign direct investment.
- Renewable energy / environmental legislation.
- Social – medial marketing.

About organizers

The primary focus of VŠTE activities is the implementation of full-time and part-time studies for professional bachelor's and master's degree programmes. The professional study programmes are supported by cooperation with regional entrepreneurs. VŠTE´s other priorities include applied research, the implementation of innovative technologies and knowledge transfer into areas that are current and crucial with regards to the regions needs. More information about VŠTE at www.vstecb.cz.

North China University of Technology (NCUT), located on the western side of Beijing, is a multi-disciplinary university that combines the natural sciences and engineering with liberal arts, economics, management and law. NCUT targets the development of qualified, high-level, application-oriented professionals with solid basic theoretical knowledge and strong practical skills and creative abilities. A quality education is the university's lifeline. Teaching is a primary focus at NCUT and the center of its activities, while the university also emphasizes the importance of research and the contribution research can make both to the educational environment and to address local and national needs. The under graduate program offers a full range of undergraduate degrees; there are also numerous graduate degree programs, as well as continuing education programs. The academic education is complemented with moral, intellectual, physical and aesthetic training in order to build well-rounded students prepared to meet the challenges of modern society. More information about NCUT at www.ncut.ciss.org.cn.

Conference report

Conference Innovative Economic Symposium 2017 took place on 19 October 2017 in the Spa Hotel Vita in České Budějovice. The thematic focus of this year was Strategic Partnership in International Trade. The whole conference started with opening speeches of doc. Ing. Marek Vochozka, MBA, Ph.D., Rector of the Institute of Technology and Business in České Budějovice (VŠTE) and prof. Zhao Jixin, Dean of the Faculty of Economics and Management of the North China University of Technology in Beijing (NCUT). The participants of the conference were academics, officials and entrepreneurs from China, Russia, and the Czech Republic. The topic One Belt, One Road, has been repeatedly mentioned as one of the promising opportunities for similar strategic partnerships at the Czech-Chinese level.

Expert guarantor

prof. Jan Váchal,
Vice-Rector of the Institute of Technology and Business (CZE)

doc. Marek Vochozka
Rector of the Institute of Technology and Business (CZE)

The Organizing Committee

The Scientific Committee

CONFERENCE ORGANISERS

Institute of Technology and Business in České Budějovice

North China University of Technology

CONFERENCE SPONSORS

Conference Innovative Economic Symposium 2017 is co-financed with the support of the South Bohemian Region.

Conference program

Innovative Economic Symposium 2017

Strategic Partnerships in International Trade

Spa Hotel Vita, October 19, 2017

10.00 – 10.10
Opening ceremony
Ing. Pavel Rousek, Ph.D.

10.10 – 10.30
Introduction

doc. Ing. Marek Vochozka, MBA, Ph.D.
Rector
The Institute of Technology and Business in České Budějovice

prof. Zhao Jixin
Dean
North China University of Technology

10.30 – 12.00
Plenary session
Host: Ing. Pavel Rousek, Ph.D.

Associate Prof. Natalia Pletneva, Ph.D.
Peter the Great St. Petersburg Polytechnic University
Issues of Ecostrategies Implementation for Tourism Enterprises in Russia

Prof. Maksim Kiselev, DSc.
Saratov Socio-Economic Institute
Two Types of Management Mistakes and How to Avoid Them

Associate Prof., Olga Kalchenko, Ph.D.
Peter the Great St. Petersburg Polytechnic University
Fuzzy sets and real options approaches for innovation-based investment projects effectiveness evaluation

Prof. Alla Yakunina, DSc.
Saratov Socio-Economic Institute
Liberalization of Russian gas exports: benefits and challenges

Mr. Li Chenguang, Ph.D.
North China University of Technology
Market opportunity and Policy support for chinese old aging industry: an application of text mining

13.00 – 14.30

Plenary session

Host: Ing. Pavel Rousek, Ph.D.

Mgr. Petr Šuleř
ČEZ Group
Impact of Crisis Communication with the Media

Tu Shu, Zhonglan Wang
North China University of Technology
Chinese Language Teaching in the Czech Republic

Ing. Marek Vokoun, Ph.D.
The Institute of Technology and Business in České Budějovice
Innovation activities of firms: the case of 8 cis waves in the Czech industry between 2001 and 2014

15.00 – 17.00

Plenary session

Host: Ing. Marek Vokoun, Ph.D.

Prof. Dr. Yury Korobov
Saratov Socio-Economic Institute
Global Banking: Transformation, Innovation & Competition

Ing. Pavel Rousek, Ph.D.
The Institute of Technology and Business in České Budějovice
The Impact of the Economic Crisis on Municipal Expenditures

Ing. Simona Hašková, Ph.D.
The Institute of Technology and Business in České Budějovice
Strategic Partnership of Independent States as a Game

Ing. Lenka Dvořáková
The Institute of Technology and Business in České Budějovice
CZ GDP Prediction via neural networks and Box-Jenkins Method

Ms. Liu Yaqing Ph.D.
North China University of Technology
Research on the Relationship between Sino-EU Trade and Economic Growth

17.00 – 18.00

Closing ceremony

Ing. Pavel Rousek, Ph.D.

Organizer VŠTE

Co-organizer

Sponsor Jihočeský kraj

Innovative Economic Symposium 2017
Strategic Partnerships in International Trade

北方工业大學

Under the auspices of Ing. Miloš Zeman, President of the Czech Republic

Motto
"The business enterprise has two–and only two–basic functions: marketing and innovation. Marketing and innovation produce results; all the rest are costs."
Peter F. Drucker

Target
The conference brings together scientists in economics, international relations, finance and marketing both from China and Europe. The conference aims to exchange knowledge and discuss issues related to market mechanisms, international business and strategic decisions.

Thematic Areas
Primary Issues
- Competitive strategies in international trade
- Entrepreneurship, international trade and economic growth
- Quantitative and qualitative methods in management decision making with a focus on international trade
- Economic, social and environmental dimensions of strategic management in international business
- Intellectual capital, knowledge and innovation in the development of international trade
- Decision-making methods and optimization econometric modeling in corporate strategy
- Consumer behavior, international marketing, marketing modeling

Alternative Issues
- Risk management and portfolio optimization
- Financial sustainability of SMEs
- Sustainability in logistics / environmental management in the supply chain
- Monetary policy / currency area theory
- Foreign direct investment
- Renewable energy / environmental legislation
- Social – medial marketing

Conference program

09:00—10:00	Participant registration
10:00—10:30	Opening ceremony
10:30—12:00	Plenary session
12:00—13:00	Lunch
13:00—14:30	Plenary session
14:30—15:00	Coffee break
15:00—17:00	Plenary session
17:00	Closing ceremony

http://ies.vstecb.cz/

19 October 2017
Spa Hotel Vita
České Budějovice, Czechia

Sponsors and Partners
Jihočeský kraj

Expert Guarantors
prof. Jan Váchal,
Vice-Rector of the Institute of Technology and Business (CZE)
doc. Marek Vochozka,
Rector of the Institute of Technology and Business (CZE)

Scientific Committee
prof. Svetlana I. Ashmarina,
Samara State University of Economics (RUS)
prof. Stefan Forlicz,
Wrocław School of Banking (POL)
dr. Simona Hašková,
Institute of Technology and Business (CZE)
prof. Leonid D. Gitelman,
Ural Federal University (RUS)
prof. Fariborz Haghighat,
Concordia University (CAN)
prof. Alois Heißenhuber,
Technical University of Munich (GER)
prof. Branislav Mičieta,
University of Žilina (SVK)
doc. Maxim A. Ponomarev,
Plechanov University (RUS)
prof. Amarjit Singh,
University of Hawai'i at Manoa (USA)
prof. Bingtao Song,
Henan University and the Chinese Economic Society (CHN)
prof. Oleksandr M. Trofymchuk,
Ukrainian National Academy of Sciences (UKR)
prof. Elena N. Veduta,
Lomonosov University in Moscow (RUS)
prof. Liu Yongxiang,
North China University of Technology (CHN)
prof. Shen Zhili,
North China University of Technology (CHN)

Organizing Committee: **doc. Marek Vochozka,** Rector of the Institute of Technology and Business in České Budějovice; **prof. Liu Yongxiang,** Dean of School of Economics and Management, NCUT; Pavel Rousek, Ph.D.; Simona Hašková, Ph.D.; Veronika Machová, MSc.; Jaromír Vrbka, MSc.; Zuzana Rowland, MSc.; Jakub Horák, Bachelor

edp sciences

SHS Web of Conferences

Statement of Peer review

In submitting conference proceedings to SHS Web of Conferences, I certify to the Publisher that I adhere to the **Policy on Publishing Integrity** of the journal in order to safeguard good scientific practice in publishing.

1. All articles have been subjected to peer review administered by the proceedings editors.
2. Reviews have been conducted by expert referees, who have been requested to provide unbiased and constructive comments aimed, whenever possible, at improving the work.
3. Proceedings editors have taken all reasonable steps to ensure the quality of the materials they publish and their decision to accept or reject a paper for publication has been based only on the merits of the work and the relevance to the journal.

Title, date and place of the conference

Innovative Economic Symposium 2017 (IES2017), 19th October 2017
Spa Hotel Vita, Skuherského 4, 37001 České Budějovice, Czech Republic

Proceedings editor(s):

- Prof. Ing. Jan Váchal, CSc.
- Doc. Ing. Marek Vochozka, MBA, Ph.D.
- Bc. Jakub Horák

Date and editor's signature

2.11 2017

Global educational trends in intellectual capital development

Svetlana Ashmarina[1], *Elena* Kandrashina[1*], *Anna* Zotova[1*]

[1]Samara State University of Economics, Russia, 443090, Samara, Sovetskoi Armii Street, 141

Abstract. Modern "post-industrial" era is distinguished by such rapid global changes and almost instantaneous copying of products and services that make it meaningless reaching the goal of long-term retention of a certain market position. The strategy ceased to be only a long-term plan and was formed as the concept of a management process which is characterized by a constant adjustment in the on-line regime. Today, the educational community is also subjected to these changes. Universities today are doing their best to attract international students. They are mobile young people (there are more than 5 million people in the world), possessing great abilities. The war for talent leads to the rapid differentiation of educational institutions. Technological innovations came with a delay of about 10 years in the sphere of education. Today, the annual growth of the online education market is 27%, the traditional - 5%. In 2016, more than 50 million people around the world studied using the largest online platforms. So the main task of this research work is to define the influence of some educational global trends on intellectual capital development in the frame of high education system and assess the readiness of Russian universities to such influence.
Key words: global educational trends, universities, e-learning

1 Introduction

The modern "post-industrial" era is distinguished by such rapid global changes and almost instantaneous copying of products and services that make it meaningless to make the goal of long-term retention of a certain market position to be the absolute one. Constant perfection of design and service accompanying the provision of products and services, becomes a condition for successful business development. The strategy ceased to be only a long-term plan and was formed as a concept of a management process characterized by a constant adjustment in the on-line regime. Today, the educational community is also subjected to these changes. Universities, educational consortia and entire countries are doing their best to attract international students. These are mobile young people (there are more than 5 million people in the world), possessing great abilities. The war for talent leads to rapid differentiation of educational institutions. The appearance in the early 2000s of international university rankings (ARWU, QS, THE) gave us the opportunity to compare universities of different countries, which was the beginning of global competition era. The

*Corresponding authors: elka@mail.ru, azotova@mail.ru

ratings have launched new mechanisms for the redistribution of resources in education: leaders strengthen their reputation, receive funding. Academic mobility is also in the best hands: according to research, masters, graduate students and professors today choose universities on the basis of international ratings. The wall has grown between high schools of the top 100 and all the others and this inequality will grow rapidly with each passing year. As a result, a country that does not have universities in a group of leaders, in the long term, 5-10 years, risks losing competition for intellectual capital and becoming an educational province [1].

2 Methods and Results

2.1. The analysis of trends stimulating education system changes and intellectual capital development

The basic changes are based on three key preconditions:

- digital revolution
- demographic trends
- fastening speed of social development

Table 1. Educational trends and their influence on high education system

Megatrends (global trends, influencing all human spheres) [2, 3]	Trends, influencing education system	First level reaction (the result, which is the output of trend influence on education system)	Second level reaction (the reaction on the first level result)
1. Information technologies development	Education virtualization	- Emergence and development of new forms and approaches to education based on modern technologies usage. - Competition growth in the sphere of virtual (distance) education.	- New approaches to distance education implementation. - Education quality decrease.
	Personification, individualization	- E-learning education technology popularization, teaching on-line.	- Growth in the gap in the quality of students education.
	Gamefication	- Growth of education projects based on gamefication elements.	- Emergence of new education programs based on gamefication.
2. Demographic trend	Demographic crisis in Russia	- The decrease in the number of students under the influence of such factors: demography, the rise of examination entrance mark, the education fee, the decrease in the places granted by federal or	- The growth in the struggle for school leavers, competition rise. - Development of education programs aimed at adults learning. - Development of academic mobility.

		regional budget.	- Exchange of principle "education for the whole life " into the principle "education through the whole life".
	Population aging	- The necessity to teach adult population to adopt some new skills and competences.	- Development of education technologies and education programs which target element is elderly people.
	Growth in the length of population's economically active life period	- Growth in number of adult students because of necessity to study some new skill and continue their labor activity.	- New education programs, qualification rise courses or some education technologies adapted for elderly people.
	Diversification	- New competences development among the students - Emergence of new courses	

In the sphere of education, technological innovations came with delay of about 10 years. The first online projects here appeared in the 1990s and were associated with the transfer of electronic materials to educational materials. True, they did not have a serious impact on the industry. In 2011, second-generation projects appeared on the market – mass open online courses (Massive Open Online Courses, or MOOC). It is believed that it was the point from which the digital revolution began in education.

Today, the annual growth of the online education market is 27%, the traditional – 5%. In 2016, more than 50 million people around the world studied using the largest online platforms (EdX, Coursera, Udacity, etc.), 800 thousand of them are Russians. In addition to private projects, a number of states launched their own national platforms - the United Kingdom, Australia, Brazil, and Russia joined to the countries in 2015. The main innovation of such platforms is the design of the courses: they consist of small video fragments for 5-10 minutes, due to which the listeners kept concentration. 2015 was a new milestone in the development of digital education. LinkedIn for $ 1.5 billion bought the project Lynda.com, which revised the training process. The basis of the program was not mini-lectures, but tutorials, short video instructions for representatives of different professions. For several years, the world's leading experts have created more than 25 thousand tutorials, and for access to them it is enough to buy a monthly, semi-annual or annual subscription. There was a new type of educational project - a kind of "intellectual fitness room", where at any time you can buy a ticket.

The development of information technology entailed a boom of educational startups. Since the beginning of the 2000s, a new class of projects has appeared in the field of education. For the first time significant financial resources were attracted not by universities with their centuries of history and brand, but young teams of entrepreneurs. Ten years ago, the global volume of venture investments in educational startups barely reached

$ 100 million, and in 2016 it exceeded $ 3 billion, which is comparable to the annual costs of education of individual countries.

Profiles of educational startups are diverse. For example, the most successful start-up, attracted more than $ 300 million, - TutorGroup - teaches English on the Internet. And the project Achieve3000, which collected a quarter of a billion dollars, develops the ability to read and understand the text. One of the most vivid examples of the new era is the Minerva program, which claims to train world leaders and innovators of the future. Its founders managed to attract more than $ 25 million at the idea stage.

All new educational projects unite one thing - they are built on modern technologies, use a synthesis of advanced developments in the field of computer intelligence, digital technologies and behavioral psychology. The best educational startups already compete with traditional universities for their investments and talents.

The demographic megatrend appears as a trend towards a change in the population on a global scale [4].

In different countries, this trend has a different focus. In countries such as China, India, Pakistan, etc. the population is growing. The opposite situation is observed in Russia.

The main reactions of Russian education system to this trend are:

- Decrease in the number of students under the influence of the demographic factor;
- Population aging (demographic aging) is the result of long-term demographic changes, changes in the nature of population reproduction, fertility, mortality their ratio, and also partly migration. The main reaction of the education system to this trend is the emergence of the need to educate people of the older generation with new skills and competences, which ultimately leads to the development of teaching technologies and educational programs whose target segment is older generation. Another social trend that manifests itself in lengthening the life expectancy of people leading an active economic life, engaged in production, participating in economic processes can be referred to as "increasing the duration of active economic life of the population." The main reaction of Russian education system to this trend is an increase in the proportion of adult students and students in the education system, due to the need for skills and competencies to continue working. This trend is also triggered by another factor associated with the new industrial revolution and the emergence of new professions.

New technologies are causing "industrial revolutions" that change the work of companies. Business dictates new requirements to people's competencies and the speed at which they are received.

There is a number of “retired professions” leaving the market. The most striking example of recent years is the massive bankruptcies of travel agencies, losing competition to services like Booking.com or Airbnb. At the same time, thanks to new technologies, the professions of the future appear, in which intellectual skills and qualities, which help to make non-standard decisions, are in demand.

Today it is already impossible to build a successful professional trajectory for years to come, but the traditional education system is based on the principle of long-term planning. The list of the most popular competences will be updated more often. If earlier a new profession arose once in 20-30 years, today it is every 3-5 years. The ability of the education system to respond quickly to new requests, to train specialists not for 5 years, but for 5 months, becomes a condition for the country's competitiveness. To this challenge, most educational institutions in the world have only to find an answer [5].

In addition, the demographic trend associated with the aging of the population in Russia also contributes to the development of online platforms. To actively use such educational products, young people often lack internal motivation and self-organization. Students who receive basic education at the first level often need that external formal shell, which is

DOI: 10.1051/shsconf/20173901001

created by the traditional approach to learning: attending classrooms, the presence of formal, and not virtual control points.

For an adult, all these external attributes of learning are no longer necessary, his main driver is the desire to get the necessary knowledge and this inner motivation makes him a more stable consumer of online courses.

All these trends, as well as the growing demand for the practicality of scientific research and the rapid commercialization of results by erasing the "science-practice" boundary, forced the RF Government to begin implementing the State Program of the Russian Federation "Development of Education" for 2013-2020, which will lead to an increase in the role of corporate training, to a closer connection of vocational education with the subjects of demand in the labor market. By 2020, all students will study according to individual curricula, which include a significant proportion of independent work using information technology.

2.2. The analysis of distance learning development in high education system of Samara region.

The most modern trend in the development of modern higher education is the promotion of distance learning and the gradual replacement of distance education by distance education. To effectively promote the distance education system, the development of information and communication technologies is vital. The authors of this work analyzed distance learning technologies in the leading universities of the Samara region (Table 1.) The results of the authors' research showed that a number of Samara higher educational institutions have already demonstrated high readiness for the projected changes and are actively implementing information and communication technologies. The largest number of areas covered by distance education is shown by the Samara State Technical University. Individual universities are still lagging far behind in promoting distance education with the help of modern technologies.

Table 2. Types of distance education technologies used in the HEIs of Samara region

The title of the university	Separate courses may be studied by distance education technologies	The ability to get distance degree at some programs	The ability to get the second degree by distance technologies
Volga region state university of telecommunication and information technologies	+	–	–
Samara region academy of Nayanova	+	+	– –
Samara University	+	– –	– –
Samara State Agricultural Academy	+	–	+
Samara State technical University	+	+	+
Samara State Transport University	+	+	– –
Samara State Medical University	+	–	–
Samara State University of Economics	+	– –	– –
Samara branch «Moscow City Pedagogical University»	+	+	–

Samara State Economic University launched the site "Virtual Campus", but so far only for students of the Master degree faculty. This site was organized on the basis of the MOODLE platform, which is a content management system (CMS), specifically designed to create qualitative online courses by teachers.

As a result of the research, the greatest activity of non-state educational institutions and branches in the promotion of their distance programs was revealed. These educational institutions quite reasonably seek to occupy that segment of the market of educational services, which is not yet being interested too much by the key universities in the region. So in the Samara branch of the Moscow City Pedagogical University, the specialized structural subdivision - the center for distance learning was allocated. Also, a program was developed to implement distance learning based on the distance education system "Prometheus".

3 Conclusion

An analysis of the results of studies on the use of information and communication technologies in higher education suggests that there are no general concepts that would allow us to capture and present a multitude of facts accumulated in the practice of education and upbringing in a single system of concepts. The analysis of practice also allows us to conclude that there is an underestimation of the possibilities of ICT in the sphere of domestic higher education. This is primarily due to the complexity and low readiness of higher education institutions to introduce and use information and communication technologies in their activities.

 DOI: 10.1051/shsconf/20173901001

The research is done in the frame of the state task of the Ministry of Education and Science of the Russian Federation №26.9402017/PC "Change management in high education system on the basis of sustainable development and interest agreement".

References

1. *The report of world economic forum Redesigning Business Value & A Roadmap to Sustainable Consupmtion,* [online], Available at: http://www3.weforum.org/docs/WEF_RedesigningBusinessValue_SustainableConsumption_Report_2010.pdf (2010)
2. J. Neisbitt, P. Aburdene, *Megatrends 2000: Ten New Directions for the 1990s*, 1st ed.,William & Morrow Company, Inc., 384 p. (1990)
3. T.A. Shakleina, A. A. Baikova, *Megatrends: Main evolution trends of global character in XXI century,* Moscow, Aspekt Press, 448 p. (2014)
4. R. R Khasiev, Forecasting of demographic situation in the Russian Federation. *Science Symbol*, 12-1, (2015)
5. G. R. Khasaev, S. I. Ahmarina, Main trends and perspectives of HEIs competitiveness improvement. *Vestnik SSEU*, **8**(118), 62-65 (2014)

Issues of Ecostrategies Implementation for Tourism Enterprises in Russia

Svetlana Bozhuk[1,*] and *Natalya* Pletneva[1]

[1]Peter the Great St. Petersburg Polytechnic University, Politechnicheskaya st., 29, St. Petersburg, 194064, Russia

Abstract. Ecological tourism in Russia has a high potential, however its extension creates an additional environmental burden. The decision of this contradiction is achieved by selection the marketing strategy of sustainable development which is adequate to the situation and cover all stakeholders involved in the producing, promotion and realization of tourist product.
Key words: tourism, analysis, ecological, sustainable development, environmental marketing

1 Introduction

In Russia, 2017 is declared as the Year of Ecology. This fact caught interest of many commercial entities to their own environmental practices. The business focused mainly on saving natural resources (non-waste technologies, energy efficiency), environment protection from harmful emissions (waste treatment, sewage disposal, air pollution control), wildlife preservation, and biological diversity. The principles of sustainable development can be realized in many ways in different business areas as each area has different options for minimizing the environmental footprint from the company's activities. Therefore, eco-marketing priorities may differ for each area, however, in any case, the consumers' concerns about environmental problems is a significant support to the company environmental activities.

A significant part of Russian consumers is interested in saving their money when paying utility bills, so a steady demand for water and energy saving household appliances can be observed.

According to many studies, the number of Russian consumers concerned about their health and beauty and, as a result, seeking to natural products consumption is constantly growing. This tendency has given rise to development of environmental marketing in such areas as organic cosmetics, food, clothing and footwear made of natural materials, environmentally friendly materials and household items.

The consumers eco-friendly attitude enables to expect interest growth in the Russian ecotourism development. In Russia, we can see many places with relatively primary nature. About 65 percent of the Russian Federation is almost unaffected by economic activity; ecosystems preserve biological productivity and biological diversity in full. To be sure,

* Corresponding author: bojuk.svetlana@yandex.ru

Russia has a high tourist and recreational potential to develop all types of outdoor recreation tourism. Thus, the interest of tour professionals for development of ecotourism is very clear.

According to Government statistics, in 2016 the total number of tourist organizations amounted to about 87000 and the total value of commercial services in the tourism sphere (including hospitality and health-wellness sectors) amounted to more than 160 million rubles.

In 2016, foreign tourists made more than 24 million trips to Russia. Primarily, they were residents of Ukraine (8,5 million), Kazakhstan (3,5 million), Finland (1,32 million), China (1,29 million), Poland (1 million).

Ecotourism issues are widely discussed in the scientific literature and at the special industry conferences. Most scientists agree and recognize the distinctive features of ecotourism such as gentle environmental impact, provision of benefits to the local population, accompanying environmental awareness and education.

Various aspects of ecotourism development on the following studies.

The work by Andrades and Dimanche [1] was devoted to problems of destination image, infrastructure development, workforce training and education, quality management, and sustainable management.

The research by Sheresheva and Kopiski [2] sheds light on the key trends and challenges with respect to leveraging the cultural and economic potential of the Russian hospitality and tourism market.

The book «Tourism in Russia: a management handbook», edited by Dimanche and Andrades Caldito [3], describes challenges to sustainable tourism development in transition economies with special references to Russia. This book demonstrates the problems of tourism management in comparing the experience of Russia and other countries, which characterize the tourism resources. Such sectors as natural and health tourism are indicated.

Frolova et al. [4] were study the relationship between the development of tourism infrastructure and growth of the tourism industry, and the stereotypes, followed by tourists when choosing destinations for their journey.

Furthermore, numbers of researches is devoted to some aspects of the tourist business development: the existing tax incentives of the various actors in the tourism sphere, the specific questions of measuring and competitiveness forecasting in terms of competitive market [5], the issues of financial security of the Russian tour operators specializing in outbound tourism (tour operators specializing in outbound tourism) [6], and in some of them carried out the Classification of Tourism Potential by use of Cluster Analysis [7].

There are many types of ecotourism such as natural, rural (agrarian), adventure (extreme), scientific (cognitive) tourism, etc. Despite the discussions on specific differences, all authors recognize the long-term benefits of development of the ecological orientation of tourism in the Russian regions, because the intensification, for example, of active or rural tourism without ecological orientation can destroy natural objects and damage the environment. In this work, the authors adhere to a broad interpretation of ecotourism as the whole set of nature-oriented travels of different range and activity.

Currently, the share of nature-oriented ecotourism is about 1 percent of the overall structure of the Russian tourist market. According to the Federal Agency for Tourism of the Russian Federation, the creation of attractive routes to natural places untouched by human activities can provide an additional tourist flow of over 0.3 mln people. The potential annual demand for rural tourism is estimated as 0.6 mln people. At the 2nd International Environmental Forum in Kaluga (September 13-14, 2016), A. Andrianov, the Executive Director of the National Agency for Sustainable Development, stated that ecotourism could become a real driver of the Russian economy. The perspective area of tourism provides rationale for study of formation features and estimation of marketing strategies efficiency

 DOI: 10.1051/shsconf/20173901002

oriented to principles of a sustainable development.

The *purpose* of this study is to analyze the main problems of formation and implementation of environmental marketing strategies for tour operators.

To achieve the purpose, the following *tasks* are set: to consider the capabilities of tour operators to adhere to the principles of sustainable development in Russia; to analyze the problems of ensuring the ecological attractiveness of a tourist product.

2 Materials and Methods

The study was based on the concept of Orsato [8] who proposed four variants of eco-marketing strategies:

- Environmental efficiency is defined by attention to the processes of reduction of the company overall costs through resource savings that allows to reduce the price for the consumer as well.
- Price leadership is also focused on the price competitive advantage, which allows obtaining a high price award from the environmental friendliness of the product itself.
- External leadership is a strategy based on differentiation of the company due to attention to the acology.
- Environmental branding differs by the product differentiation in the market as an ecological one.

To determine the ecological segments of the consumers, the classification of the *Natural Marketing Institute (NMI)* was used which involves grouping of consumers into five groups in accordance with their attitude to environmentally responsible businesses [9]:

- *Environmental fighters* (Lifestyles of Health and Sustainability – LOHAS). The main signs of behavior are activity, healthy lifestyle and desire to preserve the environment.
- *Supporters of all natural.* The main signs of behavior are taking care of their own health even at the expense of the natural environment.
- *Undecided ones.* The main sign of behavior is trendiness.
- *Conventional ones*. The main signs of behavior are pragmatism, desire to get the best value for money; they can support environmental initiatives when it is profitable.
- *Apathetic ones.* The main signs of behavior are lack of interest and awareness about social and environmental problems due to lack of time and money.

2.1 Operational hypothesis

Tour companies in Russia have access to the technologies allowing to save resources, protect environment from harmful emissions, however, these technologies are still too expensive for widespread use. Russian consumers are also not ready to provide significant support for the environmental activities of tourist companies, if this requires an increase in travel expenses. Thus, the eco-marketing priorities in this area are primarily related to saving resources allowing to reduce the companies' costs.

2.2 Study methods

Among scientific methods was chosen: systematic analysis of scientific literature, logical analysis, benchmarking of best practice in the field of sustainable development, author's study of consumer behavior.

The information for the study was obtained from 267 online application forms Google. The respondents were asked about their belonging to the environmental segment and preferences in the field of travel. Such form of the survey was chosen because of quick access to a young audience, which will determine the priorities of environmental consumption in the future.

The authors also conducted a content analysis of national parks' websites and natural attractions of World Heritage of UNESCO in Russia.

Also used secondary data of the survey conducted by VCIOM (Russian Public Opinion Research Center (VCIOM)); proceedings of the II International Ecological Forum in Kaluga in the section «Eco-tourism: status and prospects»; databases of the results of environmental certification "Ecological Union" and materials of analytical agency «TourStat».

3 Results

To determine the abilities of tour companies to adhere to the principles of sustainable development, it is necessary to clarify the composition of the goods offered on the market to consumers for exchange. The structure of the tour package (as a necessary minimum of services provided) can be considered as the basic components:

- vacation destination (tourist center),
- transport means to the vacation destination (main transport),
- accommodation and meals (accommodation services),
- movement within the vacation destination (transfer).

Thus, a full tour package of ecotourism will include protected natural areas or picturesque rural settlements as vacation destination, accommodation in ecological buildings, health food, transfer by ecological transport. Currently, just a few examples of such integrated approach to ecotourism can be found. In most cases ecotourism offers a tour to a place without civilization signs which puts in question its compliance with the requirements of sustainable development since active tourist activities can damage the environment. According to the Agency TurStat, in 2016 more than 9 million people made a cognitive journey through Russia's natural territories (national parks, sanctuaries and forest reserves). About 80% of visitors are accounted for the national parks of the Moscow region (Losiny Ostrov) and the Krasnodar Territory (Sochinsky) that indicates a significant impact on the local ecology.

However, the situation will change if the tour operator sets the task of suppliers and other stakeholders selection on the basis of conformity with the concept of sustainable development. Not only companies that provide accommodation (hotels, mini-hotels, campsites, etc.), catering services (cafe and restaurant), transportation; entertainment (including recreational facilities) and trade, but also producers of renewable energy sources, waste-processing organizations, local community and public organizations, employees, the media, etc. are considered as stakeholders.

For tour operators, the choice of the eco-marketing strategy is determined by the possibility of obtaining a competitive advantage of the company in the form of price reduction or differentiation through environmental practices (projects) or the ecological value of a tourist product.

When implementing a strategy aimed at *environmental efficiency*, the tour operators are trying to change the ecological footprint in those processes that provide cost reduction. An example of such strategies implementation is the construction of passive houses, use of paper or floor coverings from recyclable materials in hotels and restaurants, use of energy saving equipment and devices for rational water use.

Implementation of *price leadership* strategy can only be successful when working with segment of the high-income consumers considered as economically significant for the company. An example of a strategy of this kind is the Kapawi Lodge in the Amazonian rainforest offering luxury housing with minimal environmental impact. To generate

 DOI: 10.1051/shsconf/20173901002

electricity, they use solar panels, and the buildings are built in a traditional local style without nails. The cost of accommodation is $ 100 per night. The tour operator works in close partnership with the local population; this provides only positive economic and social results from the activities. In Russia such examples are not yet known.

External leadership strategies are focused on environmental policy measures implemented in public interests. The examples of this strategy are the cases of investments in non-profit environmental projects: area clean-up operations, support of social movements for local flora and fauna protection, etc.

Application of *environmental branding* strategy is associated with the differentiation of offers due to ecological value of tourism products. Primarily, the item of ecological value is the tourist center (nature reserve), and less often this is the place of residence (eco-hotels) and organic food (rural tourism).

If the traveler attraction has a natural uniqueness, its branding facilitates the formation of awareness among the general audience and makes it popular and even fashionable to visit. In order to inform tourists, information portals, mobile applications and other media products are created to provide comprehensive information in different languages and make a set of impressions, well-known all over the world.

The research of Russian tourist centers included in the UNESCO World Heritage List has shown weak efforts in the field of branding. The main source of information about tourist centers for consumers is Internet resources. Correct presentation of information about the uniqueness of a natural site not only creates awareness, but also forms expectations, fixes positive emotions, creating a sustainable image of the branded object. For example, Lake Baikal is a natural UNESCO World Heritage site, which supports several top-rated information portals (though not always in several languages) for the tourism development. From 11 Russian natural sites included in the UNESCO World Heritage List, only one does not have its own information portal (Western Caucasus), although the information is widely represented on the sites of various tourist guides. Three natural sites have web-sites with the official reserve name that is not known to all tourists. By the short UNESCO name, you can find information only on the web-sites of tourist guides. Thus, it is difficult to talk about the development of environmental branding in the tourism industry, even if the Russian UNESCO natural heritage assets are not a sample of such a strategy.

Some representatives of the tourism industry are actively positioning themselves as eco-hotels, but in fact, they can not confirm their statute with certificates or with significant environmental responsibility. The addition of "eco" rather describes the location not spoiled by civilization, than the building or lean practice.

Best practices benchmarking has shown that "green" technologies can ensure the development of ecotourism in Russia, however, they require significant and sustained investment in the tourist centers.

The situation with environmentally friendly food is the most transparent. In 2015, Russia adopted a national standard for organic and agricultural products that allows all interested companies to confirm their status. For example, in the North-West region two organic farms that are certified according to organic standard of ICEA (the Italian Institute of Ethical and Organic Certification - IACBs Equivalent EU Standard), harmonized with the European one, are under development. According to NP Ecological Union these are Alohovschina farm in Lodeinopolsky region and Zhivoeye Pole in Kirov region. In total, 14 manufacturers are certified in Russia.

Today in Russia there are a lot of examples of successful cooperation in the field of agricultural production, catering and shopping businesses. Some retail store chains are formed around itself a pool of proven farmers. For example, a retail chain *ABC of Taste* (*Azbuka Vkusa*) arranged a cooperative farm with dozens of private traders and packages milk, meat, cereals and vegetables under the brand "Nasha Marka" (Our Brand). Turnover of organic products in the ABC of Taste is kept confidential, it is known that in 2014 it grew by 21% compared with the previous year, and the company intends to expand its production. However, there are no examples of attracting such companies to the Russian tourism industry. The Russian tourism industry is still weakly interested in differentiating its offer as ecologically oriented, but gastronomic tours in Europe are well known.

There are examples of ecological construction also. "Green" construction should solve the tasks of reducing water consumption, energy consumption and heat losses, sorting and recycling of garbage. In Europe, considerable experience has been gained in the construction and successful reconstruction of buildings into passive houses, for example, in Peltosaari (Finland) after reconstruction of multistory apartment house energy consumption has been reduced in half [10].

The ambitious project of environmentally friendly city is implemented in the Chinese mountainous region of Guangxi in an area of 175 hectares. It is assumed that in the new city 30 thousand people will be settled in houses designed as vertical gardens. The buildings are equipped with air filtration and water purification systems. Electricity is provided by solar panels, travel around the city – by electric transport, and connection with other settlements – by high-speed rail [11].

In Moscow microdistrict Nikulino 2 heat pumps for hot water and the outer fencing structures with increased thermal protection were first used in Russia [12]. There are other examples of energy efficient low-height building in suburbs of major Russian cities. For example, a group of companies EcoDolie in Ekaterinburg became the first in Russia that received the World Award FIABCI Prix d'Excellence for the environmental compatibility in low-height building [13].

The activity of builders is constrained by an increase in the cost of a building built according to "green" technologies, at least 10-15% higher than conventional construction. According to experts of YIT Homes opinion, in Russia each percent of energy saving used practically increases the cost of construction by 1% too. So, "green" technologies which save 30% of electricity automatically increase the cost of construction by one third [14]. However, in the long-term cumulative costs for construction and operation of "green" buildings turn lower, and additional costs are compensated.

Now, in Russia, 10 facilities in Moscow and the Moscow region, 7 facilities in St. Petersburg, 2 facilities in Tver and the Tver region, 1 facility in Dzerzhinsk and 1 facility in Ulyanovskin, are certified in accordance with the international environmental standard for LEED buildings (The Leadership in Energy & Environmental Design). Another 50 facilities are under certification now. There are no facilities related to the tourism industry among them.

36 facilities in Moscow and the Moscow region, 8 facilities in St. Petersburg, 7 facilities in Sochi, 12 facilities in other cities of Russia are certified by the BREEAM method (Building Research Establishment's Environmental Assessment Method). And only 2 of them are hotels (located in Sochi).

In 2012, GOST R 54964-2012 "Conformity assessment. Ecological requirements for estate properties" was put into force in Russia. And in 2014, the standard of energy efficiency named the "GREEN ZOOM" was also developed. Evaluation of real property is carried out in view of the object location and environmental sustainability of land plot, transport support, water and energy efficiency, reducing harmful emissions into the atmosphere, the selection of construction materials and waste management, indoor ecology and innovation. Today 21 facilities are certified, and 7 facilities are in the process of GREEN ZOOM certification. Among them there are 8 residential complexes and a whole microdistrict (in Yekaterinburg), the objects of the tourism industry are also not represented.

The prospects of using ecological transport are becoming real. The leading car brands (Mercedes, Porsche, Aston Martin, BMW, Volvo Chevrolet, Volkswagen, Citroen, Fiat, Ford, Hyundai) presented advanced models of electric vehicles, which will be put into production. The model range offers both miniature two-seater models (for example, electric car Microlino EV costs $1,000, accelerates up to 100 km/h and is able to cover the distance of 100 km before the battery recharging), and prestigious versions (Tesla electric cars cost more than $35,000 and are able to cover the distance of more than 400 km before recharging).

The PHEV (plug-in hybrid electric vehicle) is an advance technology also. The PHEV advantages are that they can move on conventional fuel, charging their electric batteries while moving. Thus, the problem of limiting the travel distance by a battery resource is solved, the owner has the ability to charge the battery from the mains in a convenient place.

The main problems that prevent the spread of electric vehicles in Russia are their high cost and a short range of travel. In the Russian market AvtoVAZ represents one model BA3-1817 ELLADA (cost of 1.2 mln rubles) which is operated in the Krasnodar Territory. Mikhail Prokhorov at the Geneva Motor Show in 2016 introduced the concept Yo-Move electric vehicle with cost of $10,000 but the model did not enter in mass production. The task of sustainable transport creation is solved by KAMAZ jointly with Skolkovo within the framework of the "Electrochemistry 2015". However, electric buses have not yet become a universal eco-transport. It is also too early to talk about ecological trunk transport.

4 Discussion

In general, we can state that today there are technological opportunities to provide an ecological tour package in Russia. However, examples of the application of "green" technologies are differentiated and do not represent systematic practice.

The activity of introducing environmental innovations is determined by the state of environmental interest of target consumers.

The authors conducted a series of studies of potential consumers to clarify the behavior patterns and portraits of typical representatives of segments exhibiting different environmental responsibilities.

In the course of the study in the spring of 2017, in-depth interviews and an Internet survey were conducted to identify the motives for choosing a holiday package, the presence of interest and the current consumer perceptions of ecotourism for each ecological segment.

The results have demonstrated that the segment of "apathetic consumers" may show an interest in ecotourism out of curiosity, they are willing to discover something new without blowing their budget. At the same time, there is no the value of the environmental proposal for them, they are passive when choosing a tour package and have no idea about ecotourism. In everyday life, they are not ready to experience any inconvenience for solving environmental problems.

The represents of "conventional" segment are ready to travel to ecologically clean territories provided an acceptable price for the tour package. They have an idea of ecotourism from this point of view, but without going into the nuances of the natural and rural tourism. They pragmatically connect outdoor recreation with health benefits and cheapness, and believe that such places can be easily found in Russia. Little attention is paid to other aspects of the environmental situation. They are environmentally responsible only within the framework of their interests.

"Undecided ones" have the most general ideas about the relationship between human activities and environmental problems, pay attention to picturesque places for a photo session, always in search of new impressions, but the value of an environmental proposal for them is determined by fashion trends or friends' recommendations. They have no idea about the peculiarities of ecological tourism. They believe that hikes and other types of outdoor activities are suitable for this category. They are not ready to make any additional efforts to protect the environment, to give up comfort, but try to follow certain elementary rules of conduct ("do not litter", "do not make a fire in the wrong place", etc.).

"Supporters of all natural" make high demands for the place of accommodation, follow the healthy atmosphere and food, have a vacation experience in eco-hotels, but pay little attention to the consequences of human activities for the environment.

"Environmental fighters" are not only well aware of the environmental friendliness of the chosen vacation destinations, but also ready to follow in everyday life all the recommendations for solving environmental issues and to incur additional costs in order to minimize the harm to the environment not only during travel. However, the interests of travel of this segment are not limited by tours to nature reserves. For this segment, environmental conditions of accommodation are important, without reference to the tourist center. Most of them believe that ecologically clean territories can be found only abroad.

It is obvious that detailed definition of the portrait of ecological segments according to the NMI classification in tourism did not bring any surprises concerning environmental consumption, but allowed to reveal a weak awareness about certain components of ecologically oriented recreation and ecotourism opportunities in Russia. The consumers perception distortion regarding the vacation destination ecology is also noticeable.

5 Conclusions

The results of the quantitative research allow us to identify the attitude of consumers towards the environmental responsibility of the companies in the field of tourism. The contingent of respondents is mainly young people under 25 years old, and a third part of consumers are persons of 26-45 years old. Two-thirds of respondents view themselves as a middle class who can afford to buy farm products, travel at least once a year, planning a vacation with sufficient care.

Among the significant factors affecting the choice of the tour, respondents indicated low cost, visitor reviews and ecology of the vacation destination. At the same time to maintain the place ecology, most respondents do not do anything. Only 18% of respondents had experience of ecotourism, and less than half of respondents are interested in places not affected by civilization.

This confirms once again that the consumers views about the ecology of the vacation destination are highly distorted. Representation of the environmental responsibility of the tourism industry does not affect the consumer choice.

Unfortunately, a quantitative online survey does not allow us to accurately estimate the size of environmental segments, since the sample is not random, and it is difficult to distribute the obtained results to the entire population of consumers. According to researchers' opinion, about 20% of the sample can be classified as "apathetic", 19% - "undecided" and "conventional", which, in general, within the margin of error, corresponds to the research conducted by other authors. "Supporters of all natural" are represented by 35% of the respondents. "Environmental fighters" are very few in number.

It is clear that neither consumers nor the state supports the ecological orientation of the tourism industry representatives. Russian tourism industry is interested weakly to differentiate its offer as environmentally valuable yet, since it requires a significant and sustained investment. The use of "green" technologies aimed to resource economy is gradually expanding in Russia, however, it is not a systematic practice in the tourist centers. To be actually developed, the ecotourism sector requires a national program which will enable not only to expand the range of the offered tourism products, but also to take measures to preserve he heritage for future generations

The study was devoted to the problems of sustainable development of the tourism industry.

The background of the Ecotourism conception formulation was the study of the positive experience of environmental in the building sphere, transport environmental, meals environmental.

A study of the experiences of different companies has allowed us to identify the factors contributing to and hindering the development of ecotourism in Russia.

The main barrier for total implementation of the Ecotourism conception is the lack of demand from consumers for a full touristic package, which including not only the protected natural areas or picturesque rural settlements as vacation destination, but also accommodation in ecological buildings, health food, transfer by ecological transport.

That's why, in most cases ecotourism offers a tour to a place without civilization signs which puts in question its compliance with the requirements of sustainable development since active tourist activities can damage the environment.

References

1. L. Andrades, F, Dimanche, Destination competitiveness and tourism development in Russia: Issues and challenges. *Tourism management,* **62**, 360-376 (2017)
2. M. Sheresheva; J. Kopiski, The main trends, challenges and success factors in the Russian hospitality and tourism market. *Worldwide hospitality and tourism themes*, **8**(3), 260-272 (2016)
3. F. Dimanche, L. Andrades Caldito, *Tourism in Russia: a management handbook*, United Kingdom: Emerald, 448 p. (2015)
4. E.V. Frolova, O.V Rogach, E.E. Kabanova, N.V. Medvedeva, I.Y. Shimanskaya, Tendencies and prospects of tourism industry in Russia: Sociological analysis of stereotypes among population during tourism trips. *European Research Studies Journal,* **20**(2), 308-320 (2017).
5. N.N. Koroteeva, E.L Hasanov, V.A. Mushrub, V.V Bakharev, R.A Shichiyakh, The conditions of economic efficiency and competitiveness of tourism enterprises. *International Journal of Economics and Financial*, **6**(8), 71-77 (2016)
6. G.R Bakieva, L.F. Muftieva, Management analysis as means of improving the business processes at the tourist enterprise: A case study. *International Journal for Quality Research*, **10**(4), 761-768 (2016)
7. P. Chalupa, M. Prokop and J. Rux, Use of Cluster Analysis for Classification of Tourism Potential. *Littera Scripta*, **6**(2), 59-68 (2016)
8. R. Orsato, Competitive Environmental strategies: When does it pay to be green? *California Management review*, **48**(2), 131 (2006)
9. J. Grant, Green marketing. *Strategic Direction*, **24**(6), 25-27 (2008)
10. V. Bozhuk, *Sustainable development in architecture and urban planning*, (2017)
11. U. Bezborodova, *China to build the first city-forest*, (2017)
12. Official site of the group of companies Pi-ANA, *PIANA HR GROUP* [online], Available at: http://www.pianahrgroup.com/ (2017)
13. Official site of the group of companies Ecodolie, *Ecodolie* [online], Available at: http://www.ecodolie.ru/ (2017)

14. D. Truman, When in Russia will start to build an ecological house?, *Green house* [online], Available at: http://green-dom.info/3-%D0%B0%D0%BB%D1%8C%D1%82%D0%B5%D1%80%D0%BD%D0%B0%D1%82%D0%B8%D0%B2%D0%BD%D0%B0%D1%8F-%D1%8D%D0%BD%D0%B5%D1%80%D0%B3%D0%B5%D1%82%D0%B8%D0%BA%D0%B0/%D0%BA%D0%BE%D0%B3%D0%B4%D0%B0-%D0%B2-%D1%80%D0%BE%D1%81%D1%81%D0%B8%D0%B8-%D0%BD%D0%B0%D1%87%D0%BD%D1%83%D1%82-%D1%81%D1%82%D1%80%D0%BE%D0%B8%D1%82%D1%8C-%D1%8D%D0%BA%D0%BE%D0%BB%D0%BE%D0%B3%D0%B8/ (2013)

SHS Web of Conferences **39**, 01003 (2017) DOI: 10.1051/shsconf/20173901003
IES2017

The Education and Development of Company Managers in the Czech Republic and Ukraine

Zdeněk Caha[1,*]

[1]Department of Management, Faculty of Corporate Strategy, Institute of Technology and Business in České Budějovice, Okružní 10, 370 01 České Budějovice, Czech Republic

Abstract. The education and development of managers is a key component of corporate education policy because managers have a significant influence on overall company performance. The aim of the research set out in this paper was to determine, on the basis of a sample set of 607 companies from the Czech Republic and a sample set of 145 companies from Ukraine, to what extent the education of managers takes place in companies. The companies were divided into four categories according to their size (in terms of number of employees). The assumption was made that the education of managers is most widespread in economically more developed countries where companies have greater financial resources at their disposal and are more willing to invest in the education of their employees, and therefore managers. The hypothesis was therefore put forward that the education of managers is more widespread in the Czech Republic than in Ukraine and that the organization thereof in both countries is directly related to company size (i.e. the larger the company, the more widespread the education of managers). The hypotheses were confirmed.
Key words: education, development, manager, Czech Republic, Ukraine

1 Introduction

Numerous studies deal with the importance of education for managers. Managers face more and more responsibility for a continuously growing range of activities, within which they have to apply their professional and managerial skills. They are therefore not only responsible for the development of the competences of their subordinates, but also - and chiefly – their own competences so that they are able to respond in a timely and efficient manner to the changes of the external as well as the internal entrepreneurial environment [1]. Numerous studies deal with the importance of education for managers. Farahani [2] supports the necessity of further education for managers by claiming that what managers learn at universities may not be correct in a couple of years, and that managerial qualifications are not only developed through "school attendance", but they require a complex, continual development of a manager in all areas of their personality. Aragon and Valle [3] examined the importance of the education of managers for organizational performance. These authors point out the fact that although managers play a key role in a

* Corresponding author: caha@mail.vstecb.cz

company's success, most of the education and training budget is spent on employees, not managers. They nevertheless admit that the precise contribution of manager education to the company is usually difficult to detect, as the finance performance of a company is affected by numerous other factors. Akrofi [4] also deals with the relation between the development and education of managers and company performance in his works. The study is based on an analysis of empiric data obtained from 222 managers from various regions. The research proved the overall positive impact of the development and education of managers on organization performance, and implies important consequences for the effective development of managerial skills as a tool for the improvement of the effectiveness of an organization. Čech, Chormý a Skupinová [5], amongst other authors, are convinced that if managers develop their competences and skills they not only improve their own potential but also the potential of the whole company.

Managers have started to realize the importance of their education, and the responsibility they have for this, although they continue to prefer short-term one-shot courses. Notman [6], for example, states that an interest in short-term one-shot courses is growing amongst managers. On the other hand, long-term education for managers has the opposite trend, which contradicts the fact that long-term education of managers is the driving force of the economy. Gwiazdzinska-Goraj and Rudnicki [7] published a study, which was carried out in 2002-2010 and dealt with managers in agricultural companies, in which they proved that 50% of managers participated in some form of education. The research moreover brought information that documents the benefits of continuous education for managers, which resulted in a higher success rate when acquiring money from UE funds. A study carried out by Grovo, U.S., in which 500 middle managers from various industry spheres participated shows that not less than 98% of managers feel the need for more courses, particularly in the fields of professional development, conflict solving, time management and project management. The study also shows that two out of five managers are not prepared for company management and less than a half of managers are highly effective in their positions. However, 76% of the respondents claim that even inefficient managers are often very well remunerated and supported. The study also contains the interesting fact that up to 80% of managers temporarily change their behaviour and attitudes immediately after an education course, but return to old habits within half a year. 98% of the respondents had the opinion that regular educational courses for managers have a positive effect which enables companies to retain employees and maintain morale at the workplace alongside an overall satisfaction with incomes [8]. Among research carried out in the Czech Republic we can mention one piece based on a questionnaire survey carried out among 419 Czech managers in middle and top managements which produced some interesting findings. According to this research 65% of the managers were satisfied with the courses they had taken part in. Other findings dealt with a lack of educational courses, the absence of practically oriented courses, insufficient link to practice, lack of qualified trainers (a trainer's incompetence often affected the results of the courses and the motivation to further participate in educational courses amongst managers) or the fact that the price/effectiveness ratio of the courses is often unbalanced [9]. Folwarczná [10] also dealt with the development and education of managers. She carried out research amongst managers and personal. In her research, she mapped the methods of education for managers in the Czech Republic and compared them with the methods applied in more-developed countries, and the ways in which the evaluation of the effectiveness of education programmes was made. The problems with managerial training sessions and courses are actually often based on the fact that they lack specific measurable goals [8] points out. He also claims that no more than 28% of training sessions and courses for managers have specific and measurable goals. On the other hand, he points out the importance of education for managers in the example of fast growing companies that recruit new managers from among the best consultants and sales representatives. These, however, do not automatically become successful managers. For managers to become successful, it is often necessary for them to take part in suitable training. Plakhotnik [11] also agrees with this, stating that organizations should support managers through educational courses even before they are promoted to managerial positions.

The methods and frequency of education for managers differs depending on the company size and nature of business. Bager et al. [12] stress the importance of educational methods based on experience. Experience is transformed into knowledge and skills through reflection and interaction. However, they point out that managers from small and medium companies use such educational programmes to a demonstrably smaller extent than those of large companies. Limited time and the fact that managers of small and medium companies do not deal with the development of the company and the development of themselves on a long-term basis. Some studies however indicate that a lot of managers from small and medium companies have already realized this fact, which is also backed by a study by Wang and Zhou [13], who emphasise a need for the development of managers, mainly in the areas of company strategy, management, finance and law, as seen from the results of interviews with managers from small and medium companies in China.

Education programmes for managers may have various concepts which contain the individual components of education for managers. Nguyen and Hansen [14] point out that education programmes for managers should be complex, and should distinguish and stress the difference between managing (doing things correctly) and leading (doing the right things, being proactive with a vision of permanent development), which is crucial from the point of view of company development, particularly for top management. Elmuti, Minnis and Abebe [15] propose a simple three-component structure of programmes for education for managers, so that a manager masters all of the skills necessary to lead and manage employees. The third component, which contains action-based trainings (simulation and case studies), is key according to the authors. Managers can actually only achieve the necessary experience and perfect their abilities through practice. New educational technologies play and are going to play a more and more important role in education for managers. Wankel [16] points out the advantages and the unique opportunities of this kind of education particularly in the field of intercultural know-how.

2 Methodology and data

The available data was collated on the basis of responses from 607 Czech and 145 Ukrainian companies (752 companies in total). Quantitative research was conducted on the basis of a larger questionnaire survey which was conducted between June 2016 and February 2017. The aim of the survey was to obtain a statistically representative cross-sectional data set on small, medium-sized and large companies in the Czech Republic and Ukraine.The data is stored in a pvz.xlsx file (available at the author) and contains information on: country of origin (CZ and UA); company size (1-9 employees – micro-company; 10-49 employees – small company; 50-249 employees – medium-sized company; 250 and more employees – large company); actual number of employees; and the existence of education for managers (yes or no).

The following hypotheses were statistically verified:

- H1: The education of managers is more widespread in the Czech Republic than in Ukraine.
- H2: The existence of education for managers is related to company size (the larger the company, the more widespread the education of managers).

The calculations were carried out in the R system

 DOI: 10.1051/shsconf/20173901003

3 Results

Statistical verification of H1

The relative frequencies with which companies organize education for managers in Czech and Ukrainian companies were calculated in percentages. The results are presented in Table 1.

Table 1. Relative frequencies of Czech and Ukrainian companies that organize education for managers (in %)

Country	Education yes (%)	Education no (%)	Total (%)	Number of companies
CZ	48.6	51.4	100	**607**
UA	14.5	85.5	100	**145**

In the Czech Republic, 48.6% of the companies (i.e. 295 out of 607) organize education for managers. In Ukraine this stands at 14.5% of the companies (i.e. 21 out of 145). H1 is therefore valid. The general application of the hypothesis was tested by means of the Chi-square test. The zero statistical hypothesis that the proportion of companies that organize education for managers is the same in both countries was tested against the alternative hypothesis that the proportion that do is higher in the Czech Republic. The statistical test results were: X2 = 55.92; p-value = $3.78.10^{-14}$. On the basis of these results it can be concluded that the zero statistical hypothesis and the H1 hypothesis are statistically significant.

Statistical verification of H2

The relative frequencies with which companies (do not) organize education for managers were calculated in percentages according to company size. The results are presented in Table 2.

Table 2. Relative frequencies of Czech and Ukrainian companies that (do not) organize education for managers according to company size (in %)

Education	Micro-company (%)	Small company (%)	Medium-sized company (%)	Large company (%)
Yes	23.1	29	44.7	76.6
No	76.9	71	55.3	23.4
Total	100	100	100	100
Number	**169**	**210**	**219**	**154**

It is clear from the data presented above that the proportion of companies that organize education for managers increases with increasing company size (in absolute numbers: 39, 61, 98, 118 out of 169, 210, 219, 154 companies respectively). Hypothesis H2 is therefore valid. The validity can be generally proven through a logit regression model (see [1], [2] and [3]). The dependency of the proportions (absolute frequencies) of companies that organize education on the number of employees x can be modelled on the basis of the following exponential function:

$$\frac{1}{1+e^{-(b_0+b_1x)}} \tag{1}$$

where the coefficients b_0 and b_1 are estimated from the data by means of the R system. The resulting estimates were: $b_0 = - 0.851$; and $b_1 = 0.00307$. The proportions of companies that organize education (i.e. the estimated probabilities of the existence of education for managers in a company with x employees) are therefore expressed by the formula:

$$\frac{1}{1+e^{0,851-0,00307x}} \tag{2}$$

Figure 1 illustrates the graphical course of the dependency of the monitored proportions (indicated as P_vzdělání [P_education]) on the number of employees x. Table 3 shows these proportions for the selected sizes of companies.

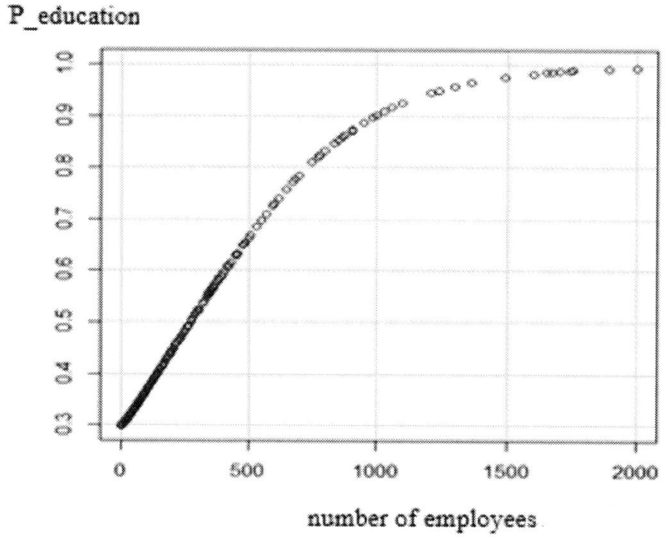

Fig. 1. Dependence of the numbers of companies that organize education for managers on the number of company employees.

Table 3. Estimated proportions of companies with education for managers according to numbers of employees

x	p
0	0.299
50	0.332
100	0.367
150	0.403
200	0.441
500	0.664
200	0.995
400	1.000

DOI: 10.1051/shsconf/20173901003

The results confirm the general validity of hypothesis H2 for the whole population of the sample sets.

4 Conclusion

The aim of the research set out in this paper was to determine, on the basis of a sample set of 607 companies from the Czech Republic and a sample set of 145 companies from Ukraine, to what extent the education of managers takes place in companies. The companies were divided into four categories according to their size (in terms of number of employees). The results of the research show that the education of managers is most widespread in economically more developed countries such as the Czech Republic, which may be the result of these companies having greater financial resources at their disposal and a greater willingness to invest in the education of their employees, and therefore managers. Two hypotheses were set. The first, that education of managers is more widespread in the Czech Republic than in Ukraine, and the second, that the organization of the education of managers in both countries is directly related to company size (the larger the company, the more widespread the education of managers). Both hypotheses were confirmed.

For follow-up research, we recommend undertaking an analysis of the areas of education for managers and the impact thereof on the economic results of companies.

References

1. M. Tyler, D. Dymock, A. Henderson, *The Critical Role of Workplace Managers in Continuing Education and Training* [online], Available at: https://www.researchgate.net/publication/303045780_The_Critical_Role_of_Workplace_Managers_in_Continuing_Education_and_Training (2016)
2. A. F. Farahani, Theory of training and development of manager's practical approach. *Proceedings of the 2nd IEEE International Conference on Emergency Management and Management Sciences*, Beijing: IEEE, 533-542 (2011)
3. I. B. Aragon, R. S. Valle, Does training managers pay off? *The International Journal of Human Resource Management*, **24**(8), 1671-1684 (2013)
4. S. Akrofi, Evaluating the effects of executive learning and development on organisational performance: implications for developing senior manager and executive capabilities. *International Journal of Training and Development*, **20**(3), 177-199 (2016)
5. P. Čech, J. Chromý, S. Skupinová, Company training of managers as a part of the human resource management in the hotel industry. *Proceedings of the International Multidisciplinary Scientific Conference on Social Sciences and Arts*. Albena: STEF92 Technology Ltd, 189-196 (2015)
6. O. Notman, The value of short-term training for Russian managers in the context of the economic crisis. *Proceedings of the 10th International Days of Statistics and Economics*. Prague: University of Economics, 1344-1352 (2016)
7. M. Gwiazdzinska-Goraj, R. Rudnicki, Agricultural education of managers of agricultural holdings in Poland in 2002-2010. *Proceedings of the International Scientific Conference on Economic Science for Rural Development*, Jelgava, 69-76 (2016)
8. E. Schimel, *Ninety-Eight Percent of U.S. Managers Want Better Management Training, Grovo Study Reveals* [online], Available at: http://www.prweb.com/releases/2016/09/prweb13719059.html (2016)

9. P. Kressová, The experience of the Czech managers with training and development courses for managers. *International Journal of Knowledge, Culture and Change Management*, **11**(4), 283-294 (2012)
10. I. Folwarczná, *Development and training of managers [Rozvoj a vzdělávání manažerů]*. Prague: Grada (2010)
11. M. S. Plakhotnik, First-time managers: increasing chances of their success before the promotion. *Industrial and Commercial Training*, **49**(3), 132-138 (2017)
12. T. E. Bager, et al., Enrollment of SME managers to growth-oriented training programs. *International Journal of Entrepreneurial Behavior*, **21**(4), 578-599 (2015)
13. Y. H. Wang, S. B. Zhou, Research on Managers' Training Needs of Small Medium-sized Enterprises in Mainland - A Case Study of Chongqing. *Proceedings of the International Conference on Human Resource, Organizational Behavior and Leadership*. Beijing, 157-160 (2009)
14. N. Nguyen, J. Ø. Hansen, Becoming a leader-manager: a matter of training and education. *Development And Learning In Organizations: An International Journal*, **30**(6), 10-12 (2016)
15. D. Elmuti, W. Minnis, M. Abebe, Does education have a role in developing leadership skills? *Management Decision*, **43**(7/8), 1018-1031 (2005)
16. CH. Wankel, Developing cross-cultural managerial skills through social media. *Journal of Organizational Change Management*, **29**(1), 116-124 (2016)

Ethical Management in Companies in the Czech Republic and Ukraine - Comparison of the Presence of a Code of Ethics

Zdeněk Caha[1,*]

[1]Department of Management, Faculty of Corporate Strategy, The Institute of Technology and Business, Okružní 517/10, 37001 České Budějovice, Czech Republic

Abstract: The aim of the study is to compare the commonness of a code of ethics as the most important ethical management tool in the business sector in two post-communist countries, namely the Czech Republic and Ukraine. The hypothesis that a code of ethics is much more widespread in the economically more developed country, which is the Czech Republic, and also the assumption that the occurrence of a code of ethics is in relation to the company size, were examined on the base of a questionnaire survey. The results definitely confirmed that a code of ethics is much more widespread in the Czech Republic than in Ukraine. The survey results have also confirmed that the commonness of a code of ethics grows with the company size. This was not confirmed in micro and small companies in Ukraine.
Key words: code of ethics, management, commonness, Ukraine, Czech Republic

1 Introduction

Ho and Oladinrin [1]; Kaptein [2] show, a code of ethics is a basic element of the ethical infrastructure of an organization. Its acceptance covers the application of formal documents, in which the explicit declaration of the values that should steer the behaviour of employees and managers and support good practices and indicators related to the core of the organization, can be found.

The first studies into codes of ethics were conducted in the USA in the 1950s and 1960s. In the 1950s, only 15% of large companies had adopted a code of ethics. This number grew relatively fast over the following decades, so that by the 1960s this number had increased to 40%, by the 1980s to 75%, and by the 1990s to 93% [3]. A Business Ethics Centre report published in 1990 contained very interesting information. It indicated that in 1986, 93% of companies were acquainted with codes of ethics, which represented a 10% increase compared to 1979 [4]. Ethics Resource Center research, which was conducted among 747 respondents [5], revealed that nearly 75% of the companies had a code of ethics in place,

* Corresponding author: caha@mail.vstecb.cz

DOI: 10.1051/shsconf/20173901004

and that this proportion was 57% among smaller companies with less than 100 employees. In 2001, in a study conducted by KPMG in South Africa, 86% of companies had a code of ethics in place [6]. In 2003, in research conducted by Deloitte [7], which involved 373 respondents, 83% of the companies in the USA had a code of ethics in place. The results of a National Business Ethics Survey published in 2005 [8], showed that 86% of companies used a code of ethics. In 2006 79% of companies used a code of ethics [9]. A global survey carried out by KPMG [10] points to the global trend for increasing numbers of companies to define and implement a code of ethics. The survey showed that in 2008 the percentage of companies with a code of ethics increased to 86%.

It is clear from the above that in last few years that a code of ethics has become a basic tool for ethical management within large and well-known companies. This premise is backed up by an informal study conducted by the Institute of Business Ethics (IBE) in 2012, which showed that a code of ethics was applied by 92% of those companies with the highest market capitalization based in Great Britain whose shares are traded on the London Stock Exchange [11]. The IBE survey [12] also revealed that in 2012, 73% of all companies used a code of ethics. The CGMA [13] conducted research in which nearly 2,500 respondents from ten countries took part. The results indicated that the proportion of companies with a code of ethics stood at 82% (and among large companies 93%). In 2016, a wide-reaching study conducted by PwC, the largest of its kind, in which 6,337 companies from 115 countries participated (of which 79 prominent companies from the Czech Republic), showed that on average 82% of companies had an officially established code of ethics [14]. In 2016, Dow Jones and Metricstream [12] jointly conducted one of the most recent surveys, which revealed that 96% of the 330 questioned companies had established a code of ethics.

The aforementioned surveys point to the growing trend towards the introduction of codes of ethics in business practice. Nonetheless, it is important to note that not all these surveys are compatible. In fact, the results were determined on the basis of numerous factors that were specific to the structure of the respondents. Although a higher proportion of large US based companies have codes of ethics, this proportion is smaller for small and medium-sized companies and companies operating in Asia. It is clear that only global studies that include large numbers of respondents from all business sectors and companies of different sizes will provide the relevant information required. Unfortunately, such surveys are challenging to realize.

A survey conducted by the Association of Chartered Certified Accountants, found that one in five executives or bookkeepers in the Czech Republic had had to deal with a serious ethical dilemma in their work. This only confirms the need to apply effective ethical management tools [15]. On the basis of smaller less representative survey conducted in 2002, in which 60 respondents participated from middle and top management in companies operating in Prague and Central Bohemia, that 35% of the examined companies had a code of ethics in place and 22% were considering preparing and applying one [16]. In contrast, Transparency International – Czech Republic and the Economic University in Prague, jointly conducted a very representative survey in 2005, which was subsequently repeated in 2006 [17]. The surveys sought to determine to what degree codes of ethics were applied by companies that operated on the Czech market. The conclusion was that in the second half of 2005 only 59 (i.e. 10.3%) of the 574 companies in the sample set had a code of ethics in place. This state of affairs was similar when the survey was repeated in the first half of 2006, whereby only 104 (i.e. 8.4%) of the 1,240 companies in the sample set had a code of ethics in place. The survey also showed that there was a relationship between company size and the presence of a code of ethics. The larger the company the greater the chance a code of ethics had been put in place. For example, in 2006, less than 5% of small companies (i.e. with less than 50 employees) had a code of ethics, whereas this percentage among large companies (i.e. with 300 or more employees) topped 24%. In 2011, less than 15% of domestic companies had a code of ethics in place [18]. According to a survey conducted by Transparency International in 2013, this percentage had increased to 46% of domestic companies. This compares poorly with the world average of 74% [19]. According to the respondents of a questionnaire survey conducted in 2015, the results of which were processed by the Financial and Economic Institute, 68% of companies had a code of ethics in place, but only 23% of companies saw the code of ethics as an obligatory set of rules [20]. According to the previously mentioned PwC survey, in which 6,337 organizations from 115 countries took part, 86% of the 79 prominent Czech companies questioned had an officially established code of ethics. Unfortunately, the low number and the size of the companies involved belies reality [11].

Despite these results, the number of companies in the Czech Republic that are establishing a code of ethics is gradually growing, although the numbers remain well below the worldwide average (which is higher than 70%). This gives the impression that the presence of a code of ethics in the Czech Republic tends to be limited to larger companies and public institutions and that this may also be dependent on market activities or economic results of companies [21, 22]. As far as Ukraine is concerned, no extensive research on the presence of codes of ethics is known to the author of this paper. However, it is assumed that the presence of codes of ethics is not as high in Ukraine as in the Czech Republic because the country has a less developed economy. It is also assumed that the presence of codes of ethics is an attribute of more developed economies, where a more cultured business environment exists.

2 Methodology of the statistical assessment

The following hypothesis was formulated and tested:

"The presence of a code of ethics is much more widespread in the Czech Republic than in Ukraine, irrespective of company size (micro-, small, medium sized and large companies), and the presence of a code of ethics increases in relation to company size".

Quantitative research was conducted on the basis of a larger questionnaire survey which was conducted between June 2016 and February 2017. The aim of the survey was to obtain a statistically representative cross-sectional data set on small, medium-sized and large companies in the Czech Republic and Ukraine. The data is stored in a pvz.xlsx file and is available at the author. The data was collected on the basis of an electronic questionnaire.The results of the questionnaire survey provided sample sets of 607 companies from the Czech Republic and 145 companies from Ukraine for the analysis. The companies in both sample sets were divided into four groups according to size: large companies (250 or more employees), medium-sized companies (50-249 employees), small companies (10-49 employees) and micro-companies (less than 10 employees). As the task was only to compare the presence of codes of ethics in companies from both countries, we combined the negative answers to the question "Does your company have a code of ethics?" into one answer "No" (the offered options were: "No, we do not miss having it" and "No, we miss having it"). Some companies did not answer this question at all. In such cases, we assigned "No" to the missing answer because it could be realistically assumed that a code of ethics was not in place in the company.

For the sample sets prepared in this way, we prepared tables of absolute and relative frequencies for the answers, as well as column graphs of relative frequencies according to both company size and country. As the absolute numbers of companies from both countries differed substantially, relative frequencies were compared i.e. the proportions of companies that have codes of ethics in place to all the companies in the sample set.

The comparisons between Czech and Ukrainian companies were conducted separately for the individual sample sets by means of a two-selection test for the differences in proportions. The first aim was to determine whether the presence of a code of ethics is more frequent in Czech companies than in Ukrainian ones. This was done by testing the hypothesis to see whether the proportion of Czech companies with a code of ethics in place is statistically significantly higher than the proportion of Ukrainian companies. From the point of view of the statistical test itself, it involved the analysis of a zero hypothesis. In other words, whether the difference between the companies with a code of ethics in the Czech Republic and in Ukraine is statistically insignificant, against the alternative that it is statistically significantly higher than 0. The tested difference was set according to the main hypothesis as the "proportion of Czech companies minus the proportion of Ukrainian companies". A one-sided interval estimation calculation was also performed on the difference in proportions to gain a concrete idea of the size of the difference. The second aim was to determine whether the proportion of companies that have a code of ethics in place grows in relation to company size. For this purpose, a two-sided interval estimation calculation was performed on the proportion of companies with a code of ethics in place according to both company size and country. The intervals are presented graphically (see Figures 1 and 2). Conclusions were subsequently drawn on the basis of the results.

All the statistical tests and calculations for the interval estimations were carried out with a confidence level of 95% i.e. at a significance level of 0.05. The outputs are presented in the form of the values of the testing statistics, p-values, lower limits of the one-sided interval estimations and both the upper and lower limits of the two-sided interval estimations. The two-sided interval estimations are presented graphically for visual comparison purposes. The frequency tables and the column graphs were prepared by means of MS Excel, the statistical tests and calculations of the interval estimations using **R** statistical software, and the graphical representation of the two-sided interval estimations using Geogebra software.

3 Results of the statistical processing

a) Absolute numbers and relative frequencies

Table 1. Presence of a code of ethics in Czech companies (absolute numbers)

Company size	Presence of a code of ethics		
	Yes	No	Total
Micro-companies	17	124	141
Small companies	42	137	179
Medium-sized companies	59	105	164
Large companies	82	40	122

Table 2. Presence of a code of ethics in Czech companies (relative frequencies)

Company size	Presence of a code of ethics	
	Yes	No
Micro-companies	12.1%	87.9%
Small companies	23.5%	76.5%
Medium-sized companies	36.0%	64.0%
Large companies	67.2%	32.8%

Table 3. Presence of a code of ethics in Ukrainian companies (absolute numbers)

Company size	Presence of a code of ethics		
	Yes	No	Total
Micro-companies	2	26	28
Small companies	2	29	31
Medium-sized companies	5	50	55
Large companies	8	23	31

Table 4. Presence of a code of ethics in Ukrainian companies (relative frequencies)

Company size	Presence of a code of ethics	
	Yes	No
Micro companies	7.1%	92.9%
Small companies	6.5%	93.5%
Medium-sized companies	9.1%	90.9%
Large companies	25.8%	74.2%

b) Presence of a code of ethics – comparison CZ and UA

Table 5. Micro-companies (relative frequencies)

	Presence of a code of ethics	
	Yes	No
Micro-companies CZ	12.1%	87.9%
Micro-companies UA	7.1%	92.9%

Table 6. Results of the statistical test and the lower limit of the interval estimation

Testing statistic Z	0.880
P - value	0.190
Lower limit of the interval estimation	- 0.043

As the p-value is higher than the significance level of 0.05, the difference in the proportions of micro-companies with a code of ethics in the Czech Republic and Ukraine is statistically insignificant. The lower limit of the interval estimation is negative, so the zero value lies inside this estimation. In spite of the fact that the deviation from zero is only 4.3%, the test results do not allow us to claim with sufficient confidence that the difference

DOI: 10.1051/shsconf/20173901004

in the proportions with regards to the presence of a code of ethics between both countries is statistically significant.

Table 7. Small companies (relative frequencies)

	Presence of a code of ethics	
	Yes	No
Small companies CZ	23.5%	76.5%
Small companies UA	6.5%	93.5%

Table 8. Results of the statistical test and the lower limit of the interval estimation

Testing statistic Z	3.132
P - value	0.00087
Lower limit of the interval estimation	0.081

In the case of small companies, the test shows a significant difference. The p-value is significantly less than the significance level of 0.05 and the lower limit of the interval estimation. This shows the extent to which the proportion of companies with a code of ethics in the Czech Republic exceeds that in Ukraine i.e. over 8% higher. As this is the lower limit, the real difference can potentially be greater still. The questionnaire survey results show 0.17 (i.e. 17 %).

Table 9. Medium-sized companies (relative frequencies)

	Presence of a code of ethics	
	Yes	No
Medium-sized companies CZ	36.0%	64.0%
Medium-sized companies UA	9.1%	90.9%

Table 10. Results of the statistical test and the lower limit of the interval estimation

Testing statistic Z	4.986
P - value	$3.078 \cdot 10^{-07}$
Lower limit of the interval estimation	0.18

In the group of medium-sized companies, the test results are even more convincing. The p-value is negligibly small; several orders below the significance level. The lower limit of the interval estimation for the difference in the proportions is 18%, while the difference in the frequencies according to the questionnaire is 26.9%. This means that the proportion of Czech companies that have a code of ethics in place is statistically significantly greater than that for Ukrainian companies.

Table 11. Large companies (relative frequencies)

	Presence of a code of ethics	
	Yes	No
Large companies CZ	67.2%	32.8%
Large companies UA	25.8%	74.2%

Table 12. Results of the statistical test and the lower limit of the interval estimation

Testing statistic Z	4.634
P - value	$1.79 \cdot 10^{-06}$
Lower limit of the interval estimation	0.267

The situation in the group of large companies is analogous to the previous case. The p-value is negligibly small; much less than the significance level. The test shows that the difference in the proportion of companies with a code of ethics is statistically significant, which is also backed up by the lower level of the interval estimation for the difference, which is 26.7%. The survey reveals that the proportion of Czech companies with a code of ethics in place is greater than that in Ukraine by no less than 41.4%.

c) Change in proportion of companies with a code of ethics in relation to company size.

Table 13. Proportions obtained from the results of the questionnaire survey

	CZ	UA
Micro-companies	12.1%	7.1%
Small companies	23.5%	6.5%
Medium-sized companies	36.0%	9.1%
Large companies	67.2%	25.8%

Table 14. Two-sided interval estimations for the proportions of companies with a code of ethics

	CZ	UA
Micro-companies	(0.077; 0.185)	(0.020; 0.226)
Small companies	(0.179; 0.302)	(0.018; 0.207)
Medium-sized companies	(0.290; 0.436)	(0.039; 0.196)
Large companies	(0.585; 0.749)	(0.137; 0.432)

 DOI: 10.1051/shsconf/20173901004

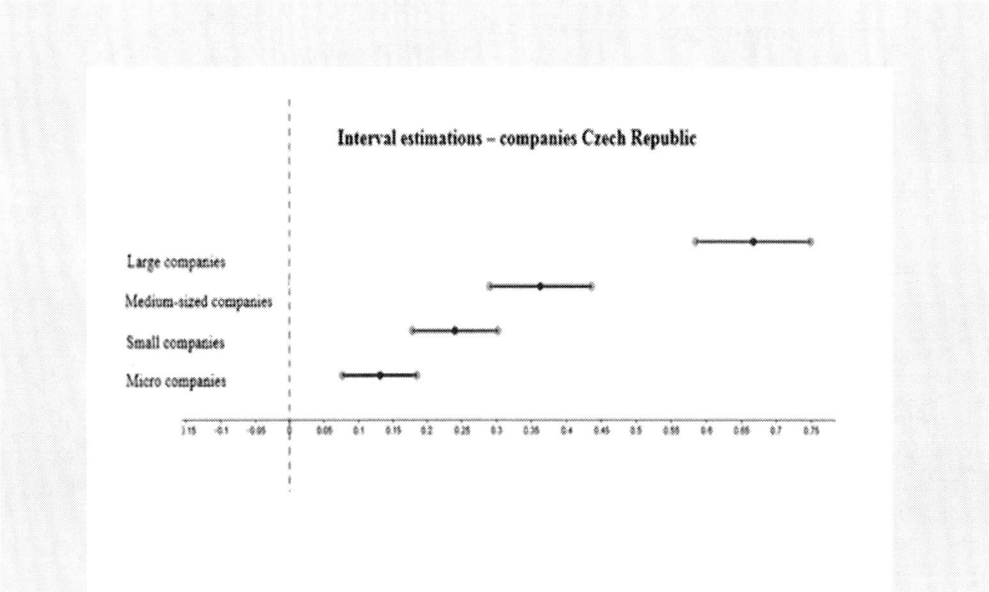

Fig. 1. Interval estimations – companies Czech Republic

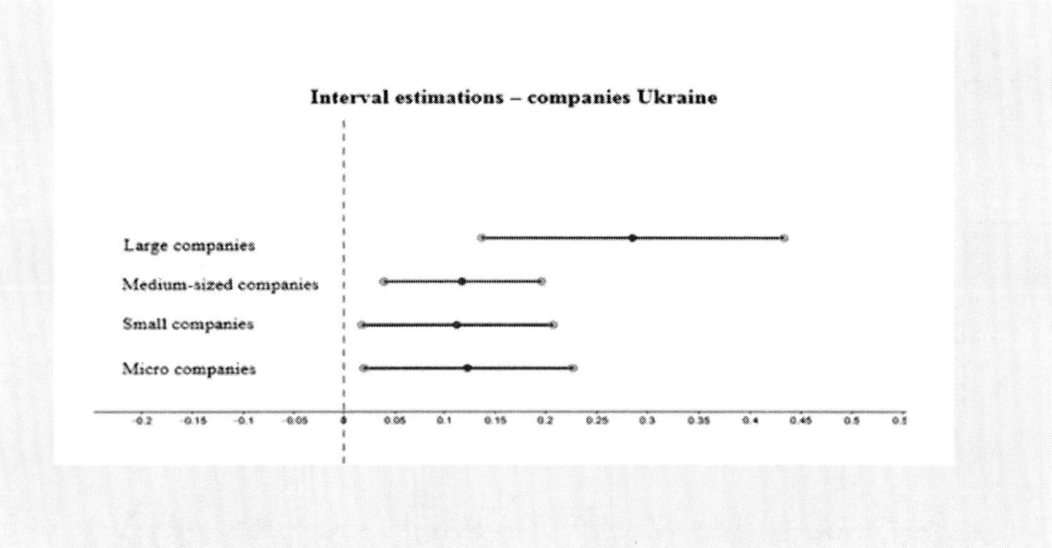

Fig. 2. Interval estimations – companies Ukraine

The results show that the proportion of Czech companies that have a code of ethics in place grows in relation to company size. The situation is clearly represented in Figure 1. The interval estimation is more to the right as the company size grows, with little overlap in the interval estimations. Furthermore, there is a significant distance between large companies and the other groups. The grey dots placed on the interval estimations lines represent the proportion of companies with codes of ethics based on the calculations made whilst processing the results of the questionnaire survey. In contrast, the growing trend is not visible among Ukrainian companies. The interval estimations for micro-companies, small companies and medium-sized companies almost completely overlap, with large companies not extending significantly beyond them (see Figure 2). This means that the presence of a code of ethics bears little to no relation to company size in Ukraine. Although a higher proportion is evident for large companies in Ukraine, this is not significant when compared to the Czech Republic.

4 Conclusion

The aim of the statistical analysis was to determine whether the proportion of companies that have a code of ethics in place is higher among Czech companies than Ukrainian ones, and whether the proportion grows in relation to company size. Overall, the results show that the presence of a code of ethics is higher among Czech companies, with the exception of micro-companies, where no difference was proven. Among Czech companies, the number of those that have a code of ethics in place grows in relation to company size, with the most significant jump identified among the group of large companies.

In the case of Ukrainian companies, the proportion of those that have a code of ethics in place is so small that it practically does not grow in relation to company size. The only group where the proportion of companies with a code of ethics in place is higher is among the group of large companies. However, when compared to the situation among Czech companies, the difference is incomparably less significant.

5 Discussion

Our research has confirmed the assumption that the presence of a code of ethics is lower in Ukraine than in the Czech Republic, which in turn confirms that codes of ethics are attributable to more developed economies, where a more cultured business environment exists. Generally speaking, this also confirms that the presence of a code of ethics in both countries is the greatest among large and medium-sized companies.

It is the opinion of the author that further research is required into the relationship between the level of development of a country's economy and the presence of a code of ethics within companies, and specifically in relation to company size.

References

1. C. M. F. Ho, O.T. Oladinrin, *Evaluation of ethical codes implementation - a fuzzy approach. Facilities* [online], Available from: http://www.emeraldinsight.com/doi/abs/10.1108/F-04-2015-0015 (2016)
2. M. Kaptein. The Effectiveness of Ethics Programs: The Role of Scope, Composition, and Sequence. *Journal of Business Ethics* [online], Available from: http://link.springer.com/article/10.1007/s10551-014-2296-3 (2015)
3. M. Štička et al., *Korupce a protikorupční politika ve veřejné správě.* Praha: Transparency International - Česká republika, (2008)

4. S. P. Rogers, J. M. Swales, We the People? An Analysis of the Dana Corporation Policies Document. *The Journal of Business Communication*, **27**(3), 293-313 (1990)
5. L. B. Harman, *Ethical challenges in the management of health information*. New York: Aspen Publishers, (2001)
6. KPMG, *Ethics Survey – 2001: Ethics in Practice* [online], Available at: https://www.gov.za/sites/default/files/ethics_0.pdf (2002)
7. Deloitte, *Business Ethics and Compliance in the Sarbanes-Oxley Era* [online], Available at: https://dart.deloitte.com/resource/1/ed57017b-3f7e-11e6-95db-4df6a2b33f51 (2004)
8. Ethics resource center, *National business ethics survey: How employees view ethics in their organizations (1994-2005)*. Washington DC, USA (2005)
9. LRN, *The impact of codes of conduct on corporate culture: Measuring the immeasurable* [online], Available at: http://www.acc.com/_cs_upload/vl/membersonly/Article/1327042_1.pdf (2006)
10. KPMG, *Business Codes of the Global 200: Their Prevalence, Content and Embedding* [online], Available at: http://www.ethicsmanagement.info/content/Business%20codes%20Fortune%20200.pdf (2008)
11. IBE, *Employee Views of Ethics at Work. Institute of Business Ethics*. British Survey. [online], Available at: http://www.ibe.org.uk/userfiles/euethicsatwork2012.pdf (2012)
12. CGMA, *Managing responsible business 2015 edition: The ethical challenges organizations must navigate to succeed in a connected world*. Durham: Chartered Global Management Accountants (2015)
13. Pricewaterhousecoopers, *Worldwide Economic Crime Survey - Report for the Czech Republic [Celosvětový průzkum hospodářské kriminality - zpráva za Českou republiku]* [online], Available at: https://www.pwc.com/cz/cs/hospodarska-kriminalita/assets/pdf/global-economic-crime-survey-2016-cz.pdf (2016)
14. Dow Jones & Metricstream, *Global anti-corruption survey results 2016* [online], Available at: https://dow-jones-2097.docs.contently.com/v/2016-dow-jones-and-metricstream-global-anti-corruption-survey (2016)
15. Transparency International – Czech Republic, *Every fifth financial manager in Czech has solved an ethical dilemma in his job [Každý pátý finanční manažer v česku řešil v zaměstnání etické dilema]* [online], Available at: https://www.transparency.cz/kazdy-paty-financni-manazer-v-cesku-resil-v-zamestnani-eticke-dilema/ (2015)
16. Z. Džbánková, Etické aspekty ekonomického prostředí v ČR v procesu připojování k EU. *Marathon*, **49**(6), 3-19 (2013)
17. Transparency International – Czech Republic, *Ethics code application survey [Průzkum aplikace etických kodexů]* [online], Available at: http://transint.xred.cz/doc/ve_pruzkum_aplikace_etickych_kodexu.pdf (2006)
18. Z. Dytrt, *Managerial Ethics in Questions and Answers: Is morale and success in business together? [Manažerská etika v otázkách a odpovědích: jde morálka a úspěch v byznysu dohromady?]*, Brno: Computer Press (2011)
19. Neovlivní.cz. The state will protect the company from prosecution, but they are lagging behind in the struggle with competitors [Stát ochrání firmy před stíháním, ty přitom v boji s konkurencí zaostávají]. In: *Neovlivní.cz* [online], Available at:

http://neovlivni.cz/stat-ochrani-firmy-pred-stihanim-ty-pritom-v-boji-s-korupci-zaostavaji/ (2015)

20. Finanční a Ekonomický Institut, Only in less than a quarter of companies follow an ethical code [Jen v necelé čtvrtině firem se řídí etickým kodexem]. In: *Finanční a ekonomický institut* [online], Available at: https://faei.cz/jen-v-necele-ctvrtine-firem-se-ridi-etickym-kodexem/ (2015)
21. Z. Caha, J. Urban, *A Code of Ethics as an Organizational Management Tool and its Use in the Czech Republic*. Lüdenscheid, Germany: RAM-Verlag, (2017)
22. Z. Caha, M. Vokoun, The impact of the presence of a code of ethics on the economic prosperity of a company. In: *Dvouletý,O., M. Lukeš a j. Mísař. Innovation management, entrepreneurship and sustainability* 2017. Praha: Oeconomica (2017)

CZ GDP Prediction via neural networks and Box-Jenkins Method

Lenka Dvořáková[1,*]

[1]The Institute of Technology and Business in České Budějovice, School of Expertness and Valuation, Okružní 10, 370 01 České Budějovice, Czech Republic

Abstract. Economic indicators are nowadays ones of the most observed, their development does not only serve for comparing individual countries among each other but also show how the given country is prospering. That is why economists are trying to predict also the future development of these indicators via different statistical instruments. Neural networks or Box-Jenkins Method, able to predict future development based on data from the past, are one of the many instruments. The aim of this contribution is to find CZ GDP prediction per individual quarters using neural networks and Box-Jenkins Method, compare them mutually, and evaluate which of them is better.
Key words: GDP, prediction, neural networks, Box-Jenkins Method

1 Introduction

The development of economy is a constantly observed factor, its past proves how the specific country was prospering in the past, and some institutions are trying to predict individual economic indicators, such as Ministry of Finance in the CZ, Czech National Bank, both are trying to predict the future annual indicators such as GDP, interest rate, inflation and unemployment. The Czech Statistical Office (CSO), Czech Banking Association (CBA) or Commercial Bank's Prognosis with the next year's view deal with prognoses. The GDP indicator is one of the most observed economic indicators, and it rightly deserves this amount of attention because its amount influences the economic level of a given country.

The effort to predict a possible development of the GDP indicator using mathematic models is obvious in articles by such economists as Kriz [1]; Markovic, Petkovic, Nikolic and Milovancevic [2]; Abdelaal and Mohamed [3].

Another possibility are predictions of future GDP via Holt-Winters static tool smoothing based on the division of individual provinces, and not from the perspective of a national scale, as stated by Zhao, Liu and Cao [4]. Ways to predict economic indicators are also demonstrated by Jerabkova's article [5], who has, using the Box- Jenkins Method, predicted the unemployment development in the Czech Republic. Klimek [6] focuses on the seasonality of economic and financial indicators with the option of using Box-Jenkins Method as a useful tool for predicting this data. Vochozka [7], which his research was to compare the prediction of the GDP development in common prices within the US until

* Corresponding author: dvorakova@mail.vstecb.cz

2024 performed through neural networks and standard time series, when statistical regression has proven significantly better results of the future USA GDP development in common prices than neural networks have. Sokolov-Mladenovic, Milovancevic, Mladenovic and Alizamir [8] focus in their study was to predict the economic growth with help artificial neuron network with extreme learning machine of based on trade in services, exports of goods and services, imports of goods and services, trade and merchandise trade on the economic growth. Other authors Stevanovic, Stevanovic, Vujicic and Gajic [9] estimate gross domestic product (GDP) based on electricity estimation by artificial neural network (ANN) when their results it can be concluded that the ANN model with extreme learning method could produce the acceptable prediction of the GDP based on the electricity utilization.

With regard to the option of using both neural networks and Box-Jenkins Method for prediction, certain authors are interested in comparing these models and their prediction accuracy of the future values. Ho, Xie a Goh [10] compare these methods from both short-term and long-term views in the area of predicting repairable system failure. Yip, Fan a Chiang [11] also compare both models to predict maintenance of construction equipment costs. Gabor and Dorgo [12] have decided to compare statistical models for the prognosis of the return of an enterprise dealing with production and export in the wood industry. Or a prediction according to Lam and Oshodi [13], who state that fluctuations in the building industry influence the building industry and economy negatively, thus it is necessary to develop reliable and exact prediction models to carry out efficient strategies to mitigate the impact of these fluctuations.

Neural networks and Box-Jenkins Method are statistical programmes via which the future development of economic values may be predicted. The aim of this contribution is to compare neural networks to Box-Jenkins Method and, through the results state which of the methods is more suitable for predicting quarterly CZ GDP data for the future.

2 Materials and methods

The basic data set of the Gross Domestic Product development (GDP) for the quarterly period from 1995 until the first quarter of 2017 comes from the webpages of the Czech Statistical Office. To prepare the data set, MS Excel will be used. To process the data, the DELL Statistica software, version 12 will be used to predict neural networks as well as Box-Jenkins Method.

2.1 Neural Networks

First, linear regression will be carried out. Subsequently, automated neural networks will be used for regression purposes. Then, regression will be carried out using neural structures. We will generate multi-layer perceptron networks, and basic radial function neural networks. Time will be the independent variable. GDP per individual quarters will be determined as the dependent variable. Time series will be divided into training sets, testing, and validational sets, in the following ratio: 70%, 15%, and 15%. A delay in time series will be equal to 1 and 1000 neural networks will be generated, out of which 5, proving the best results, will be preserved. In case of a multi-layer perceptron network (MLP) at least two neurons will be set, 8 hidden neurons maximum. In case of a radial basic function (RBF) the hidden layer will consist of at least 14 neurons, 19 neurons maximum. For a multilayer perceptron network's purposes we will recognize these distributional functions in both the hidden layer and in the output layer:

- Linear;
- Logistical;
- Atanh;

DOI: 10.1051/shsconf/20173901005

- Exponential;
- Sinus.

Other settings will be left as default (according to the AN – Automated Networking tool).

2.2 Box-Jenkins method

At first, time series analysis will be carried out. Subsequently, the ARIMA model and autocorrelation function will be used. Time will be the independent variable. The GDP development per individual quarters will become the dependent variable. The ARIMA model parameters will be the following:

- p – Autoregressional will be set at 4,
- q – the Moving Average will be set at 4,
- the Seasonal Shift will be also set at 4.

The estimate method will be approximate with the option of a maximal number of 50 iterations, required accuracy of 0.00010 and the maximal amount of iterations for back-casting of 10. The amount of backups per a variable will be 20. Autocorrelation will contain standard errors of white noise, and in partial auto-correlation the p-value will be highlighted for 0.050, amount of moves equal to 15.

In the conclusion, we will compare the results of neural networks and the ARIMA model. The comparison will be carried out via the residue analysis regarding the evaluator´s perspective and experience.

3 Results and Discussion

3.1 Neural Networks

In Graphics No. 1, there is the output data of training, testing and validation performances representing assymerical results, however this division is given by the seasonality of the CZ GDP data observed, per individual quarters.

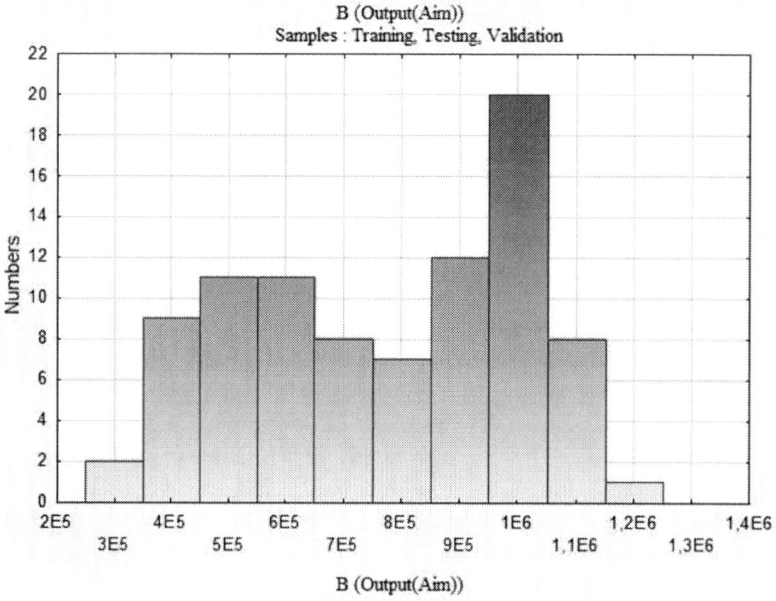

Fig. 1. GDP CZ Output Aim

Table No. 1 called 'The Overview of GDP Neural Networks' shows five of the best neural networks for prediction. The results of training, testing, and validation data have a very high-quality corresponding value of performances. Nevertheless, data performances have been very similar, that why it was necessary to add the results of residues and absolute residues up to evaluate which one of them is closest to reality and that there is a prerequisite of the most exact prediction to it. All five of the best networks may be included among the three-layer network with a forward signal spreading, and learning with a lecturer, known as the Radial Basic Function (RBF). The first RBF 1-19-1 network has got 1 input of 19 hidden layers and 1 output layer, the add-up of the residues is -681262, and the add-up of absolute residues is 4209248.1. The second RBF 1-15-1 network has 1 input of 15 hidden layers and 1 output layer, the add-up of residues is 494839, and the add-up of absolute residues is 4622729.8. The third RBF 1-17-1 network has 1 input of 17 hidden layers and 1 output layer, the add-up of residues is 448907, and the add-up of absolute residues is 4617032.8. The fourth RBF 1-15-1 network has 1 input of 15 hidden layers and 1 output layer, the add-up of residues is -135730, and the add-up of absolute residues is 3636381.1. The fifth RBF 1-16-1 network has 1 input of 16 hidden layers and 1 output layer, the add-up of residues is 208485, and the add-up of absolute residues is 3697098.9. The residue results thus prove that the RBF 1-15-1 network, which is the most suitable one for predicting time series, has the smallest residue reaching 0 and the lowest value of absolute residues.

Table 1. An Overview of GDP Neural Networks

An Overview of Active Networks (data HDP)											
Index	Name of Network	Train. Perf.	Test. Perf.	Valid. Perf.	Train. Error	Testing Error	Valid. Error	Training Algorithm	Error Funct.	Hidden Layer Activation.	Input Activation Funct.
1	RBF 1-19-1	0.971287	0.9391	0.98778	1.40E+09	2.70E+09	2.17E+09	RBFT	Quarter Sum	Gauss	Identity
2	RBF 1-15-1	0.957353	0.94728	0.98543	2.17E+09	2.22E+09	1.64E+09	RBFT	Quarter Sum	Gauss	Identity
3	RBF 1-17-1	0.957808	0.94642	0.98764	2.15E+09	2.07E+09	1.07E+09	RBFT	Quarter Sum	Gauss	Identity
4	RBF 1-15-1	0.982321	0.95041	0.98623	8.64E+08	2.25E+09	2.28E+09	RBFT	Quarter Sum	Gauss	Identity
5	RBF 1-16-1	0.975471	0.96121	0.98593	1.20E+09	1.38E+09	1.40E+09	RBFT	Quarter Sum	Gauss	Identity

Time series prediction for CZ GDP is visible in Graphics No. 2, where a curve reaching the blue-marked letter B (CZ GDP development) is being looked for. Five of the best neural networks for future prediction have been identified from the previous Table No. 1, and from the residue and absolute residue calculations we already know that the best option is offered by the fourth RBF 1-15-1 network, displayed in black colour in the Graphics.

 DOI: 10.1051/shsconf/20173901005

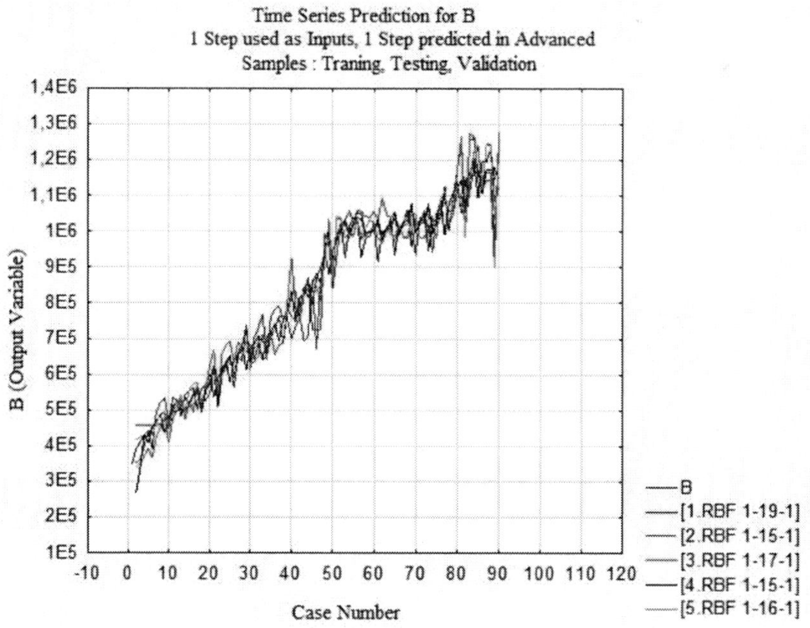

Fig. 2. Time Series Prediction

3.2 Box-Jenkins Method

Graphics No. 2 called ′ CZ GDP Histogram′ clearly proves that resulting values are not copying Gauss ordinary curve, but the data appears as assymetrical. It may be stated that this display is given by the seasonality of data analysed.

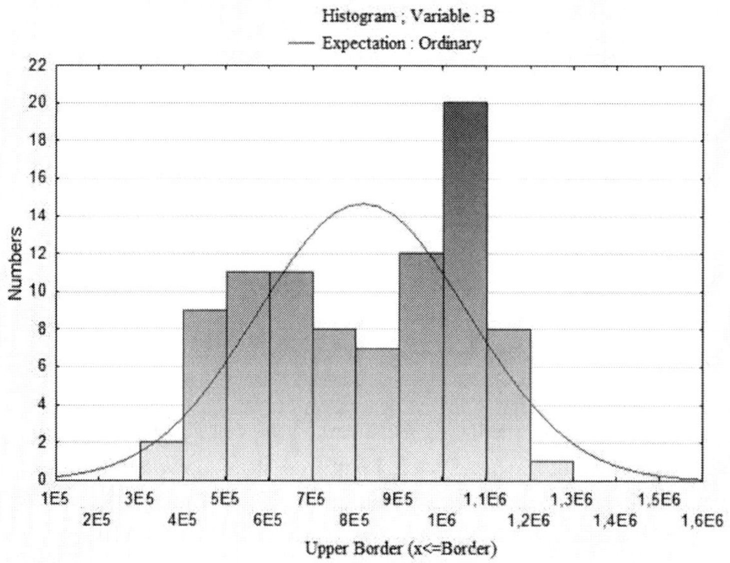

Fig. 3. GDP CZ Histogram

Table No. 2 called 'Descriptive Statistics of CZ GDP ' shows how many cases have been processed to evaluate the future CZ GDP prediction using the Box-Jenkins Method. Moreover, the average value, equal to CZK Million 815, 777.1, the standard deviation of CZK Million 241, 840.5, the minimum of CZK Million 349 683 and the maximum of CZK Million 1,229, 963 have been calculated.

Table 2. Descriptive GDP CZ Statistics

Variable	Descriptive Statistics (data_HDP)						
	Average	Stand. Dev.	Minimum	Maximum	First Case	Last Case	N
B	815777.1	241840.5	349683.0	1229963	1.000000	89.00000	89.00000

Via values presented in Graphics No. 4 we can see that the Series proves an obvious exponential trend as the series variability is not constant. Ordinary probability is relatively distanced from the expected values (the red growing curve), mostly at the beginning and at the end of the CZ GDP observed period. It may be assumed that the resulting prediction of future CZ GDP will not be responding to the real future assumption.

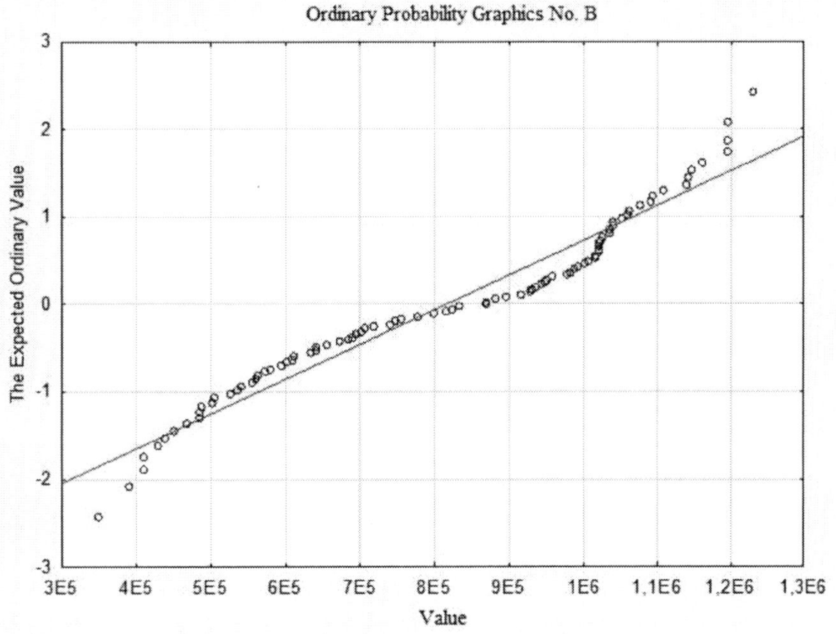

Fig. 4. Ordinary Probability

The values presented in Graphics No. 5 'Ordinary Probability without Trend' are really very distanced from the expected value (red decreasing curve), the deviation from the expected value tells us that prediction values of the CZ GDP period observed will not be responding to the real future development.

DOI: 10.1051/shsconf/20173901005

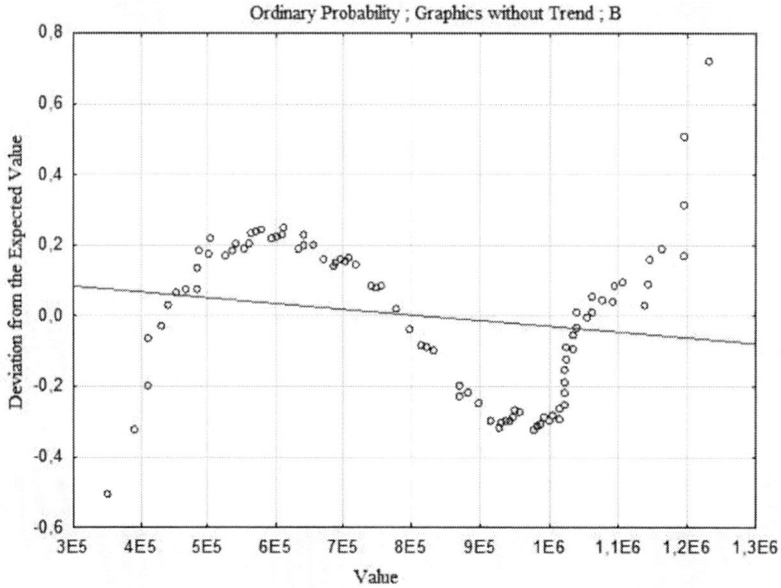

Fig. 5. Ordinary Probability without Trend

In the p1 - p4 a q1 - q4 parameter table they are highlighted because they point at the moving average (p=0) and at the autoregression model (q=0), other parameters are so-called ′positive′. The given data proves that the p2 and p3 parameters are characterised as moving averages, while q1 and q3 are classified as autoregression models. Due to comparing both methods, even here the ARIMA model residues have been added up, having the quality of 21820482,7. The residue result reaches high values, due to which it is impossible to predict the future CZ GDP values exactly.

Table 3. Parameter Results

Paramet.	Input: B (data_HDP) Transformation: None Model: (4,0,4) PČ Resid. = 8081E7					
	Param.	Asympt. SmCh	Asympt. t(81)	p	Lower 95% spol	Upper 95% spol
p(1)	0.296555	0.303906	9.758109E-01	0.332064	-0.30812	0,901233
p(2)	0.967356	0.146480	6.604034E+00	0.000000	0.67591	1.258804
p(3)	0.466198	0.144058	3.236191E+00	0.001755	0.17957	0.752827
p(4)	-0.876520	0.526549	-1.664649E+00	0.099847	-1.92419	0.171148
q(1)	-0.919758	0.196606	-4.678182E+00	0.000011	-1.31094	-0.528574
q(2)	0.225150	0.154531	1.456994E+00	0.148984	-0.08232	0.532618
q(3)	1.014520	0.000000	2.471890E+13	0.000000	1.01452	1.014520
q(4)	0.247900	0.409295	6.056770E-01	0.546423	-0.56647	1.062268

 DOI: 10.1051/shsconf/20173901005

4 Conclusion

The aim of this contribution has been to compare the methods and evaluating time series, and based on the results of a future prediction to compare whether neural networks or the Box-Jenkins Method are more suitable to be used to predict the development of CZ GDP per individual quarters.

Only RBF networks have reached the 5 of the best neural networks, these networks were really similar, and that was why it was necessary to add up the individual residues to identify the best one. The fourth RBF 1-15-1 network was, based on the residue add-up evaluated, from the perspective of performance, as the best one, it had the lowest residue value, equal to -135730 reaching 0, and the lowest absolute residue value equal to 3636381.1.

The results of the Box-Jenkins Method prove that only a half of parameters responds to moving averages or autoregression models. Specifically, they are the p2, p3, q1 a q3 parameters. The p1, p4, q2 a q4 parameters are so-called positive parameters. To compare them to neural networks it was necessary to add up the residues the value of which is 21820482.7. The resulting comparison of residue values unambiguously proves that the Box-Jenkins Method proves much higher values than neural networks. Neural network performance thus may be understood as much more exact values than using Box-Jenkins Method for future prediction of individual CZ GDP quarters.

References

1. R. Kriz, Nonlinear prediction of the GDP growth rate in the globalized world. *Proceedings of the 16th International Scientific Conference of Globalization and its Cocio-Economic Consequences*, (2016)
2. D. Markovic, D. Petkovic, V. Nikolic, M. Milovancevic, Soft computing prediction of economic growth based in science and technology factor. *A: Statistical Mechanics and its Applications*, **465**, 217-220 (2017)
3. M. M. A. Abdelaal, S. F. S. Mohamed, Statistical model of Egyptian economic growth prediction. *Advances and Applications in Statistics*, **47**, 225-246 (2015)
4. N. Zhao, Y. Liu, G. Cao, E. L. Samson, J. Zhang, Forecasting China's GDP at the pixel level using nighttime lights time series and population images. *Giscience & Remote Sensing*, **54**(3), 407-425 (2017)
5. V. Jerabkova, Unemployment in the Czech Republic and its predictions based on the B ox-Jenkins methodology. Proceedings of the 12th International Scientific Conference on Applications of Mathematics and Statistics in Economy, 189-196 (2009)
6. P. Klimek, Practical use of the BoxJenkins methodology for seasonal financial data prediction. *7th International Scientific Conference on Finance and Performance of Firms in Science, Education and Practice,* 598-611 (2015)
7. M. Vochozka, Practical Comparison of Results of Statistic Regression Analysis and Neural Network Regression Analysis. *Littera Scripta*, **9**(2), 156-168 (2016)
8. S. Sokolov-Mladenovic, M. Milovancevic, I. Mladenovic, M. Alizamir, Economic growth forecasting by artificial neural network with extreme learning machine based on trade, import and export parameters. *Computers in Human Behavior*, **65**, 43-45 (2016)

9. M. Stevanovic, S. Vujicic A. M. Gajic, Gross domestic product estimation based on electricity utilization by artificial neural network. *Physica A: Statistical Mechanics and its Applications*, **489**, 28-31 (2018)
10. S. L. Ho, M. Xie a T. N. Goh, A comparative study of neural network and Box-Jenkins ARIMA modeling in time series prediction. *Computers & Industrial Engineering*, **42**(2-4), 371-375 (2002)
11. H. Yip, H. Fan, Y. Chiang, Predicting the maintenance cost of construction equipment: Comparison between general regression neural network and Box-Jenkins time series models. *Automation in Construction,* **38**, 30-38 (2014)
12. M. R. Gabor, L. A. Dorgo, Neural networks versus Box-Jenkins method for turnover forecasting: a case study on the romanian organisation. *Transformation in Business & Economics,* **16**(1), 187-210 (2017)
13. K. C. Lam, O. S. OshodI, Forecasting construction output: a comparison of artificial neural network and Box-Jenkins model. *Engineering, Construction and Architectural Management,* **23**(3), 302-322 (2016)

Is the size of an enterprise in the construction industry an important prerequisite for its success?

Bořivoj Groda[1*]

[1]Institute of Technology and Business in České Budějovice, School of Expertness and Valuation, Okružní 517/10, 370 01 České Budějovice

Abstract. The construction industry can currently be ranked among the most important sectors of the national economy, with a large number of companies operating within it. These undertakings differ from each other by subject of activity, place of activity, chosen strategy, the legal form of the enterprise or the size of the enterprise. This paper aims to determine what impact the size of a construction company has on its future success, with asset returns and return on equity being considered success indicators. A data sample of 3139 construction companies from between 2011 and 2015 was used for analysis. As a result, the predetermined hypotheses have been verified: a larger enterprise generates a higher ROE and a larger enterprise generates a larger ROA. It was found that the size of the company greatly affects its liquidity.
Key words: enterprise size, construction industry, ROA, ROE, liquidity

1 Introduction

The construction industry plays a vital role in the national economy and its long-term socio-economic development. Its key role lies primarily in the development of housing, facilities and the infrastructure necessary for national social and economic development [1]. According to Naderpajouh, Choi and Hastak [2], the construction industry has its own specificities, primarily in the demand for capital, individuality, manual difficulty and the exceptionalism of requirements.

According to Chan and Owusu [3], the construction industry is the sector of the national economy that concerns itself with land preparation and the construction, adjustments and repair of buildings and other estates. The construction industry is thus an economic field that enables many functions such as social functions (sports, health care, housing, etc.), energy, transport, agricultural and industrial production, etc. [4].

This paper focuses on company size, which is considered to be a very important and basic trait of the enterprise. In order to determine the optimum size of a company, we can indicate the speed and range of growth that is ideal for a particular business [5]. Company size depends on many external and internal factors. Most companies are seeking to develop their activity [6]. The growth of any factor, whether revenue, profits, the number of

* Corresponding author: groda@mail.vstecb.cz

employees or the size of facilities, is essential for almost all enterprises. Kim, Lin and Chen [7] note that for many companies competing in rapidly changing industries, expansion (production capacity, geographical presence, market share, etc.) may also be necessary for continuation.

At present, small and medium-sized enterprises are considered to be major drivers of industrial structure growth in many countries, and such sectors are very important for many developing countries seeking to revitalize their economic structure [8]. Vochozka and Mulač [9] report that the small and medium-sized enterprise sector plays a very important role in the economy, especially in job creation, as it is a factor of social stability and is conducive to the development of certain regions, cities and municipalities. This business area can therefore be labelled as the driving force in the sphere of business, growth, innovation and competitiveness. Barhatov and Belova [10] note that under the current economic conditions, the development of small and medium-sized enterprises is not possible without some financial support. National small and medium-sized enterprises have a particularly acute need for state aid. Special funds, corporations, or agencies are established [11].

Small and large enterprises differ from one another considerably. The advantage of large businesses is the possibility of career growth for employees. Large companies offer their devoted and capable employees a career to build. In addition, we can mention advantages such as benefits for employees, higher financial valuation, etc. The disadvantage of large companies lies in their decision-making thanks to a complex organizational structure and more difficult response to changes in the environment, especially in the market [12]. According to De Toni and Nassimbeni [13], small and medium-sized enterprises are often family businesses run by entrepreneurs who worked as factory employees in companies engaged in the same industry at the beginning of their careers. Therefore, they bring with them a considerable amount of technical experience that allows them to think about realizable innovative products and processes. On the other hand, these entrepreneurs suffer from a lack of managerial competencies. We can also mention lower wages as a disadvantage of small and medium-sized enterprises, therefore the total area of financing. It is also important to acknowledge that work intensity is higher in these businesses [14]. The advantage of small and medium-sized enterprises is their speed in response to a changing external environment, such as changes in demand or market [15]. Of course, as far as decision-making is concerned, small businesses are far more flexible and faster in the area of decision-making, because of the clearer and simpler organizational structure. Employee relationships are more confidential than in large businesses, which helps to build high-quality teamwork [16].

Rizos et al. [17] argue that small and medium sized enterprises are increasingly aware of the benefits of improving resource efficiency, such as saving material costs, creating competitive advantages and accessing new markets. At the same time, however, these businesses are aware of their problems stemming of lack of finance and managerial skills. Grando and Belvedere [18], however, argue that small and medium-sized enterprises perform worse than large enterprises, due to the lack of human and financial resources that prevent small and medium-sized enterprises from adopting new technological solutions and innovative management practices needed to improve overall performance.

Therefore, the aim of the paper is to determine whether the size of an enterprise indicates its future success. We can verify several hypotheses as to which variable will be considered an indicator of success. In this article, we look at the enterprise through the eyes of its owners. We will consider return on equity and return on assets as indicators of success.

DOI: 10.1051/shsconf/20173901006

The hypotheses are thus:

1. A larger enterprise generates a higher ROE[1].
2. A larger enterprise generates a higher ROA[2].

2 Data and methods

Data from a sample of construction enterprises from the Albertina database will be used in the analysis. The sample of businesses will consist of micro, small, medium and large enterprises. The size of the business will be determined by the number of employees in the enterprise. The total set consists of 3139 records from the period of 2011 to 2015. The representation of enterprises in the data file for individual years by size is evident in Table 1.

Table 1. Division of enterprises in sample surveyed.

	Year					Total sum
Quantity	2011	2012	2013	2014	2015	
0	51	60	64	49	52	276
3	104	96	87	102	94	483
7	26	27	19	23	24	119
15	110	114	118	119	114	575
23	23	37	34	32	36	162
38	65	47	45	47	49	253
75	133	130	139	132	123	657
150	60	63	50	56	65	294
225	6	7	11	12	12	48
375	37	35	34	39	36	181
750	13	13	10	6	7	49
1250	2	2	4	3	3	14
1750	4	4	2	2	1	13
2250	1	1	1	2	1	6
2750					1	1
3500	1	1	2	2		6
4500	1	1				2
Total sum	637	638	620	626	618	3139

The division corresponds to the actual representation of enterprises in the different size segments of the construction industry in the years 2011 to 2015. This is a selection equal to one twentieth of all enterprises operating in the construction sector in the observed period. Figure 1 summarizes the number of enterprises in the sample throughout the observed period.

[1]ROE = Return on Equity
[2]ROA = Return on Assets

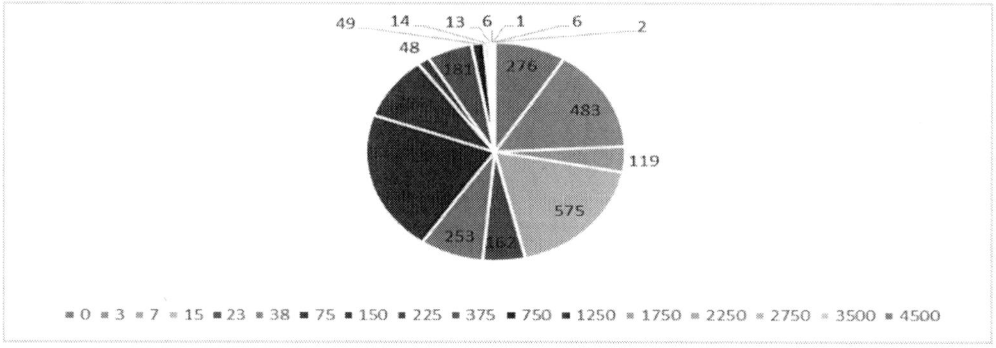

Fig. 1. Total division of enterprises by size.

Using graphical methods, we will investigate the ROE and ROA dependence on enterprise size, as predicted by the hypotheses. It can be assumed that the value of both indicators will increase with growth in enterprise size.

In addition, we will examine other indicators in which the size of the company can be reflected – indicators of liquidity or possibly of the company's indebtedness.

3 Results

Figure 2 shows the pattern of return on equity as dependent on the size of the enterprise

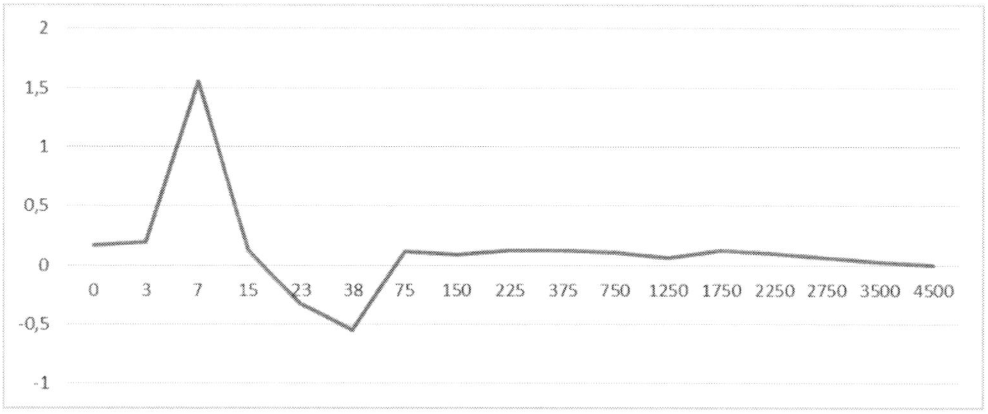

Fig. 2. Average ROE of the sample of enterprises surveyed.

The figure shows a large deviation in enterprises with seven employees, with a ROE of over 1.5. The ROE then drops to negative values. It reaches the lowest value for enterprises with 38 employees (less than – 0.5), then subsequently grows and maintains positive values for the largest enterprises. Figure 3 shows why fluctuations have occurred on both sides.

DOI: 10.1051/shsconf/20173901006

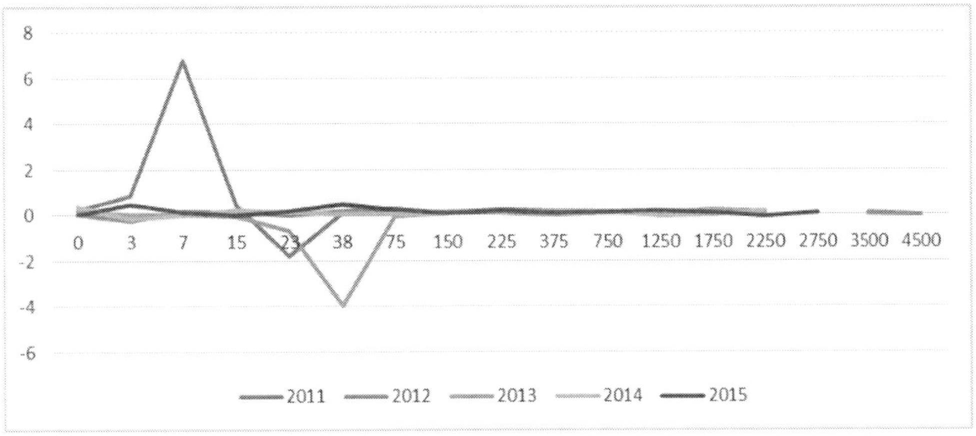

Fig. 3. The average ROE of enterprises in individual years.

The deviation in enterprises with seven employees was caused by results from the year 2011, when the ROE of these enterprises reached an average of approximately 7. In the case of enterprises with 38 employees, the fluctuation was caused by an average ROE of almost -4 in the year 2013. Under other conditions, the value of ROE stayed slightly above zero throughout the observed period. Thus, if we disregard the extreme values, we can state that hypothesis number one has not been fulfilled. The size of an enterprise does not affect the ROE value.

The circumstances are significantly more complex in the case of ROA. The entire situation for the set of enterprises surveyed is illustrated in picture 4.

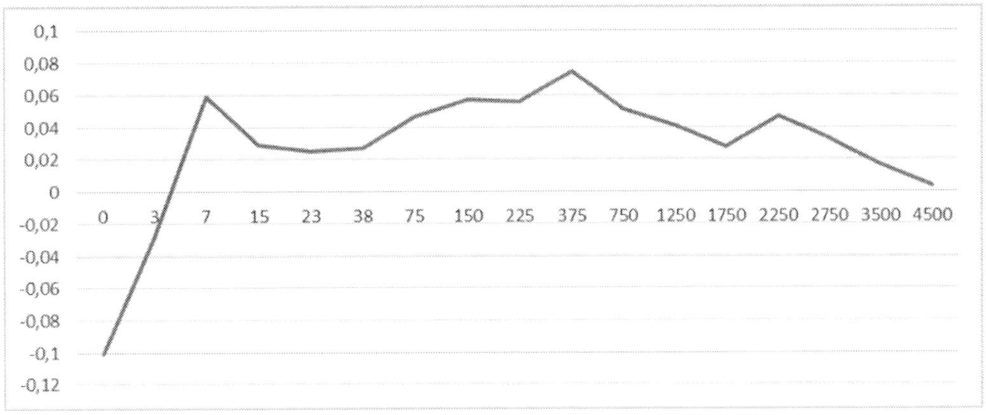

Fig. 4. Average ROA for the set of enterprises surveyed.

The lowest value of ROA is evident in companies with no employees. Specifically, it reaches 0.1, then starts rising. The first peak is reached by businesses with seven employees. Further in its course, it falls to reach its peak for businesses with 375 employees. Then it begins declining, while, however, maintaining positive ROA values. Overall, it ranges from - 0.1 to just below 0.8.

The progress over individual years is the subject of figure 5. The interval increases to values from - 0.3 to 1.3. Fluctuations, however, are basically minimal, just as with the average ROA (Figure 4). It should be noted that the development of ROA appears to be more stable and predictable in larger companies for individual years of the surveyed period.

Deviations can only be found in small and medium-sized enterprises (up to 38 employees), they keep decreasing further down the interval with results appearing quite similar.

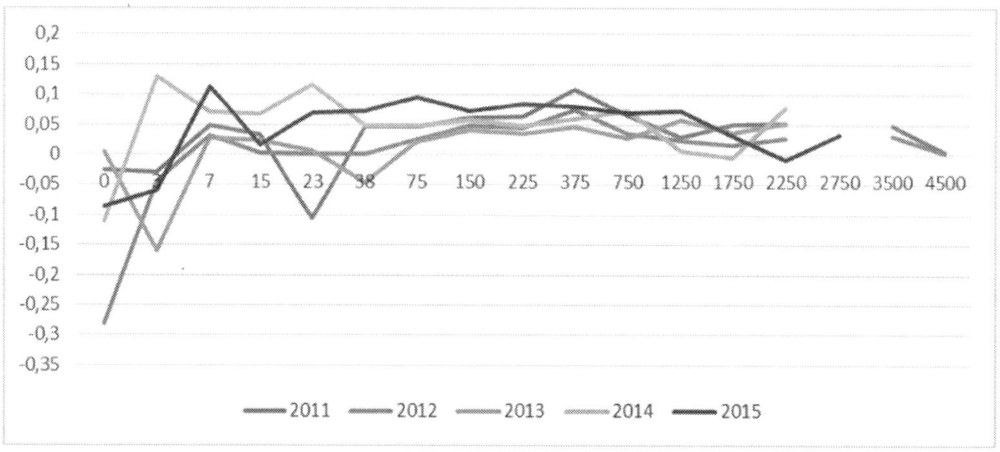

Fig. 5. Average ROA of enterprises during individual years.

Even these results are insufficient to prove that the size of an enterprise could have a positive effect on the ROA.

Of course, one can think that the relationship between enterprise size and some other indicators would be more prominent. As already indicated in the methods of the article, we will focus our attention on the company's indebtedness (i.e. the ability to utilize the resources and potential of the company). The company's indebtedness with respect to the size of the company can be seen in figure 6.

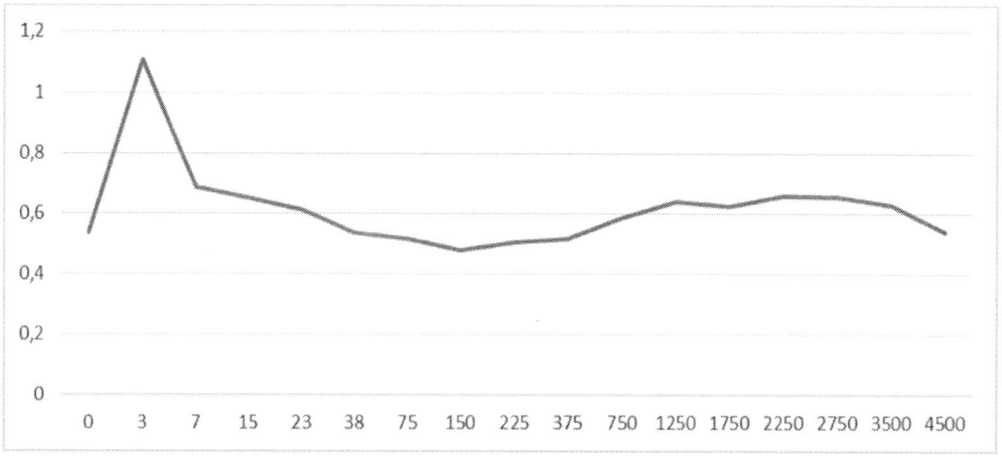

Fig. 6. Average indebtedness of enterprises in respect to their size.

The figure shows a fluctuation in enterprises employing 3 workers, where the average indebtedness is more than 1.1. This means that an average enterprise of this size is over-indebted, and if so in the long-term, it should end its activity. Otherwise, it can be stated that the average indebtedness for all sizes of construction enterprises is approximately comparable, ranging from 0.5 to 0.65. Again, we can say that the size of the enterprise is not relevant.

If we consider indebtedness in the context of the ROE value, the progress of indebtedness only confirms the progress of the ROE. It is obvious that the companies are not concerned with the effects of financial leverage, or more precisely, its positive effects cannot be demonstrated. In general, it can be assumed that large businesses would be more indebted, generating higher ROEs than other businesses. However, this cannot be confirmed.

Let us examine, however, the relationship between the size of the company and its liquidity. Figure 7 provides the cash flow of an enterprise in accordance to its size.

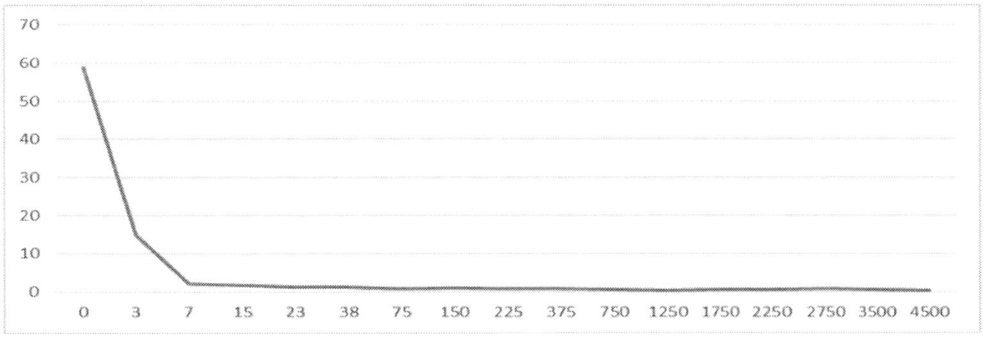

Fig. 7. Average financial liquidity of enterprises by size.

It is clear from the graph that small businesses with a total of seven employees deal with financial liquidity very inefficiently. They hold large stocks of funds. In general, we can say that they fear risk.

Figure 8 represents the flow of instant and total liquidity of a business in respect to its size.

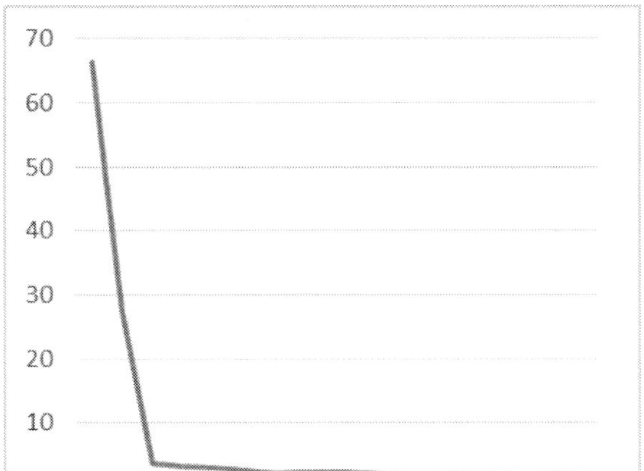

Fig. 8. Average liquidity.

The progress of instant and total liquidity is identical to the progress of financial liquidity. Small businesses are concerned about possible insolvency. Therefore, they manage money very inefficiently. By contrast, large businesses hold a minimum money stock and are therefore more efficient. However, not even this aspect of management has been reflected in the ROE and ROA results.

4 Conclusion

The aim of the article was to determine whether the size of an enterprise indicates its future success. Two hypotheses were set to support the goal of the paper:

1. A larger enterprise generates a higher ROE.
2. A larger enterprise generates a higher ROA.

The hypothesis number 1, a larger enterprise generates a higher ROE, has been disproven. We can state that the size of a company in the construction industry did not have an effect on the ROE.

The hypothesis number 2, a larger enterprise generates a higher ROA, has also been disproven. The size of a company in the construction industry did not affect the ROA.

The aim of the article has been met. In the context of result discussion, it was found that the size of the company had a major impact on the development of the company's liquidity. Small enterprises, or more specifically micro-enterprises with up to 7 employees, have an aversion to risk, keep a large stock of money and thus behave inefficiently. By contrast, large corporations manage their liquidity very carefully and therefore behave responsibly. However, not even very precise liquidity management has been reflected in higher ROE and ROA values.

References

1. Z. Rowland, J. Vrbka, Optimization of a company's property structure aiming at maximization of its profit using neural networks with the example of a set of construction companies. *Mathematical Modelling in Economics* [Математичне моделювання в економіці], **7**(3-4) (2016)
2. N. Naderpajouh, J. Choi, M. Hastak, Exploratory Framework for Application of Analytics in the Construction Industry. *Journal of Management in Engineering,* **32**(2), (2016)
3. P. C. Chan, E. K. Owusu, Corruption Forms in the Construction Industry: Literature Review. *Journal of Construction Engineering and Management*, **143**(8), (2017)
4. P. Šuleř, Optimizing the capital structure of the company to maximize its profits by using neural networks on the example of building companies. *Mathematical Modelling in Economics* [Математичне моделювання в економіці], **7**(3-4), (2016)
5. P. Wei, L. Xu, B. Zeng, Corporate hedging, firm focus and firm size: the case of REITs. *Managerial Finance,* **43**(3), (2017)
6. D. C. Joaquin,a N. Khanna, Investment timing decisions under threat of potential competition: Why firm size matters. *The Quarterly Review of Economics and Finance,* **41**(1), (2001)
7. D. Kim, S.-Ch. Lin, T.-C. Chen, Financial structure, firm size and industry growth. *International Review of Economics,* **41,** (2016)
8. H. Esmailpour, M .R.Tavallai, Innovative marketing and competitive advantage of small and medium-sized businesses according to the entrepreneurial orientation and Features. *International Journal of Advanced Biotechnology and Research*, **7**(3), 1130-1143 (2016)

9. M. Vochozka, M, P. Mulač, Financial management. In *Business economics* [Finanční řízení. In. *Podniková ekonomika*]. 1st ed. Prague: Grada (2012)
10. V. Barhatov, I. Belova, Infrastructure of state support of small and medium-sized business in Russia. *Proceedings of the 16th International Scientific Conference on Economic and Social Development – The Legal Challenges of Modern World*, 508-517 (2016)
11. M. Vochozka, M. Psárska, Factors supporting growth of added value, performance and competitiveness of SMEs and selected EU countries. *Innovation Management Entrepreneurship and Corporate Sustainability*, 756-767 (2016)
12. G. S. Dangayach, S. G. Deshmukh, Manufacturing strategy. *International Journal of Operations,* **21**(7), (2001)
13. A. De Toni, G. Nassimbeni, Small and medium district enterprises and the new product development challenge. *International Journal of Operations*, **23**(6), (2003)
14. A. Kasych, M. Vochozka, Theoretical and methodical principles of managing enterprise sustainable development. *Marketing and Management of Innovations*, (2), (2017)
15. M. Vochozka, Z. Rowland, The evaluation and prediction of the viability of construction enterprises. *Littera Scripta*, **8**(1), (2015)
16. M. Starbek, J. Grum, Concurrent engineering in small companies. *International Journal of Machine Tools and Manufacture*, **42**(3), (2002)
17. V. Rizos, A. Behrens, W. V. Gaast, et al, Implementation of Circular Economy Business Models by Small and Medium-Sized Enterprises (SMEs): Barriers and Enablers. *Sustainability,* **8**(11), 1212-1230 (2016)
18. A. Grando, A, V. Belvedere, District's manufacturing performances: A comparison among large, small to medium sized and district enterprises. *International Journal of Production Economics,* **104**(1), (2006)

Prediction of stock price developments using the Box-Jenkins method

Bořivoj Groda[1], *Jaromír* Vrbka[1*]

[1]Institute of Technology and Business, School of Expertness and Valuation, Okružní 517/10, 37001 České Budějovice, Czech Republic

Abstract: Stock prices develop in a non-linear way. Naturally, the stock price prediction is one of the most important issues at stock markets. Therefore, a variety of methods and technologies is devoted to the prediction of these prices. The present article predicts the future development of the stock price of ČEZ, a. s., on the Prague Stock Exchange using the ARIMA method - the Box-Jenkins method. The analysis employs the final price of the last trading day in a given month, from February 2012 to September 2017. The data come from the Prague Stock Exchange database. Statistica software is used for processing the data, namely advanced time series prediction methods, the ARIMA tool, and autocorrelation functions. First, the current stock development of ČEZ, a.s., was graphically evaluated, and this was followed by a stock price prediction for the next 60 days in which the shares would be traded. Lastly, the prediction residues were analysed. It was confirmed that the calculation was done correctly, but with little accuracy. The conclusion is an assertion that the Box-Jenkins method is not a suitable tool for prediction.
Key words: Box-Jenkins, ARIMA, prediction, shares

1 Introduction

Successful forecasting of future share prices development can bring a considerable profit for investors. The effective market hypothesis indicates that share prices reflect all information currently available, and that the changes in prices which are not based on newly discovered information are unpredictable [1]. However, some experts disagree with this statement, and therefore there are a number of methods and technologies associated with this concept, which enable to gain information on future development of share prices [2].

The basic characteristics of share prices are sensitivity, stationarity and asymmetric volatility. Investors naturally want to be able to forecast the share price movement, despite of the fact that it is a stochastic process. The development of share prices is a non-linear and dynamic process. There are a number of macroeconomic, industry and company factors which may have influence on share prices. These include for example global indices of share prices, overall economic activity, exchange rate, interest rate etc. [3]. Share prices

* Corresponding author: vrbka@mail.vstecb.cz

forecasting is thus very useful and attracts interest of researchers and investors, who carry out subjective investment judgements on the basis of objective technical indicators [4]. Expected share price is one of the important issues of stock market research. Accurate prediction of share prices, which is the basis for deciding on financial investment, is probably the biggest challenge for capital investment [5]. As referred to above, currently there are a number of applications and new technologies for share prices forecasting, such as neural networks, the ARIMA method etc. [6]. In this contribution, the ARIMA method (Box-Jenkins) is used for analysing and share prices forecasting.

ARIMA, i.e. a model of Autoregressive integrated moving average, is one of the most widely used models in time series forecasting. At present, the ARIMA model is often used for its unique predictive capability also for share prices development forecasting [7]. Frances [8] favours this statement claiming that the ARIMA models are the most frequently used models for predicting time series that can be stationary transformation. ARIMA can therefore enable identification of time series characteristics as well as predicting their behaviour in the future. Lags of differenced series appearing in the forecasting equation are called Autoregressive terms, while lags of the forecast errors are called moving average terms. As a general rule, a time series which needs to be differenced to be made stationary is said to be an integrated version of a stationary series [9]. Random variable which is a time series is considered stationary if its statistic characteristics are constant over time [10]. The models are used mainly for short-term forecasts, when there are no data for explanatory variables, or in case the model has poor predictive power [11].

Forecasting future stock market values on the basis of both past and current data series is thus one of the most required financial applications. The objective of the contribution is to forecast the future development of ČEZ share prices on the Prague Stock Market using Box-Jenkins method.

2 Data and methods

ČEZ, a.s. is one of the most important business entities in the Czech Republic. Its operations, structure, vision and business activities are characterized as follows [12]: "ČEZ group is an integrated energy cluster operating in a number of Central and South-Eastern countries and Turkey, with headquarters in the Czech Republic. Its core business consists of production, distribution, trading and sale in the field of electricity and heat, trading and sale of natural gas and coal mining. ČEZ group currently employs nearly 27,000 employees.

The most significant shareholder of the parent company ČEZ, a.s., is the Czech Republic with a holding in the capital of the company (as of June 14, 2017) of nearly 70%. The shares of ČEZ, a. s., are traded on Prague and Warsaw Share Exchange, where they are a part of the PX and WIG-CEE share exchange indices.

ČEZ mission is to provide safe, reliable and positive energy to customers as well as the whole company. Its aim is to bring innovations to meet energy needs and thus contribute to a better quality of life. Its strategy reflects the fundamental transformation of European energy market. ČEZ wants to operate its energy assets in the most efficient way and to adapt to the growing share of decentralized and non-emission production. Another priority is to offer a wide range of products and services for the customers, along with selling electricity and gas. The third priority is to invest actively in prospective energy assets with a focus on the Central European region and in supporting modern technologies at an early stage of their development.

In the Czech Republic, the ČEZ group are active in coal mining, generation and distribution of electricity and heat, trading in electricity and other commodities, sell electricity, heat and natural gas to end-customers and provide other services. The

 DOI: 10.1051/shsconf/20173901007

production portfolio consists of nuclear, coal, gas, water, photovoltaic, wind and biogas sources."

Data on share prices between February 2012 and September 2017 are available. Specifically, the data on the final price at the end of the last trading day in a month will be used. In total, 69 records will be available for analysis. The data come from the Prague Stock Exchange database.

For the data processing, DELL Statistica V12 will be used. Advanced methods of Time series/ ion will be applied. Subsequently, the tools of ARIMA and autocorrelation function will be selected.

As a target variable, share prices will be selected. Next, we will specify 20 variable backups. We will assume seasonality of data at the level of each month of the year. The parameter of seasonal variation will be set to 12. Parameter p (Autoregressive constant) as well as parameter q (moving average) are set at level 1. The number of movements will be 15, the p-value for sharpening is 0.05.

First, we will review graphically the current development of ČEZ share prices. Next we will focus on forecasting the value of ČEZ shares. The development of prices will be predicted for the following 60 days on which the shares will be traded. Finally, residual stock will be analysed.

3 Results

ČEZ share prices development in the monitored period can be seen in Figure 1.

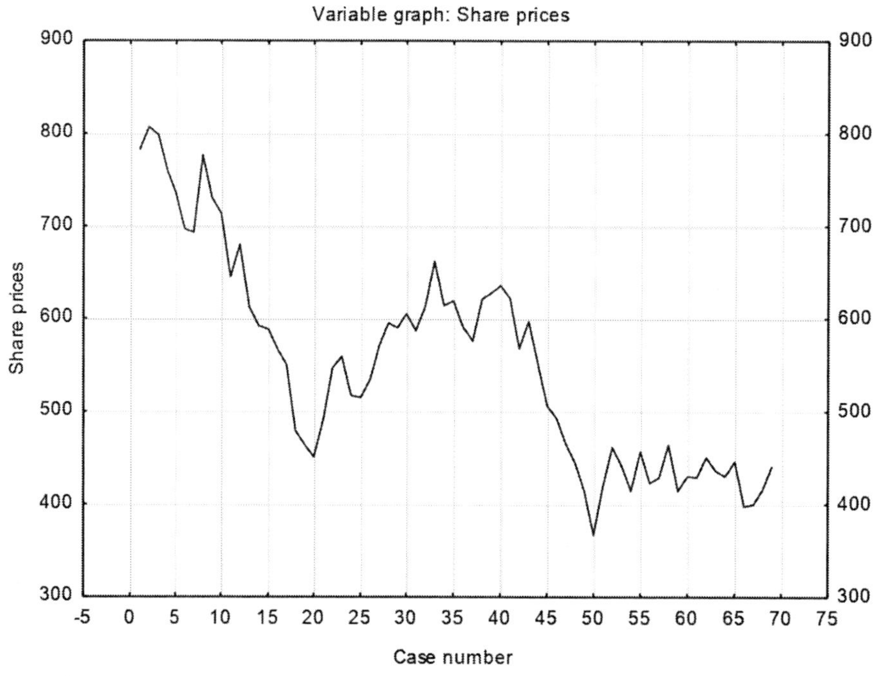

Figure 1 ČEZ share prices between February 2012 and September 2017

ČEZ share prices reached the peak at the beginning of the monitored period. At that time, the price was above 800 CZK per share. Subsequently, the price fell below 700 CZK per share. Then the share price grew to about 780 CZK and immediately decreased to 450 CZK per share. Approximately in the middle of the monitored period, share price increased

DOI: 10.1051/shsconf/20173901007

to more than 650 CZK. In February 2016, the price was between 400 (or more precisely, 399.9) and 450 (or more precisely, 455.5) CZK per share.

The figure shows there is no regular movement of share prices. A question therefore arises, whether the Box-Jenkins method, which is based on seasonal cycles of time series, is the right tool for forecasting ČEZ share prices.

This could be better seen from the graph in Figure 2.

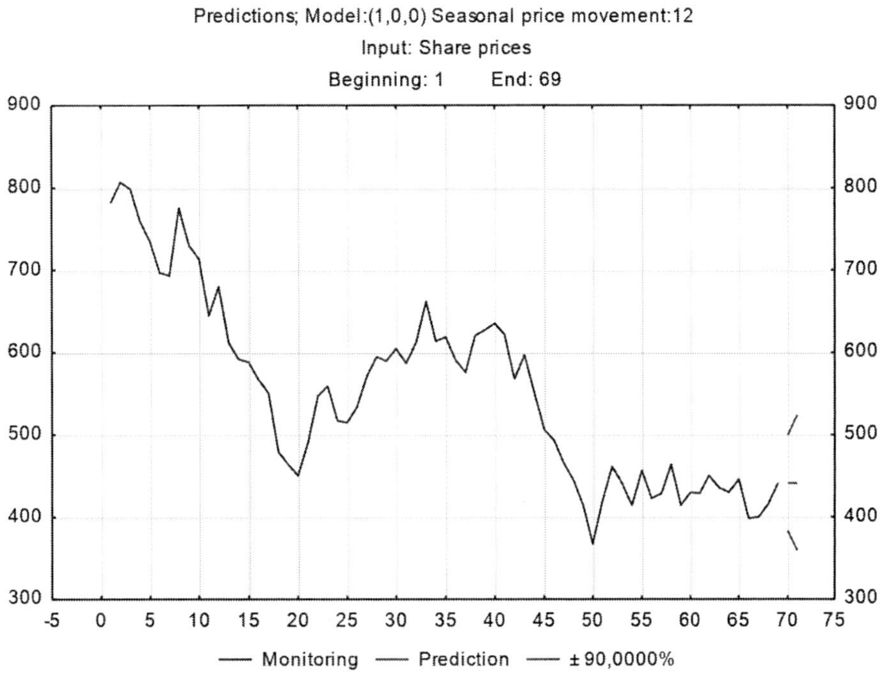

Figure 2. Prediction of future development of ČEZ share prices

The graph in Figure 2 illustrates the development of share prices in the monitored period. In addition, it also shows forecasts for the following two months of trading on the Prague Stock Exchange.

The actual forecast values are listed in Table 1

Table 1 Forecasting values of the future ČEZ share prices development.

Case number	Forecasts; Model:(1,0,0) Seasonal movement:12 (Development of share prices - ČEZ) Input: Share price Beginning : 1 End : 69			
	Forecast	Lower 90.0000%	High end 90.0000%	Standard error
70	441.3000	383.1765	499.4235	34.84797
71	441.3001	359.1010	523.4991	49.28247

DOI: 10.1051/shsconf/20173901007

The figure 2 as well as the table 1, shows that the forecast for both following months is almost the same. The values differ by 0.0001 CZK. However, the Box-Jenkins method indicates the pessimistic and optimistic variant, or lower-end and high-end limit of the forecast interval. The limit is, however, set at the level of +- 10 %. Therefore, this leads to situation, when in October 2017, the share price is assumed to be in the interval of ⟨383.1765; 499.4235⟩. The difference between the high-end and lower-end limit is 116 CZK. Similarly, in November 2017, the share price is supposed to be in the interval of ⟨359.1010; 523.4991⟩. In this month, the share price is thus predicted with an inaccuracy of more than 164 CZK. In November, we are thus working on the interval that is close to half of the lower predicted price. The question is whether it is possible to consider such inaccuracy adequate and whether it is possible to work with it in practice. We will therefore try to analyse the forecast residuals (that is, the difference between predicted and real price of ČEZ shares) and find out whether better results can be achieved using the Box-Jenkins method.

Figure 3 shows graphical illustration of residuals.

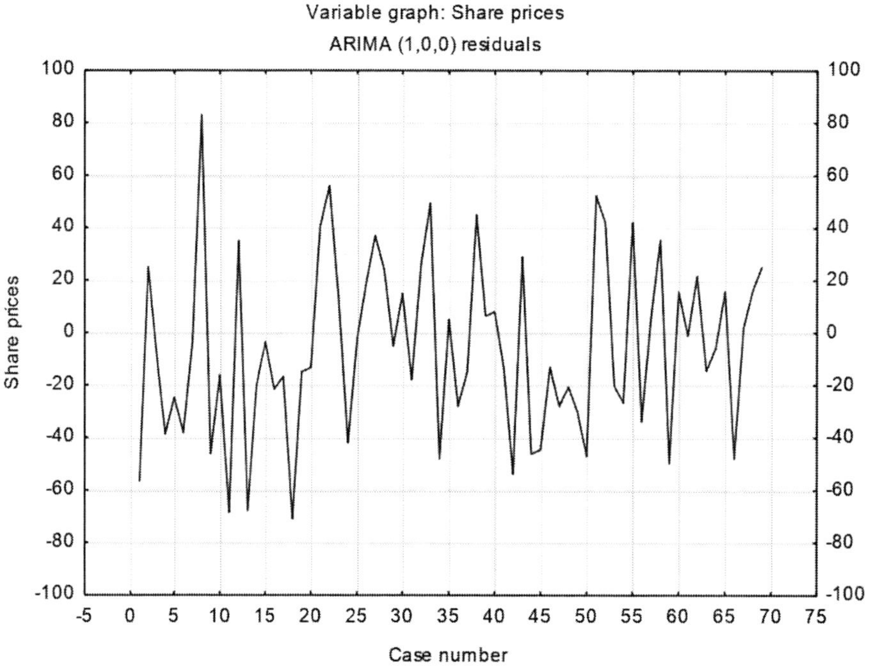

Figure 3. Residuals of share prices forecasts

The shape of the curve connecting residuals shows that the differences of predicted and real share prices development are rather big and the model suggested is not optimal. Two types of results could be achieved. We could either obtain a relatively precise model (with a minimal differences within the interval of the high-end and low-end limit), whose accuracy would be relatively low, or a model with a wider range between lower-end and high-end limit. However, share prices will most likely move within the predicted interval. In our case, the second alternative has been chosen, that is, wider range of interval with a high probability that we will move within the given interval.

Figure 4 illustrates this situation.

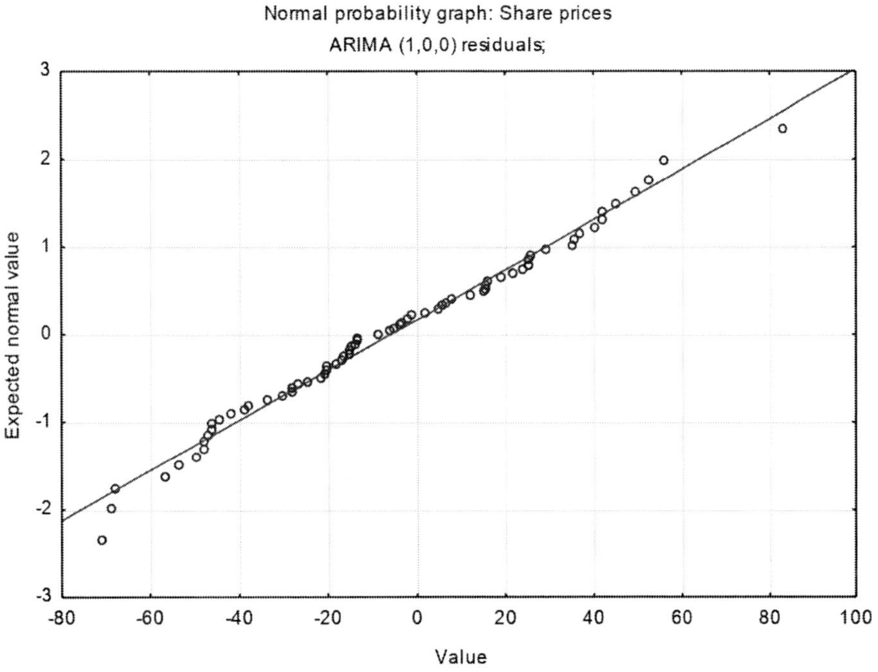

Figure 4. Normal probability of share prices

The figure shows the residuals fitted with the normal probability curve. Figure 5 shows histogram of residuals.

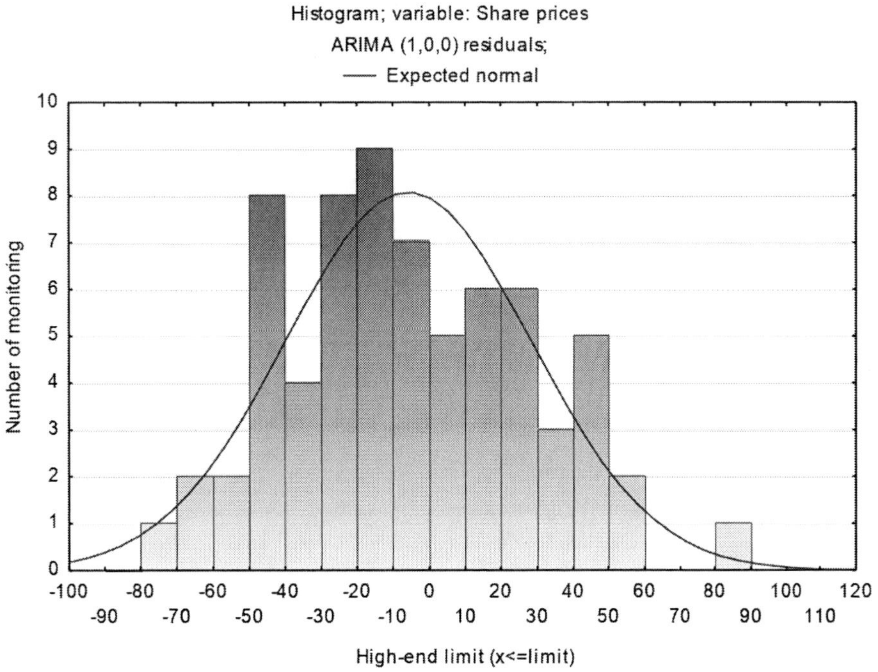

Figure 5. Histogram of residuals

DOI: 10.1051/shsconf/20173901007

The figure shows normal residuals distribution (replicating the Gaussian curve). However, the peak of the curve is at – 10 CZ. The Gaussian curve is shifted to the left, to negative values. Most probably, the sum of the residuals will be negative. This is, however, not a major problem. The most important thing is the overall number of residuals and their relative high values. This, in fact, confirms the accuracy of a poorly precise result.

Figure 6 shows autocorrelation functions of residuals.

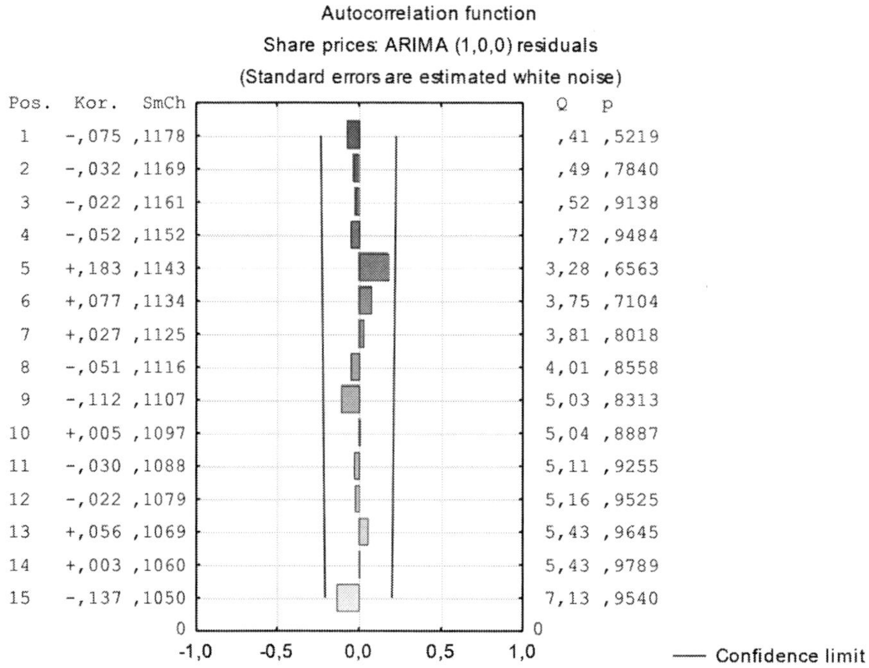

Figure 6. Autocorrelation function of residuals

Figure 6 shows a wide range of possible occurrence of the correct value. The figure clearly shows standard errors, correlations and especially confidence limits marked in red. It may therefore be concluded, that the result has been calculated correctly and that in October and November 2017, the ČEZ share price will be within the calculated interval.

4 Conclusion

The objective of the contribution is to forecast future development of ČEZ share prices on the Prague Stock Market using the Box-Jenkins method.

First of all, the data series was analysed. Next, forecasts of future ČEZ share prices development were calculated. The low-end and high-end limits of share prices forecasts for October and November were identified. However, the results indicate that there is a significant difference between low-end and high-end limits. Subsequently, the forecast residuals were analysed. Mainly due to autocorrelation functions of residuals and thus also elimination of white noise it was confirmed, that the calculation was carried out correctly, but with low accuracy. The question thus arises, whether the Box-Jenkins method is the right forecasting tool. It would therefore be convenient to validate the results by means of other method (with the same data set).

The objective of the contribution was achieved.

References

1. D. Walterová, M. Vochozka, Š. Bendová, The Subprime Mortgage and Bank Crisis in the Financial Market. *Littera Scripta,* **2**(1), 95-108 (2009)
2. X. Li, H. Zhao, K. Zheng, S. Sun, A DFS Model for Forecasting Stock Price. *Proceedings of the 2nd Workshop on Advanced Research and Technology in Industry Applications*, Dalian, China, **81**, 1678-1683 (2016)
3. Y. Ch. Wang, T. N. Nguyen, Forecasting Stock Prices for an Emerging Market: A Case of Vietnam. *Proceedings of the 2nd International Conference on Finance and Economics*, Ho Chi Minh City, Vietnam, 190-200 (2015)
4. Ch. Wu, P. Luo, Y. Li, K. Chen, Stock Price Forecasting: Hybrid Model of Artificial Intelligent Methods. *Engineering Economics,* **26**(1), (2015)
5. M. F. Anaghi, Y. Norouzi, A Model for Stock Price Forecasting Based on ARMA Systems. *Proceedings of the 2nd International Conference on Advances in Computational Tools for Engineering Applications,* Zouk Mosbeh, Lebanon, 265-268 (2012)
6. Y. S. Lu, J.L. Zhang, Forecasting stock price by SVMs regression. *Proceedings of the 11th International Conference on Artificial Intelligence: Methodology, Systems, and Applications*, Varna, Bulgaria, **3192**, 295-303 (2004)
7. P. F. Pai, Ch. S. Lin, A hybrid ARIMA and support vector machines model in stock price forecasting. *Omega – International Journal of Management Science*, **33**(6), 497-505 (2005)
8. P. H. B. F Franses, The Econometric Modelling of Financial Time Series. *International Journal of Forecasting*, **16**(3), 426-427 (2000)
9. A. B. Sánchez, C. Ordóñez, F. S. Lasheras, F. De Cos Juez and J. Roca-Pardiñas, Forecasting SO 2 Pollution Incidents by means of Elman Artificial Neural Networks and ARIMA Models. *Abstract and Applied Analysis*, (2013)
10. J. Junttila, Structural breaks, ARIMA model and Finnish inflation forecasts. *International Journal of Forecasting*, **17**(2), 203-230 (2001)
11. G. Mélard, J. M. Pasteels, Automatic ARIMA modeling including interventions, using time series expert software. *International Journal of Forecasting*, **16**(4), 497-508 (2000)
12. ČEZ, 2017, *ČEZ, a.s. – about company* [online], Available at: https://www.cez.cz/cs/o-spolecnosti/cez/profil-spolecnosti.html (2017)

Strategic Partnership of Independent States as a Game

Simona Hašková[1,*]

[1]The Institute of Technology and Business, School of Expertness and Valuation, Okružní 517/10, 37001 České Budějovice, Czech Republic

Abstract. The paper analyses the basic strategies of the international trade of two countries, analogous to the strategies of the market behaviour of two companies in the duopoly. The two-stage methodological approach was applied to analyse a problem. The first stage includes the formulation of an analytical tool in the form of an international trade model based on the model of comparative advantage whose outputs are transformed into the language of the game theory apparatus and presented by a payoff matrix. Based on the payoff matrix, the relevant analyses are then performed in the second stage. In analogy with market reality, two elementary game strategies were analysed – the strategy of cooperation and the myopic strategy. It has been shown that the cooperation based on the „cartel" model of two firms is to a certain extent an adequate description of the behaviour model of two co-trading countries provided that the input assumptions of game theory are not significantly violated. The article explains why the „foreign trade cartel" is economically beneficial to all citizens of the trading countries unlike the cartel of companies operating within the same market, where the companies take key advantage from cooperation at the expense of consumer welfare.
Key words: international trade, cartel, strategic partnership, game theory

1 Introduction

Game theory, as a discipline of applied mathematics, analyses a wide range of conflicting (game) decision situations that can occur anywhere where conflicts of interest between stakeholders occur. The theoretical models analyse these conflicting situations, and based on the mathematical model of the conflict and with the help of calculations they try to find optimal strategies for the participants (players) of the strategic game. The basis of the analysis is to determine which strategy is the best answer to the strategy chosen by other players in terms of a particular criterion. If the criterion is the wealth maximization then the optimal response is a strategy that gives the player maximum benefit with regard to the chosen strategy of other player / players (e.g. [1] or [2]). The use of the conclusions of the game theory in practice is problematic, mainly because of the inconsistency between the model's initial assumptions (individualism, rationality and interdependence of participants in the conflict [3]), which co-determinate how things should evolve in contrast to

* Corresponding author: haskovas@post.cz

phenomena occurring in society, which deny axioms of the game theory (bounded rationality [4], collectivism [5]).

If any normative theory in social sciences deviates from realistic assumptions, it necessarily converges to utopia [6]. Is this the case of game theory? The irrational behaviour of participants in decision-making situations in the theory of games has been studied by a number of authors. In Camerer, Ho and Chong [7] it was demonstrated on a choice of a number between 0 and 100 according to predefined rules (the player who determines the number closest to 2 / 3 of the average of numbers chosen by all players wins the cash prize; in the case of more than one winner, the prize will be shared equally) that in the real world, a clear game theory strategy (on the basis of which all participants would choose the number 0, and therefore the best strategy is to choose this number too) would not lead to the prize because the numerical average of the answers of the test groups ranged between 20 and 30. Other empirical research has shown that people are not economic maximizers but that they take into account the interests of others and are sensitive to standards of cooperation and justice; specifically, a robust test published in Jensen, Call and Tomasello [8] has shown that people will reject the proposed distribution of unexpected money gain at the expense of themselves if they consider it unfair (according to the theory of games, however the proposer will distribute the amount except the ratio 0 : 100 the optimal answer is to accept the offer, as „something is always better than nothing").

There are many other games, however, in which the players are very close to the state of equilibrium. One example is a stylized game of decision about entering a market in which n players simultaneously decide whether to enter the market with a known market demand d where d < n. Players prefer to enter the market if a total of d or fewer players enter the game; in the case n > d they do not enter the market. The theory of market equilibrium states that the total number of players entering the market will be close to demand d, which corresponds to the real state of affairs in most markets (the equilibrium has also been achieved in one-off experiments, when for example, past experience that could lead to better results did not have the chance to be applied), see [9].

If theory delivers the optimal solution, why is its application in reality often unsuccessful? In Kahneman [10] a response was offered: by analysing two modes of thinking (logical and intuitive), a conclusion was come to that people are often unable to consider things in a deeper way and to think in a constructive and deductive way; they prefer rather to act intuitively and impulsively, which in the absence of experience in the given field can lead to irrational decisions. The solution that is offered is to lead individuals to consistency in thinking, not to rely by experience on unsubstantiated intuition and to learn from mistakes [11]. Another answer suggests that understanding the theory of games as a guide to the application of logical laws in an effort to maximize one's own benefits is incorrect and needs to be reassessed with respect to psychological realism [9], whose essence is that people behave psycho-logically rather than logically (this is based on peoples´ wrong premise who assume that the others think similarly to themselves).

We have presented examples of individual decision-making in the field of game theory and the causes of the failure of its application, in which the style of thinking and the subjective aspects of individual psychology play a significant role. In the field of strategic decision-making of companies maximizing the profit, subjective aspects are eliminated in favour of maximizing the wealth of the firm owner, while the applied style of thinking is consistent with the desired goal and to a great extent corresponds to the principles of game theory [12].

Analogously with the strategic behaviour of companies, in the following it will be assessed whether game theory has its application in the strategic decision making of two countries as partners in foreign trade. For this purpose the impact of the cooperative and non-cooperative behaviour of partner countries on the development of the wealth of economies measured by the change in gross domestic product will be examined. The input macroeconomic model for the assessment of game strategies is the analysis of the comparative advantages resulting from the cooperation between two states. The purpose is to identify the prerequisites for functional cooperation of strategic partners in the spirit of game theory.

2 Methodological approach to the analysis of the cooperative behaviour of two countries

From the point of view of microeconomic theory, the cooperative duopoly is a market structure consisting of two producers who, on the basis of a collusive agreement on the quantity produced or price, divide the market. The fact that the duopoly agreement is unlawful and therefore legally unenforceable causes the duopolists to monitor their behaviour in order to detect a possible breach of the agreement and act accordingly [13]. This dependency property distinguishes duopoly (which is a special elemental form of a general oligopoly) from other market forms.

From a macroeconomic perspective, the duopoly market structure can be likened to an international cooperation agreement of two states, whose behaviour depends on compliance with conditions of the partnership, except that the partnership agreement between states is a legal treaty. In order to analyse the strategic behaviour of the two participants in international trade, it is necessary to adopt simplifying assumptions to construct a macroeconomic model of the consequences of the utilization of comparative advantages in the co-operation of states, whose conclusions will then be presented by a payoff matrix of the respective game:

- There are two countries involved in international trade. The engagement of other states will not be taken into account in the analysis;
- The states produce and trade two goods - food and textiles;
- The cost of acquirement of a standard unit of resources for producing these goods is the same in both countries;
- Production reaches the limit of their consumption possibilities (at the production possibility frontier curve (PPF)) in both countries and gross domestic product (GDP) is measured by the consumption method;
- In the different productions the same factor of production is equally productive in the given country;
- Two strategies at the level of international trade are taken into account - the strategy of cooperation (strict observance of the contractually agreed exchange ratio between the units of traded goods) and the myopic strategy (the attempt of the country to, at least in the short term, deflect the agreed exchange ratio to its advantage);
- The goal of states is to maximize the wealth measured by gross domestic product.

3 Model of functioning of comparative advantage

Let us consider two different sized and advanced closed economies, the economy of country A and the economy of country B, differing in the amount of standard resource units available (i.e. the combinations of land, labour and capital). Each of them produces two goods, food (F) and textiles (T), with the maximum number of units of these goods (UF and UT) that can be produced from one standard unit of resources as shown in Table 1

Table 1. Output per unit of resources in countries A and B prior to the commencement of international trade

Economy (country)	A	B
Food (the number of UF)	240	80
Textiles (the number of UT)	60	40

From Table 1 we see that the economy of country A is more efficient than the economy of country B in the production of both goods. It is 3 times more productive in food, but only 1.5 times in textiles. Therefore, it is comparatively better in food production. Similarly, although country B is worse in the production of both goods, it is less bad in textile production than in food; therefore it has a comparative advantage in textiles.

Let us define

- Cz … as a price of a standard resource unit in world currency (e.g. in $);
- F_A, F_B … as the number of UF produced from a standard resource unit in countries A, B;
- T_A, T_B … as the number of UT produced from a standard resource unit in countries A, B;
- $Cf_A = Cz / 240$, $Cf_B = Cz / 80$, $Ct_A = Cz / 60$, $Ct_B = Cz / 40$ … as prices of UF and UT in countries A, B.

Let us neglect the fact (in accordance with the fifth assumption in section 2) that in various productions the same production factors may be differently productive as a result of the economical transfer from one production to another, which shapes the production possibility frontier curve (PPF) as a concave curve. The production possibility frontier curves of countries A, B at full employment of a standard unit of resources in the plane (F, T) can be thus graphically approximated by the line segments $Cz = Cf_A \cdot F_A + Ct_A \cdot T_A$, $F_A \geq 0$, $T_A \geq 0$, or respectively $Cz = Cf_B \cdot F_B + Ct_B \cdot T_B$, $F_B \geq 0$, $T_B \geq 0$ with coordinates endpoints (0, 60), (240, 0) and (0, 40) (80, 0) – see Figure 1. Each combination of quantities (F_A, T_A), or respectively (F_B, T_B) located on these line segments produces GDP = Cz, which means 100 % of the possible output (e.g. if $F_B = 20$, then $T_B = 30$, hence $Cf_B\ 20 + Ct_B \cdot 30 = Cz$).

Let us assume that countries A and B will conclude a strategic partnership in which their economies will open to each other; each of them will devote to the production of the goods in which it gains a comparative advantage (i.e. country A will devote to food production, country B the textile production). These goods will be exchanged in a relaton 1 UT = 3 UF. This international exchange relation is mutually beneficial as it ensures that what one wants to import the other wants to export:

- the imported goods are cheaper than domestic (the food producer in country A pays for imported 1 UT only 3 UF, while 1 UT from domestic production costs 4 UF, since $Ct_A = 4 \cdot Cf_A$; similarly, the textile producer in country B pays for imported 1 UF only UT / 3, whereas 1 UF produced domestically costs UT / 2 as $Cf_B = Ct_B / 2$);
- the export yield is higher than from domestic sales (in country A the export revenue of 1 UF equals $1\ UF = UT / 3 = Ct_A / 3 = (Cz / 60) / 3 = Cz / 180$, the yield of the sales of 1 UF in the domestic market reaches $Cf_A = Cz / 240$, in country B the export yield of 1 UT equals $1\ UT = 3 \cdot UF = 3 \cdot Cf_B = 3 \cdot Cz / 80 = Cz / 26.6$, whereas the yield on the sale of 1 UT domestically is $Ct_B = Cz / 40$);
- Each country spends less resources on a given volume of imported goods than if it would be produced domestically (on imports of 1 UT country A spends 1 / 80 of a standard resource unit, since $1\ UT = 3 \cdot UF = 3 \cdot Cf_A = 3 \cdot Cz / 240 = Cz / 80$, whereas domestic production of 1 UT requires 1 / 60 of a standard resource unit,

since $Ct_A = Cz / 60$; country B spends on imports of 1 UF 1 / 120 of the standard resource unit, as 1 UF = UT / 3 = $Ct_B / 3 = (Cz / 40) / 3 = Cz / 120$, whereas home production of 1 UF requires 1 / 80 of the standard resource unit that has to be spent, since $Cf_B = Cz / 80$).

As a result of the opening of two economies to each other and the commencement of mutual trade, the above Table 1 changes as follows:

Table 2. Output per unit of resources in countries A and B after the initiation of international trade

Economy (country)	A	B
Food (the number of UF)	240	120
Textiles (the number of UT)	80	40

The value of F_B has increased from the original 80 UF to 120 UF by means of producing T_B = 40 UT from a standard resource unit in country B, which were then exchanged in exchange ratio 1 UT = 3 UF with country A for F_B = 120 UF. Similarly, T_A value has changed from the original 60 UT to 80 UT. From a standard resource unit it was first produced F_A = 240 UF in country A, which were then exchanged with country B for $T_A = F_A / 3$ = 240 UF / 3 = 80 UT.

The new frontier of production possibilities of the full standard resource unit employment thus became line segments in the plane (F, T) with the endpoints of the coordinates (0, 80), (240, 0) and (0, 40), (120, 0). Each combination of (F_A, T_A) located at PPF of country A then produces GDP in the amount of the price 80 UT in country A, i.e. 80 Ct_A = 80 Cz / 60 = 1.33 Cz. Compared with the closed economy, GDP in country A increased by one third. Analogically, any combination of the amount (F_B, T_B) lying at PPF of country B produces GDP in the amount of the price 120 UF in country B, i.e. 120 Cf_B = 120 Cz / 80 = 1.5 Cz. Compared with the closed economy, GDP has grown by half in country B (see the shift of the frontiers of production possibilities of country A and B in Fig. 1).

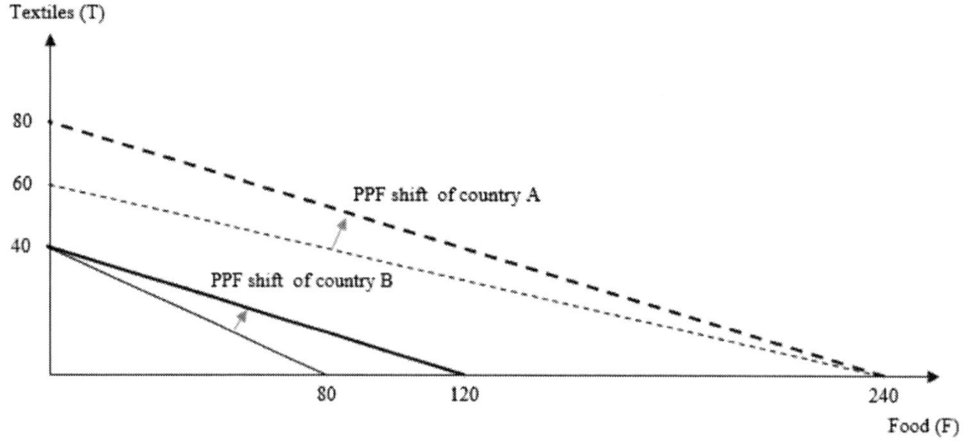

Fig. 1. The benefits of countries A and B from international trade illustrated as PPF shift

4 Results: International trade as a game with a non-zero sum

DOI: 10.1051/shsconf/20173901008

A game of non-zero sum is a game in which everyone can win if they cooperate. In the above-mentioned game of two players this was the case. Both countries agreed on an optimal exchange ratio lying in the middle between the opportunity costs of 1 UT (or 1 UF) in countries A and B from the perspective of the theory of comparative advantage. These are 4 UF in country A and 2 UF in country B (or respectively ¼ UT in country A and ½ UT in country B). Therefore, the exchange ratio of 1 UT = 3 UF has been chosen. From the viewpoint of the comparative advantage theory this is optimal as the less-developed economy B grows faster than the more developed economy A, as a result of which the desired conversion occurs.

However, mutually beneficial is every exchange ratio UT: UF in the interval (2, 4). The closer to 2, the relatively more advantageous (more profitable) the trade will be for country A than country B and vice versa. In the case of the exchange ratio of 1 UT = $\sqrt{8}$ UF ≈ 2.83 UF (this follows from the equality of ratios 40 · x · Cz / 80 = 240 / x · Cz / 60, where x signs for the exchange price) the original GDP grows by 41 % in both economies after initiation of a trade. In the case of 1 UT = 2 UF, the GDP of economy A doubles, while nothing changes in economy B compared to the state before opening to the trade. In the case of 1 UT = 4 UF, the GDP of economy B doubles, whereas nothing is changed in economy A. This leaves the space for competitive behaviour between countries to drag on the size of the exchange rate, whose possible (limit) results are recorded in the payoff matrix in Fig. 2.

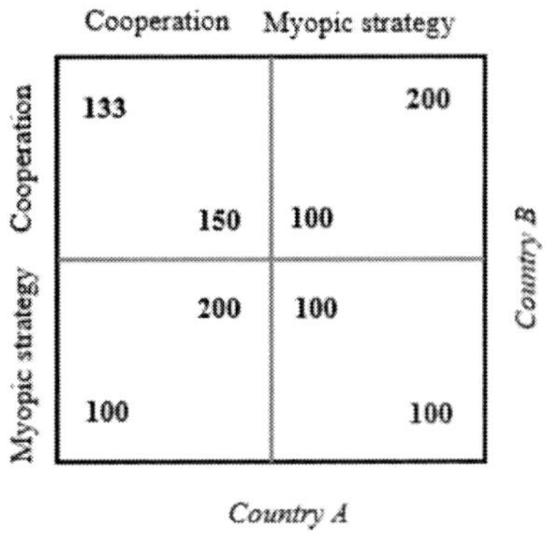

Fig. 2 Payoff matrix of benefits from foreign trade expressed in GDP growth (in %) of countries A and B

The payoff matrix exhibits this case: each of the countries can choose two strategies – cooperation and limit myopic strategy. In the upper left quadrant both countries cooperate. We have focused on the clash of cooperation strategies of both countries so far, the consequence of which was a shift of PPF that increased the initial GDP from 100 % of the potential of originally closed economies of both countries to a percentage GDP growth of open economies (133 % in country A, 150 % in country B) – see Fig. 1.

The essence of myopic (short-sighted) strategies lies in the fact that the country by deliberately generating trade deficits tries to deflect the exchange ratio to one's own benefit resulting in a drop of the GDP of the trade partner in behalf of own GDP growth. It can be assumed that the partner will not tolerate it and applies subsequently myopic conflicting strategy; in the case that the countries do not conclude a deal on any exchange rate once again, they end up in the bottom right-hand quadrant where the economies will close to each other in terms of the foreign trade – they will fall into trap of Nash equilibrium.

5 Discussion

As the practice has shown, many companies with the intention to maximize their profits orient their strategies in order to gain market shares of their competitors. In Armstrong and Collopy [14] it was documented that such an effort may, however, damage the company's profitability. Their research has shown that out of twenty US companies the competitive firms have achieved lower returns on investment and the probability of their survival was lower than that of the companies that were not competitive. In the case of a limited number of „players" in the market who produce not substitutable or badly substitutable goods, a strategic market-sharing agreement, which is not legal in the economically developed countries, is the solution to prevent losses resulting from competitive market struggles (see [15], who analyses the post-cartel price during litigation). Detecting collusive agreements is difficult because collusive arrangements are usually hidden [16]. Unlike the collusive agreements of companies, whose strategies result in an increase in the socially undesirable deadweight loss, the collusive agreement of foreign trade partners is beneficial for the national economies in the long run.

This is illustrated by the above-described simple analysis of the payoff matrix of the contribution of two countries from foreign trade, where the GDP growth (%) of country A is recorded in the marginal parts of the payoff matrix, the GDP growth (%) of country B at the inner angles of the matrix (see Fig. 2). In the case of the cooperating countries in foreign trade, the GDP grew by 33 %, or respectively 50 % (see the left upper quadrant) compared to the state before opening of economies to each other. The myopic competitive strategy (right upper and left bottom quadrant) brings a temporary advantage of GDP growth by 100 % to country A, or respectively to country B. The temporality of the advantage can be inferred from the fact that the legally enforceable contracts are a part of strategic partnership in foreign trade practice that severely disadvantage or penalize non-cooperating behaviour – in our case the myopic behaviour. Penalty sanctions can ultimately lead the affected country to the withdrawal from the treaty, by means of which the economies find themselves in a pre-existing state before entering into trade cooperation – see the bottom quadrant on the right the results of which correspond to the undesired Nash equilibrium. To prevent the trap of Nash equilibrium means to observe the terms of contractual cooperation and not to behave myopically.

6 Conclusive summary

The paper analysed two basic strategies in foreign trade – cooperation and myopic strategy in analogy with the strategic behaviour of companies in duopoly. In the first phase of the analysis a model of comparative advantages of two countries was constructed based on an agreement on mutual exchange of two goods (food and textiles) at a mutually advantageous exchange rate. The outputs of this model were transformed into a game theory apparatus – the payoff matrix; the benefit of the mutual opening of economies was quantified by the percentage increase in GDP. The matrix showed the optimal strategic choice of the two countries – the orientation towards cooperation, which in the long run provides the countries with the desired benefits from international trade.

A unilateral violation of the cooperation by the application of a myopic strategy would very likely lead the „affected" country to the application of contractual sanctions contained in a binding trade agreement, which, unlike cartel cooperation of firms, is a legal document. In the extreme case, a breach of contract may lead to an end of cooperation resulting in a drop of GDP to the original level of closed economies – the trap of Nash equilibrium. In order to avoid the trap means to consistently follow the partners' agreement. Unlike cartel cooperation of firms, from which companies benefit at the expense of consumers and whose existence is associated with a socially undesirable occurrence of deadweight loss, two countries′ tight cooperation enables, under certain circumstances shown in the paper, to increase the wealth of nations as a whole.

References

1. R. A. McCain, *Game Theory*, Edward Elgar Publishing, Drexel University, Philadephia, (2010)
2. D. M. Frankel, S. Morris, A. Pauzner, Equilibrium selection in global games with strategic complementarities. *Journal of Economic Theory,* **108**(1), 1-44 (2003)
3. H. Singh, Introduction to game theory and its application in electric power markets, *IEEE Computer Applications* in Power, **12**(4), 18-20 (1999)
4. D. Kahneman, A perspective on judgment and choice: Mapping bounded rationality. *American psychologist*, **58**(9), 697-720 (2003)
5. M. B. Brewer, Y. R. Chen, Are Collectives in Collectivism? Toward Conceptual Clarification of Individualism and Collectivism. *Psychological review,* **114**(1), 133-151 (2007)
6. W. J. Samuels, J. E. Biddle, J. B. Davis (eds), *A companion to the history of economic thought*, Blackwell Publishing Ltd, John Wiley & Sons (2008)
7. C. Camerer, T., J. Ho, J. Chong, A Cognitive Hierarchy Model of Games. *Quarterly Journal of Economics*, **119**(3), 861-898 (2004)
8. K. Jensen, J. Call, M. Tomasello, Chimpanzees are rational maximizers in an ultimatum game. *Science*, **318**, 107-109 (2007)
9. C. F. Camerer, *Behavioral Game Theory: Experiments in Strategic Interaction*, Princeton, NJ: Princeton University Press (2011)
10. D. Kahneman, Maps of Bounded Rationality: Psychology for Behavioral Economics. *The American economic review,* **93**(5), 1449-1475 (2003)
11. S. Hašková, E. Babulak, M. Vochozka, Intellectual Capital Optimisation in Microeconomic Sphere. *Proceedings of the 8th European Conference on Intellectual Capital: ECIC2016*, p. 112-118 (2016)
12. M. J. Mazzeo. Product choice and oligopoly market structure. *RAND Journal of Economics*, **33**(2), 221-242 (2002)
13. P. Billand, C. Bravard, S. Chakrabarti, S. Sarangi, A note on networks of collaboration in multi-market oligopolies. *The Annals of Regional Science*, **53**(2), 325-335 (2014)
14. J. S. Armstrong, F. Collopy, Competitor Orientation: Effects of Objectives and Information on Managerial Decisions and Profitability, *Journal of marketing research,* **33**(2), 188-199 (1996)
15. J. E. Harrington, Post-cartel pricing during litigation. *The Journal of Industrial Economics,* **52**(4), 517-533 (2004)
16. R. H. Porter, Detecting Collusion. *Review of Industrial Organization,* **26**(2), 147-167 (2005)

Fuzzy sets and real options approaches for innovation-based investment projects effectiveness evaluation

Olga A. Kalchenko[1,*]

[1]Peter the Great St.Petersburg Polytechnic University, Institute of Industrial Management, Economics and Trade, 195251 Polytechnicheskaya, Saint-Petersburg, Russia

Abstract. The paper focuses on the specifics of innovation-based investment projects effectiveness evaluation (IIPEE). When using traditional approaches, errors occur in evaluation, since innovation-based investment projects are long-term with high uncertainty. Fuzzy sets and real options approaches have long been used as alternative for IIPEE. The novelty of this paper is in the joint use of these two approaches with the consideration of the wide variety of project implementation scenarios. Consequently, it reasonably seems more efficient use of fuzzy sets approach together with Weighted Average Polynomial Option Pricing Model for IIPEE.
Key words: fuzzy sets, innovation, investment, effectiveness, evaluation

1 Introduction

Innovative changes are essential proactive actions to prevent future negative situations of dynamic environment.

Innovation-based investment projects differ from traditional investment projects by the following features:

- High degree of uncertainty – very often cash flows have irregular nature, i.e. there is no probability distribution of key parameters.
- Problems in strategic effect valuation that can have more qualitative than quantitative character.
- More qualitative than quantitative information on the project, as well as unpredictable risks that have to be taken into consideration.

In case of innovation investment projects, efficient instruments that would allow managing high risks and future uncertainty are urgently needed. Today so far Discounting Cash Flow Method (DCF) with main criterion Net Present Value (NPV) is the most widespread analytical tool for the investments effectiveness evaluation due to calculation simplicity and practical accessibility of the approach.

However, this approach has fundamental inaccuracies concerning IIPEE:

- Ideal market conditions requirements;

* Corresponding author: o.kaltchenko@mail.ru

 DOI: 10.1051/shsconf/20173901009

- Investor's flexibility and unpredictable qualitative information flow disregard. DCF method does not take into account any new unexpected market information that arises during the project lifetime. Changes in legislation, sudden competitor's actions, a new technology development, exact experimental results, and others are examples of opportunities and threats that may change investor's behaviour and strategy. Investors can force projects or stop them relying on the new market information [1];
- Inadequate risk calculation. Risk calculation in the denominator by the cumulative discounting rate does not solve the problem of considering high risk in innovation-driven projects. Incorrectness arises by the reason of decreasing value to the present moment not only for cash in-flows, but also for out-flows. This situation is associated with the fact that discounting to the present value gives correct values only if the cash flows sequence is standard [2].

2 Fuzzy sets approach for IIPEE

Fuzzy sets approach allows taking into consideration the value of management flexibility, it is proposed to use a slightly modified fuzzy-interval method [3].

The main advantages of fuzzy sets approach are the possibilities:
- to implement multi-criteria evaluation, including both qualitative and quantitative indicators;
- to operate data, the values of which cannot be set explicitly;
- to formalize in a unified form and to use all available heterogeneous information (deterministic, interval, statistical, linguistic), which increases reliability and quality of decisions in selecting innovation projects;
- to take into account the uncertainty of scientific and technical developments;
- to save time on finding exact values of variables and writing equations that describe them (when the decision maker performs the forecast for the initial parameters, it is not required to generate point probabilistic assessment, it is sufficient to specify the calculated range of the predicted parameters);
- to form a sufficiently complete set of possible scenarios of innovation-based investment project development;
- to present the expected development effectiveness not as a single-valued indicator, but as a certain range of values with its expectations distribution, characterized by the membership function of the corresponding fuzzy number. Weighted complete set of expectations allows to get an integrated measure of negative results expectations during the development implementation and to assess the risk level.

According to the fuzzy sets approach [3], the basic parameters of an innovation-based investment project can be evaluated in the form of triangular fuzzy sets (minimum, maximum, and most expected values) and, for each level of confidence (from 0 to 1), construct interval evaluation of net present value (NPV):

$$[NPV_1, NPV_2] = \left[- l_2 + \sum_{i=l}^{N} \frac{\Delta V_{il}}{(1+r_{i2})^i} + \frac{C_l}{(1+r_{N+1,2})^{N+1}} - l_l + \sum_{i=l}^{N} \frac{\Delta V_{i2}}{(1+r_{il})^i} + \frac{C_2}{(1+r_{N+1,1})^{N+1}}\right] \tag{1}$$

where $[I_1, I_2]$ - the range of possible values of the initial investment volume, N - the number of planned intervals (periods, stages) of the investment process, $[\Delta V_{i1}, \Delta V_{i2}]$ - interval working balance of receipts and payments in the i-th period, $[r_{i1}, ri_2]$ - interval discount rate, $[C_1, C_2]$ - liquidation value of the net assets, established during the investment process, (N+1) - interval, not related to the life of the project, but the time of completion of the settlement of all innovation project participants [4].

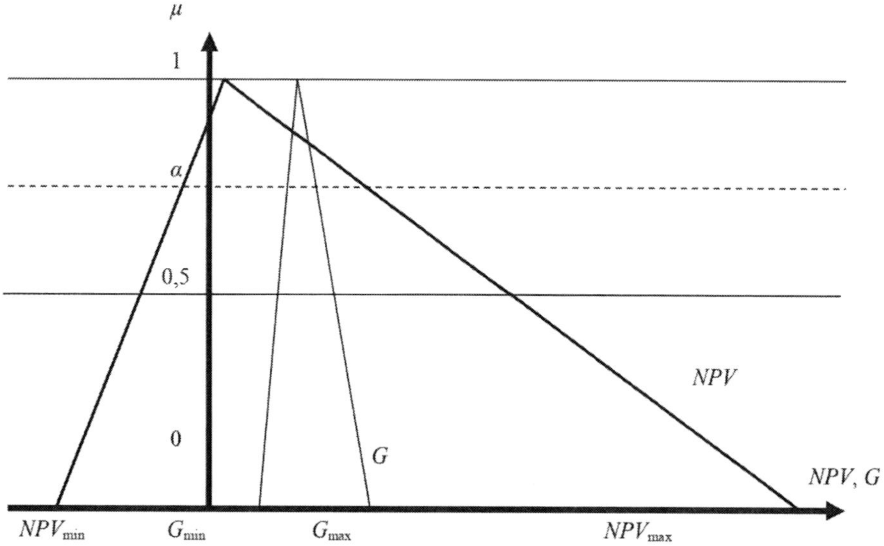

Fig. 1. NPV fuzzy interval evaluation [4]

The project is considered effective when the NPV is more than a certain G level (in the common case G = 0). This parameter is also a subject of fuzzy sets evaluation. G indicator is a return requirement for investments.

Given an acceptable level of sampling at the interval α [0, 1], it is possible to reconstruct the resulting fuzzy number NPV by approximating its membership function μ broken curve on interval points. Often, it is possible to bring the NPV to the triangular-term, being limited by calculations of the main points of the outcome data fuzzy numbers. As the performance phases of the project are realized, further information results, allowing refinement and better adjustment of the NPV fuzzy evaluation.

After each stage, as additional information becomes available for the analysis of the innovation project and for possible uncertainty reduction, the fuzzy profile of the project is rebuilt and the risk is recalculated.

3 Real options approach for IIPEE

Basic real options - those that have the greatest impact on the implementation of the project, which received the most important decisions to minimize losses and give companies management important information.

In our earlier studies [4], we wrote about the possibility of applying real options approach for IIPEE:

- Practical implementation of classical probabilistic analysis (multi-scenario "decision tree") to the investment evaluation using real options approach with a large number of control points is quite problematic, both due to the complexity of probabilities estimation, and due to the lack of a clear methodology for NPV changes indicators measurement;
- Application of Cox-Ross-Rubinstein Binomial Option Pricing Model (BOPM) [5]. BOPM is based strictly on the binomial changes assumption. For innovation and investment

projects evaluation is limited, because the future uncertainty is reduced only to two states: positive and negative. However, researchers still have to accept it.

- Scope of Black-Scholes Option Pricing Model (BSOPM) [6] for real investments evaluation is possible when the project success is strongly related to the price of an actively traded liquid asset. For example, the price of oil or other popular resource, when the volatility statistics can meet the requirements. BSOPM is based on the continuous time assumption. This implies a possibility to sell or buy a share at any moment like at the Stock Exchange. BSOPM is based on the replicated portfolio assumption. Innovation market is developing, and there are no statistics on it. An expert evaluation of volatility, taking into account external and internal company risks, is required.

The following conclusions were drawn: these approaches are based either on subjective probabilities or tied to market indicators, therefore, the possibilities of their application to IIPEE are limited. Innovation-based investment projects are characterized by significant uncertainty, large number of decision points and weak predictability of environmental parameters, therefore none of these methods of applying the real options approach can be used to evaluate effectiveness. However, the real options approach provides an opportunity to develop an algorithm for decision-making in projects implementation.

We proposed to apply real options approach for IIPEE in two ways:

1) quantitative interpretation of an option and determination of its present value, either on the basis of outcomes probabilities estimation, or as a link to some liquid replicable underlying asset;

2) option is a mechanism for constructing decision algorithms, a qualitative understanding of options as a resource of flexibility.

Innovation-based investment projects can be considered real options, because, in the initial stages the shortage and unreliability of financial information occurs. Such projects can be stopped with a much smaller financial loss. The consideration is aimed at identifying their additional features and capabilities by:

- varying project parameters during implementation;
- quantitative assessment of the available options and their inclusion in the project cost.

As previously mentioned, real options approach allows overcoming the mentioned problems:

1. investors' flexibility valuation;
2. risk calculation in the numerator by the scenarios (or decision) tree.

The idea of Real Option Valuation (ROV) was taken from the stock market where investor's opportunity, but not the obligation to sell or to buy an asset, is valued.

S. Myers was the first, who implemented option finance technique to the real investment [7]. Nowadays the leading specialists in the real option valuation significantly contributed to this area [1, 8-10].

E. Rogova and A. Yarygin suppose that the plenty of researches devoted to the ROV method may be divided into two parts [2]:

1. strong mathematic papers that sometimes do not give the clear way of using results in practice [11, 12];
2. papers devoted to method popularization [13].

On the basis of these models pitfalls, E. Rogova and A. Yarygin developed Weighted Average Polynomial Option Pricing Model (WAPOPM). Since BOPM does not need to estimate volatility parameter (σ) and is based on the discrete time assumption, it is considered as a basic method in their research. The aim of their research was to construct a decision tree with any possible complex structure (any time-intervals between the project's stages, and any amount of the scenarios at each stage) [2].

Each specific innovation-based investment project leads to corresponding decision tree.

Instead of searching the formula for real option value at the initial moment of time in all possible cases, they propose a unique analytical algorithm of calculation in any subtree (part of the tree constructed from parent node and its children nodes).

After real option value calculation at the leafs, ROV at the parent node of these leafs is calculated. The goal is the ROV evaluation at the root.

An innovation-based investment project is considered to have three possible scenarios: successful; non-profitable and breakeven; detrimental.

In case of two possible ways, BOPM algorithm is used, which is based on the equal portfolio value assumption regardless of the way (basic asset price change). The portfolio consists of basic asset, risk-free obligations, and an option on them. Cox-Ross-Rubinstein's remark is [5]: "from either the hedging or complete markets approaches, it should be clear that three-state or trinomial stock price movement will not lead to an option pricing formula based solely on arbitrage considerations."

In the work of E. Rogova and A. Yarygin valuation for call-option and for put-option is proposed.

Strategic NPV of innovation-based investment project equals to:

Expanded (strategic) NPV=direct (passive NPV)+Real Options Value (flexibility value) (2)

Their unique analytical calculations algorithm in any subtree includes:

1. Technological input data identification (a decision tree which reflects particularities of the innovation-based investment project, time-intervals between the project's stages).
2. Financial input data identification (risk-free rate, basic asset price at the initial moment of time, option exercise price).
3. Parameters that reflect change in basic asset price identification (fuzzy sets approach that was proposed earlier in our research is applicable, as innovation-based investment project are often characterized by poor statistic data).
4. Leafs ROV calculation.
5. Root ROV evaluation by iteratively using WAPOPM in all subtrees.

The valuation of the real options in innovation-based investment projects with extended uncertainty is an important practical problem.

WAPOPM:

- takes into account investors' flexibility and it calculates investment risk by the scenarios tree (decision tree);
- does not need an estimation of the volatility parameter and it is based upon the discrete time assumption;
- is based on the polynomial changes;
- can be easily used in practice together with fuzzy sets approach.

4 Conclusions

In our previous research we proposed and approved the use of fuzzy sets approach together with real options as quantitative assessment and project parameters variation for IIPEE [4].

Presently, according to recent development [2], it reasonably seems more efficient use of fuzzy sets approach together with WAPOPM for IIPEE.

Fuzzy sets and real options approaches for IIPEE allow constant critical analysis and the consideration of alternative parameters in the internal and external environment. It becomes possible not to evaluate the entire project at once, but divide it into several stages as investor opportunities are acquired. Fuzzy sets and real options approaches allow evaluating the efficiency and choosing between alternatives without using other methods, by considering the relationship between risk and opportunities that companies face.

 DOI: 10.1051/shsconf/20173901009

It becomes possible to correct innovation-based investment projects during their realization. The ability to manage innovation-based investment projects is achieved by the joint use of fuzzy sets and real options approaches.

Fuzzy sets and real options approaches for IIPEE were used and approved at three different sized industrial companies, situated in one of the most innovative region of Russia – Saint-Petersburg and Leningrad region. The first is a medium-sized company, JSC "Eco-energetika," which is the developer and owner of advanced Russian technology in the field of energy saving. The second is a medium-sized agro-industrial company of Leningrad region. The third is a large-sized organization, JSC "NPP" Radar MMS," which is a world leader in the field of electronics, special and civil systems, precise instruments, and special software.

It is very important and interesting to scrutinize in further calculations improvement on the basis of a software product, for example Matlab Fuzzy Logic Toolbox.

Author is thankful to Mr. Prof. Dr. Vasiliy R. Okorokov for his guidance and valuable comments. All errors are the sole responsibility of the author.

References

1. V. Antikarov, T. Copeland, *Real options: a practitioner's guide,* Texere (2001)
2. E. Rogova, A. Yarygin, Valuing innovative 4G (LTE) technology with real options approach. *The XXIV ISPIM Conference Proceedings* (2013)
3. A. O. Nedosekin, The applications of fuzzy sets theory to problems of finance control. *Audit and financial analysis*, **2**, 15-18 (2000)
4. V. R. Okorokov, O. A. Kalchenko, An innovative project evaluation technique under conditions of information uncertainty. *International Journal of Business and Information,* **10**, 180-197 (2015)
5. J. Cox, S. Ross, M. Rubinstein, Option pricing: A simplified approach. *Journal of Financial Economics,* **7**(3), 229-263 (1979)
6. F. Black, M. Scholes, The pricing of options and corporate liabilities. *The Journal of Political Economy*, **81**(3), 637-654 (1973)
7. S. Myers, Determinants of corporate borrowing. *Journal of Financial Economics,* **5**(2), 147-175 (1977)
8. L. Trigeorgis, *Strategic Investment,* Princeton University Press (2004)
9. A. Damodaran, *The Promise and Peril of Real Options.* Working paper, Stern School of business (1999)
10. J. C. Hull, Options, *Futures and Other derivatives,* Prentice Hall (2002)
11. S. M. Turnbull, *Option Valuation,* Dryden Press (1987)
12. P. Wilmott, S. Howison, J. Dewynne, *The Mathematics of Financial Derivatives,* Cambridge Univ. Press (1995)
13. K. J. Leslie, M. P. Michaels, The real power of real options. *The McKinsey Quarterly,* (3), 5-22 (1997)
14. H. Smit, L. Trigeorgis, *Real Options: Examples and principles of valuation and strategy,* Oxford University Press (2003)

Motion to innovation: Brand value sources have (not) changed over time

Jana Kliestikova[1,*] and *Maria* Kovacova[1]

[1]University of Zilina, Faculty of Operation and Economics of Transport and Communications, Department of Economics, Univerzitna 1, 010 26 Zilina, Slovak Republic

Abstract. Innovation drives the expansion of economies in a global dimension. This is also the reason why contemporary researches indicates the trend of incorporation of the innovation to the strategic concept of brand value building and managing. It has been proven, that innovation is relevant source of brand value perceived by consumers. Such a trend has been established in reaction to the growing importance of brand value for competitive advantage creation in a global perspective. Currently published scientific contributions mainly highlight the importance of brand innovation in selected sectors of national economies abstracting from innovation perceptions due to the national socio-psychological profile. This is the reason why universally applicable theory of innovation in scope of brand value building and managing is missing. The lack of the theory can be removed by identification of importance of innovation attributes as brand value sources in context of market specifics. So, the aim of this paper is to provide such an identification and to verify existence of divergences between "foreign" theory and "domestic" practice. To do that, we use questionnaire, selection analysis and cluster analysis. We detect specifics of brand value perception focusing on innovation and its attributes comparing theory and reality of Slovak environment.
Key words: innovation, brand value, cluster analysis, selection analysis

1 Introduction

The concept of brand value building and management changes dynamically despite practical experience of brands which have been well known over the world has shown that this part of managerial theory should be applied selectively and without time pressure. It is not an exception that insufficient consideration of relevant brand and market specifics has led to the phenomena opposite to the desired – to the decrease of brand value [1]. Nowadays, brand innovation is such a trend. It is because although theory recommends it as a possible source of subjectively perceived brand value, the longstanding practice of brands proves that more than changes, consistency is connected with brand value growth [2]. Possible reason of this fact is, that brand value is very fragile and each inappropriate intervention often raises unpredictable fluctuations in it. So, it is very important to: 1) detect strength of influence of innovation phenomena on branding practice worldwide; 2)

* Corresponding author: jana.kliestikova@fpedas.uniza.sk

analyse possibilities of general implementation of new theories in conditions of brand and market specifics; 3) identify those attribute of innovations which are sources of subjectively perceived brand value by costumers in selected market (either applying sectoral or regional approach) and 4) compare the importance of traditional vs. newly formulated brand value sources across products and markets.

2 Literature review

The idea of innovation as a significant source of brand value is relatively new despite the issue of brand and innovation penetration has been discussed intermittently in a scientific literature [3]. Up to now, innovation in scope of its branding implications has been analysed mainly partially, without coherent theory creation and highlighting different points of view.

Thus, for example Aaker as a "god father" of brand theory states that not brands should be innovated but on the contrary, innovations should be branded. This perspective is ideologically significantly different from modern theory of brand and innovation penetrance. Mentioned theory is based on the assumption that brand strategy can be crucial to the success of innovation, especially in the long run.

There are times when a company literally needs to label innovation or lose it. Without a successful branding strategy, innovation may be short-lived - scattered on a confused, scattered impact market - or become another forgotten internal initiative. In such cases the brand can do everything. The mark to emphasize does not simply mean putting the name and logo on innovation. This means that the brand is an integral part of a coherent strategy that supports actively managed and adequately funded branding programs [4].

So, brand is not an addressee of the innovative managerial activities but only a tool of ensuring the market acceptance of such activities.

But innovation is not always perceived without restrictions as positive. There are indications that a company needs to consider consumer perceptions of the company as a whole, and not just new products and technologies, and take into account a functional-cognitive perspective as well as consumer emotions and experiences [5].

The extension of this theory has been made later investigating the question if it is possible to affect negatively perceived value by exploring how consumers view and evaluate brands following an innovation failure. It has been proven that innovation failures are more detrimental to high-equity brands that have preannounced the innovation and to low-equity brands that do not receive supportive word-of-mouth from an opinion leader after the failures occur [6].

It means that the correlation between perceived brand perception and innovative activities exists. The theory of innovation as relevant brand value source in its actual meaning starts to be created later, when the influence of company innovativeness and product innovativeness on the components of customer value, is analysed [7].

According to this, provided research aims to examine the impact of product innovation attributes (complexity, relative advantage, compatibility, trialability and observability) on brand equity, and to detect whether these attributes exert a different effect on low-versus high-equity brands. Overall, it was found that innovation attributes have an effect on brand equity, and this effect differs between low- dand high-equity brands, with a low-equity brand being benefited more than a high-equity brand from perceptions towards a product's innovation attributes. Additionally, it was found that the impact of complexity and relative advantage on brand equity increases when consumer innovativeness increases in the case of a high-equity brand. However, no significant difference was found between low- and high-equity brands regarding the proposed moderating effect of innovativeness [8].

In the same time, the theory of the need of regional specifics taking into account is the imperative of successful incorporation of innovation into the process of brand value building and management [9].

This is one of two main evolutionary directions of innovation perception in scope of brand value creation. The second one is based on the sectoral approach and specifics of innovation perception when sector is a variable [10].

These two approaches have been originally developed separately, but not long time later, there it is possible to detect their connection.

At the beginning, this connection led to the development of so called negative theory of innovation as a significant source of brand value. There has been developed and tested a model of critical brand innovation factors by examining key factors influencing firm-level brands' innovation and increased market performance. Adapting both organizational elements and market response characteristics, the model integrated four key variables in China's industrial service markets: innovation, internationalization, market orientation, and organizational learning. Findings provided a foundation for understanding how firms improve their innovation and subsequent market performance in an emerging and dynamic market. The study demonstrated that when brands are more innovative, their performance increases - brand innovation plays a fully mediating role on the effects of market orientation and organizational learning to market performance, but has no mediating effect on internationalization and market performance. A lack of innovation reduces market performance even when internationalization, market orientation, and organizational learning are present [11].

Later, the negative approach has been conversed to the positive approach. Its basis consists in identification of innovation attributes leading to perceived brand value optimization across sectors and markets. According to mentioned, it has been analysed how innovation impacts brand perception of Brazilian consumers of food products and processed beverages, in addition to identifying brands considered innovative in this sector as well. To do so, it has been provided content analysis and it have been defined ten categories of innovation attributes: 1) entrepreneurship; 2) marketing and communication; 3) product mix; 4) practicality and functionality; 5) packaging and design; 6) consumer relations; 7) management; 8) social and environmental sustainability; 9) technology and R & D; 10) new + new experience. It has been found out that, in general, the innovative brands have a mixing process or combination among the ten characteristics, with emphasis on three: entrepreneurship; marketing and communications and product mix. Overall, the study showed that innovative brands are creative; confident and bold; present interesting advertisements; are affordable; value quality of their products and seek to satisfy desires, often still unknown by consumers [12].

Some of detected innovation attributes have been discussed separately. These are: 1) entrepreneurship; 2) marketing and communication; 3) product mix and practicality and functionality; 4) technology and R & D.

The study focused on entrepreneurship is the first study engaging in the missing link regarding retail innovation and branding by providing a brand-driven process to systematically develop retail format innovation projects. The so called brand-driven retail format innovation locks into anthropological research findings where cultural meanings are considered as the main source for the construction of brand identities whereby the new retail format is transformed around brand-derived touchpoint experiences [13].

Marketing and communication attribute have been developed mainly in scope of partial marketing communication activities such an events. Provided study contributed to the body of knowledge regarding event management and corporate branding and shed light on future research to explore the initiative and benefit of pushing forward event innovation [14].

When product mix and practicality and functionality are discussed, there is only a little space to find shades of difference between these two categories of innovation attributes. Existing results suggest that consumer perceptions of product-description usefulness, technology fluidity, product usefulness, product-use ease, consumer-review trustworthiness, consumer-review usefulness, user ratings of consumer reviews and retail-platform trustworthiness have a direct or indirect effect on their attitude and purchase intention toward the technology product [15].

Technology and R & D are analysed in a study which combines the signaling theory and dynamic marketing capabilities perspective to investigate the mediating role of product innovation in the influence of R & D expenditure and brand equity on marketing performance. Overall, the research provides fresh insights into the process by which R & D expenditure and brand equity affect product innovation and marketing performance in highly competitive product categories [16].

As it is declared above, the theory of innovation in scope of brand value building and management is one of the key progressive concepts of contemporary branding. On the one hand it seems to be well developed, but on the other hand the fragmentation of this theoretical approach will be a possible source of later managerial misunderstanding when it will be implemented without restrictions.

It is because the importance of national socio- psychological profile has been proven but following researches elaborate the general theory of innovation attributes without taking into account the specifics which exists across sectors and markets. Such a specific in its very rough features is the status and perception of quality in the process of brand value building. According to previous research, there are significant discrepancies between practical implications of countries where brand is identified with quality and countries where brand is identified with image and social status [17,18, 19, 20].

Slovak Republic is typical country where qualitative aspects of brand are highlighted.

In general, we can say that this trend dominates in former post-soviet countries. The reason is that previous influence of the principles of centrally planned economy has distorted the market and the consequences of such a state have been shown in such a way. Due to the mentioned, the undesirable spiral mechanism can be started – domestic companies apply inconvenient patterns of brand value building and management – brand value decreases – companies rather do not build and manage value of theirs brands – brands loss their competitive potential in comparison with foreign competitors and the market deforms – survive only strong foreign brands applying their national branding mechanisms – the impression of so called "good practice" is created – domestic companies apply inconvenient patterns of brand value building and management and the circle can start again.

To minimize the probability of occurrence of such a negative phenomenon, we fill the gap in the theory of innovation attributes perceptions form the point of view of brand value building and managing in regard to specifics of Slovak market environment. So, we prove, that the selective approach to the issue of innovations should be applied to achieve really effective results.

3 Methodology and data

Basic data set which is needed to achieve set goal of the paper is not fully available. The newest data about innovation activities of companies published by Statistical Office of Slovak Republic is from year 2014. In order to obtain reliable and valid data, it is used also brand value ranking from this year.

The lack of recency of analysed data is eliminated by the methodology of the research where these data are used only as a framework platform for initial situational analysis. The methodology is the following: 1) identification of the innovation practice in Slovak Republic; 2) identification of the perception of innovation as a source of brand value by Slovak consumers; 3) identification of the position of innovation in group of classical brand value sources; 4) identification of innovation attributes with highest impact on brand value.

The identification of the innovation practice in Slovak Republic can be realised by the situational analysis provided on the basis of analysis of the structure of innovation in Slovak Republic (see Tab. 1).

Table 1. Structure of innovation in Slovak Republic in 2014

Type of innovation		2014		
		All companies	Industry	Services
Companies with innovation	All companies	2 432	1 250	1 182
	Small	1 615	661	954
	Medium	594	414	180
	Large	223	175	48
Technological innovation	All companies	1 549	867	682
	Small	951	412	539
	Medium	405	301	104
	Large	193	154	39
Successful innovation	All companies	1 399	777	622
	Small	848	366	482
	Medium	363	260	103
	Large	188	151	37
Product innovation	All companies	412	257	155
	Small	259	148	111
	Medium	108	70	38
	Large	45	39	6
Process innovation	All companies	439	235	204
	Small	301	131	170
	Medium	96	66	30
	Large	42	38	4
Product and process innovation	All companies	549	286	263
	Small	289	88	201
	Medium	160	124	36
	Large	100	74	26
Incomplete and (or) stopped innovation	All companies	150	90	60
	Small	103	46	57
	Medium	42	41	1
	Large	5	3	2
Non technological innovation	All companies	883	382	501
	Small	664	249	415
	Medium	190	113	77
	Large	30	21	9
Companies without innovation	All companies	5 214	2 596	2 618
	Small	4 057	1 867	2 190
	Medium	973	609	364
	Large	184	120	64

Simple descriptive analysis of these data is adequate platform for following selection analysis realised in scope of Superbrands ranking 2014 (see Tab. 2). The criteria for its realization are: 1) company size; 2) innovative character; 3) type of innovation and 4) presence of brand in later ranking.

 DOI: 10.1051/shsconf/20173901010

According to the results of selection analysis, it is possible to formulate hypothesis which will be verified in scope of realised questionnaire survey.

Table 2. Superbrands 2014 ranking in Slovak Republic

2014 ranking		2017 ranking	Innovative character
Brand	Sector		
Acidko	food and agriculture		
AVIS	transport	X	X
Baumit	construction		X
BMW	automotive		X
Budiš	food and agriculture		
COOP	retail	X	
decodom	furniture		X
DEMA	sport equipment	X	X
dm drogerie markt	retail	X	X
eiffell optic	retail		
Elesko	food and agriculture	X	
Eurolines	transport		X
Fernet Stock Citrus	food and agriculture		
Husqvarna	construction		X
Liga proti rakovine	non-profit sector		
Mary Kay	cosmetics and drugstore	X	X
McCain	food and agriculture		
Mecom	food and agriculture	X	
Melife Amslico	insurance	X	
NAY	retail	X	
Nicolaus Extra Vodka	food and agriculture	X	
OLO	waste liquidation		X
Pokec.sk	on-line paltform		
Považský cukor	food and agriculture		
Raciol	food and agriculture	X	
Radisson Hotel Carlton	traveling	X	
Rajo	food and agriculture	X	
Satur	traveling	X	
Slovak Telecom	telecommunication		X
Slovnaft	petrochemical		X
SND	media and culture		
Starmedia	media and culture	X	
Sweet family	food and agriculture	X	
topreality.sk	on-line paltform	X	
Yves Rocher	cosmetics and drugstore		
ZIPP	construction		X
zlavy.sme.sk	on-line paltform		X

The basic research problem of realised questionnaire survey was the insufficient knowledge about impact of innovation on perceived brand value in specific conditions of Slovak republic. The survey was conducted from July to September 2017. We used a standardized method of the direct questioning. As a recommended tool of this survey it was chosen a semi-structured written questionnaire. A basic set of surveyed respondents was formed of Slovak citizens older than 15 years (depending on the size of the basic set, the survey sample was 384 respondents). That age limit was set as one of the essential prerequisite for autonomous buying decision-making, according to a valid Slovak law.

In accordance above mentioned theoretical approaches to the solved research problem and knowledge about Slovak consumer's specifics, the hypotheses of the research were set:

- Hypothesis H1: The presence of innovation has an impact on perceived brand value.
- Hypothesis H2: The character of sector has an impact on brand innovation.

Hypotheses were tested using binomial test which uses the binomial distribution to decide if the outcome of an experiment in which we count the number of times when one of two alternatives has occurred. The following characteristics were used for the hypothesis testing (1):

$$t = \frac{|\bar{x} - \mu|}{\sigma} \Rightarrow \frac{x - np}{\sqrt{npq}} = \frac{\frac{x}{n} - \frac{np}{n}}{\frac{\sqrt{npq}}{n}} = \frac{\tilde{p} - p}{\sqrt{\frac{pq}{n}}} \tag{1}$$

Where:

- x the frequency of the trait,
- n the frequency of observed phenomena,
- p predicted probability of the phenomenon incidence,
- q probability of alternative phenomenon frequency,
- $\tilde{p}$ statistical probability of the phenomenon incidence.

Hypothesis H1: After the equation (1) substitution, the value of testing characteristic was calculated (7,325). The critical testing characteristic was calculated through the Microsoft Excel, using the function NORMINV (0,05; 0; 1). The resulting value was -1,421 (t_k). To accept the null hypothesis (H0) which is the object of verification, the "t" value has to be minor than "t_k" value. As the "t_k" value is minor than "t" value (7,325 > -1,421), the null hypothesis (H0) is rejected and we accept the alternative hypothesis (H1) at the significance level of 0, 05.

Hypothesis H2: After the equation (1) substitution, the value of testing characteristic was calculated (6,249). The critical testing characteristic was calculated through the Microsoft Excel, using the function NORMINV (0,05; 0; 1). The resulting value was -1,565 (t_k). To accept the null hypothesis (H0) which is the object of verification, the "t" value has to be minor than "t_k" value. As the "t_k" value is minor than "t" value (6,249 > -1,565), the null hypothesis (H0) is rejected and we accept the alternative hypothesis (H1) at the significance level of 0, 05.

So, we found out that the innovation is a relevant factor of brand value. Its importance varies according to the sector. But this is a very general statement and its specification is vital. Possible way how identify relevant innovation attributes is to provide more detailed analysis. Basic method for processing of secondary data is cluster analysis. The objective of this analysis consists in clustering of the objects into the groups with condition that two objects from the one cluster were more similar each other than two objects from different cluster. The cluster analysis consists of following steps 1) selection of input database; 2) selection of parameters; 3) name of the object; 4) choice of clustering technique; 5) choice of using distance; 6) placing in clusters; 7) choice of distances in Dendrogram; 8) selection of appropriate techniques for creation of Dendrogram; 9) an explanation of the best Dendrogram similarities of objects and variables. Input for clustering in terms of data analysis is data matrix. Data matrix contains variables (in our case the attributes of

DOI: 10.1051/shsconf/20173901010

innovation) in *m* columns and *n* rows (in our case brands) on which these characters are measured. Every object it means the row of data matrix (1), that is, $x_i^T = (x_{i,1}, \ldots x_{i,j}, \ldots x_{i,m})$ is characterized by its characteristics. Tab. 3 provides a basic set of data which we need for realizing the cluster analysis of socio-cultural profiles OECD countries; this table is consist of 1) E is entrepreneurship; 2) M&C is marketing and communication; 3) PM is product mix; 4) P&F is practicality and functionality; 5) P&D is packaging and design; 6) CR is consumer relations; 7) M is management; 8) S&ES is social and environmental sustainability; 9) T&R&D is technology and R & D; 10) N is new + new experience. All the data included in the table were obtained through the realised questionnaire survey. These data express the average value of innovation attribute importance in percentage.

Table 3. Set of data used for cluster analysis

Brand	Attribute of innovation									
	E	M&C	PM	P&F	P&D	CR	M	S&ES	T&R&D	N
AVIS	5	8	12	16	22	3	11	16	4	3
Baumit	9	1	10	12	13	19	9	15	8	4
BMW	1	19	17	16	14	7	5	7	11	3
decodom	2	10	18	21	10	8	6	13	9	3
DEMA	4	13	15	14	17	11	6	4	11	5
dm	3	20	19	10	8	15	5	14	5	1
Eurolines	6	22	17	12	9	13	6	12	2	1
Husqvarna	5	11	15	20	13	9	4	5	14	4
Mary Kay	4	16	18	21	7	21	4	6	1	2
OLO	2	1	10	23	4	3	5	27	22	3
Slovak Telecom	3	19	22	9	7	16	5	6	11	2
Slovnaft	2	15	16	13	6	9	3	26	9	1
ZIPP	1	26	14	11	6	10	7	12	7	6
zlavy.sme.sk	2	23	19	8	9	18	6	7	5	3

As a clustering method we used "nearest neighbour method" which is based on the minimum distance. It is that this method finds two objects which are separated by the shortest distance and placed them into the cluster. Another cluster is formed by adding a third nearest neighbour. This process is repeated until all objects are in a common cluster. The distance between two clusters is defined as the shortest distance of any point in the cluster to any point in another cluster.

4 Results and discussion

In accordance with results of hypotheses testing, we can conclude that 1) the presence of innovation has an impact on perceived brand value and 2) the character of sector has an impact on brand innovation. So, we declared the importance of innovation in scope of brand value building and management in Slovak republic as well as the heterogeneity of its appearance across sectors. To detect the relevance of innovation attributes, chosen valuable brands (according to the criteria of their innovative character) were grouped into clusters based on their innovation attribute. This distribution is the following:

Cluster 1: Avis, Baumit
Cluster 2: dm – drogerie markt, Eurolines, Mary Kay, Slovak Telecom, ZIPP, zlavy.sme.sk
Cluster 3: OLO

Cluster 4: BMW, decodom, DEMA, Husqvarna
Cluster 5: Slovnaft

Fig. 1 graphically shows these clusters in more detail

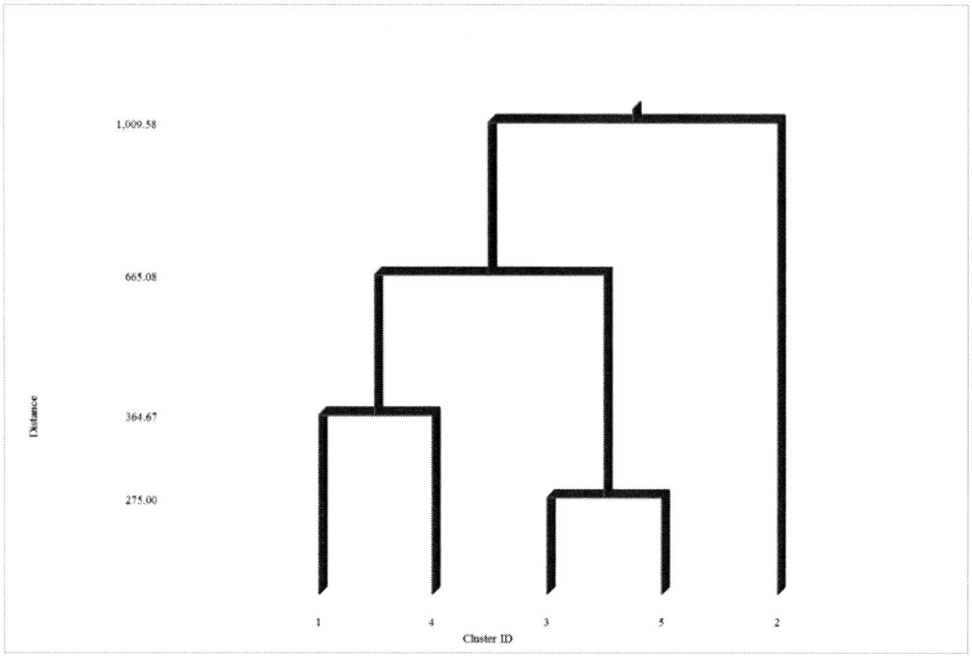

Fig. 1. Cluster Dendrogram of most valuable brands based on their innovation attribute

We can state that despite the general diversity of analysed brands, the greatest degree of convergence is between clusters 3 and 5. So, the greatest degree of inter-group similarity occurs between OLO and Slovnaft. It means that innovation attribute relevant in scope of OLO brand value could be implemented without significant changes in case of brand Slovnaft and vice versa. This is caused by the proximity of these clusters when the prospective of vertical axis is taking into account. On the contrary, the greatest degree of divergence is between these clusters (3 and 5) and cluster 2 where dm – drogerie markt, Eurolines, Mary Kay, Slovak Telecom, ZIPP, zlavy.sme.sk were clustered. Possible reason of such a situation is really different brand value nature what implicates also convergences between innovation attributes as prospective brand value source.

While grouping of some brands into the cluster could be considered to be unambiguous in relative to their sectoral character (industrial brands in cluster 4), cluster analysis also found some surprising results. Originally, we supposed that brands with innovative character which declared their position in Superbrands 2017 ranking (Avis, DEMA, dm drogerie markt, Mary Kay), will be grouped in very convergent clusters and so we will be able to detect relevant dominant innovative attributes in scope of brand value perceived by Slovak consumers. We thought so because of results of initial situational analysis where it has been observed that majority of companies operating in Slovak market provide simple combined product and process innovation. But the results indicate real divergences between valuable brands across sectors. So, sectors where brand value is built effectively through the adequate innovation attributes are: 1) transport (AVIS); 2) industry (sport equipment – DEMA and cosmetics and drugstore – Mary Kay) and 3) retail (dm – drogerie markt). For these sectors the relevant innovation attribute in scope of brand value building and management are the following: 1) transport - packaging and design, practicality and functionality and social and environmental sustainability; 2) industry - practicality and functionality, product mix and consumer relations and 3) retail - marketing and communication, product mix and consumer relations.

5 Conclusion

In general, current managerial trends highlight innovation as a possible relevant source of brand value. This motion is observable also in Slovak market environment. We found out that the presence of innovation has an impact on perceived brand value as well as that the character of sector has an impact on brand innovation. At the same time, we detected sectors where innovation attributes are implemented effectively and have a positive impact on brand value. In these sectors, brands which have been identified as valuable should serve as patterns of innovation usage in scope of branding. But there are also many sectors where the position and relevance of innovation in the scale of relevant brand value sources has not been proved. So, there is still very challenging to found the key to the innovation effect exploitation when brand value is built and managed.

The research leading to these results has received funding from the project titled "Integrated model of management support for building and managing the brand value in the specific conditions of the Slovak Republic" in the frame of the programme of Slovak Research and Development Agency under the grant agreement number APVV-15-0505.

References

1. S. Banerjee, V. Yadav, and S Ch. Banerjee, A social marketing framework for innovation and technology adoption: The case of agricultural extension in India. *Journal of Self-Governance and Management Economics*, **4**(2), 63-82 (2016)
2. V. Markova, P. Lesnikova, Strategic level of sustainable development of enterprises. *Economical and managerial spectrum [Ekonomicko-manažerské spektrum]*, **9**(1), 9-15 (2015)
3. B. Lowe, F. Alpert, Forecasting consumer perception on innovativeness. *Technovation*, **45-46**, 1-14 (2015)
4. D. Aaker, Innovation: Brand it or lose it. *California Management Review*, **50**(1), 8-24 (2007)
5. W. Kunz, B. Schmitt, A. Meyer, How does perceived firm innovativeness affect the consumer? *Journal of Business Research*, **64**(8), 816-822 (2011)
6. S.L. Liao, C.C.J. Cheng, Brand equity and the exacerbating factors of product innovation failure evaluations: A communication effect perspective. *Journal of Business Research*, **67**(1), 2919-2925 (2014)
7. J. Kim, K.H. Kim,T.C. Garrett, H. Jung, The contributions of firm innovativeness to customer value in purchasing behavior. *Journal of Product Innovation Management*, **32**(2), 201-213 (2015)
8. S. Norskov, P. Chrysochou, M. Milenkova, The impact of product innovation attributes on brand equity. *Journal of Consumer Marketing*, **32**(4), 245-254 (2015)
9. C.Y.Lin, Conceptualizing and measuring consumer perceptions of retailer innovativeness in Taiwan, *Journal of Retailing and Consumer Services*, **24**, 33-41 (2015)

10. R. Panda, D. Kapoor, Relationship between information systems integration, innovation and consumer – based commitment practices for knowledge sharing in creating power brands. *Market-Tržište,* **29**(1), 59-74 (2017)
11. B. Nguyen, X.Y. Yu, T.C. Melewar, S. Gupta, Critical brand innovation factors (CBIF): Understanding innovation and market performance in the Chinese hightech service industry. *Journal of Business Research,* **69**(7), 2471-2479 (2016)
12. CH.P. da Silva, V.M.F. de Paula, V.A.F. da Paula, Innovative brands: How consumers perceive innovation. *Revista Brasileira de Marketing,* **16**(1), 83-97 (2017)
13. G. Botschen, P.K. Wegerer, Brand-driven retail format innovation: A conceptual framework. *International Journal of Retail and Distribution Management,* **45**(7-8), 874-891 (2017)
14. F.N.X. Yang, S.X.C. Tan, Event innovation induced corporate branding. *International Journal of Contemporary Hospitality Management,* **29**(3), 862-882 (2017)
15. T.Y Wu, C.A. Lin, Predicting the effects of eWOM and online brand messaging: Source trust, bandwagon effect and innovation adoption factors, *Telematics and Informatics*, **34**(2), 470-480 (2017)
16. P. Sharma, N.S. Davcik, K.G. Pillai, Product innovation as a mediator in the impact of R&D expenditure and brand equity on marketing performance. *Journal of Business Research,* **69**(12), 5662-5669 (2016)
17. M. Nadanyiova, The quality mark SK and its impact on the shopping behavior of Slovak consumers. *Procedia Economics and Finance,* **23**, 1509-1514 (2015)
18. J. Greblikaite, W. Sroka, R. Daugeliene et al., Cultural integration and cross-cultural management challenges in the Central European Countries: Lithuania and Poland. *European Integration Studies,* **10**, 29-43 (2016)
19. K. Janoskova, A. Krizanova, Comparison of selected internationally recognized brand valuation methods. *Oeconomia Copernicana,* **8**(1), 99-110 (2017)
20. G. H. Popescu, Does economic growth bring about increased hapiness? Journal of Self-Governance and Management Economics, **4**(4), 27-33 (2016)

Metrics for assessing retailers based on consumer perception

Anastasii Klimin[1,*], *Dmitrii* Tikhonov[1], and *Aleksei* Efimov[1]

[1]Peter the Great Saint Petersburg Polytechnic University, Institute of Industrial Management, Economics and Trade, Graduate School of Business Technologies. 195251, Saint Petersburg, Russian Federation

Abstract. The article suggests a new look at trading platforms, which is called "metrics." Metrics are a way to look at the point of sale in a large part from the buyer's side. The buyer enters the store and make buying decision based on those factors that the seller often does not consider, or considers in part, because "does not see" them, since he is not a buyer. The article proposes the classification of retailers, metrics and a methodology for their determination, presents the results of an audit of retailers in St. Petersburg on the proposed methodology.
Key words: metrics, assessing, retailers, consumer perception

1 Introduction

The objective of the article to propose a new methodology for the classification of retailers, based on the Maslow pyramid. The authors offer 9 metrics for estimations of retailers, which allow to classify the shop to one of five categories, according to Maslow pyramid. The results of study of retailers of St. Petersburg on the offered methodology are given. In the conclusion of the article authors indicate problems in the application of the proposed methodology and show directions for its development and improvement.

Most classifications of retailers are based on size, range, price level, parameters of the layout of goods in the point of sale, location of stores and other parameters [1-4]. Practically there are no classifications of stores, based on motivational and psychological characteristics of buyers. This study proposes a classification of retail stores based on the Maslow pyramid [5], considering the traditional classification criteria, which are listed above. This classification can be applied not only to FMCG stores, but also, for example, to clothing stores, other products of the B2C markets. In addition, it can be extended to the classification of any sales tools, which we call "trading platforms", for example, to glossy real estate magazines (apartment, houses, etc.) or Internet sites about restaurants.

Retailers are divided into five categories corresponding to the levels of the Maslow pyramid and one additional, which is called "no category". The higher the declared category, the higher the impression of customers about the store, where customer can satisfy different levels of its needs, that is, to receive ever deeper satisfaction from the

* Corresponding author: klimin@kafedrapik.ru

DOI: 10.1051/shsconf/20173901011

purchase according to the Maslow pyramid. In Table 1 you can see the proposed classification.

Table 1. Classification of retailers according to the Maslow pyramid

Number of Maslow pyramid level – category of retailer	Name of Maslow pyramid level	Type of retailer category
0	No category	Street sellers
1	Physiological needs	Tent, pavilion.
2	Safety needs	Convenience store, shop near the house.
3	Social belonging	Network retail, corporate system
4	Esteem	Specialized store
5	Self-actualization	Boutique

Briefly describe the categories of stores.

1. Tent, pavilion - "Physiological needs", "Survival"

These small shops sell goods of daily demand, such as vegetables, milk, bread, packaged products. The advantage of these points of sell is that they have high mobility, thanks to which it is possible to place a tent or pavilion near potential buyers, and reduce the time for the purchase of products. This is the lowest level in the pyramid, where the consumer buy, when "very hungry". In such places, there is a minimum of comfort, convenience and assortment.

2. Convenience store, shop near the house - "Safety".

The pavilion does not guarantee safety, the level of sanitation there can be quite low, there is often no place for buyers (a shopping hall), the goods are given away from a window or through a counter to the street. The store near the house is not a network retail, it is targeted for buyers who live near the store. In most cases, this marketplace is located within a radius of 500 meters from the house, that is, "within walking distance", which is undoubtedly a significant advantage for buyers. Unlike the tents and the pavilions, they have a trading hall, some standards of sanitation and hygiene are observed.

3. Network retail, corporate system - "Social belonging".

Network retailers - the most common in modern conditions, the form of trade in the FMCG markets. In the opinion of the authors, it corresponds to the third level of the Maslow pyramid, social belonging. Visitors to network stores select them in accordance with their social status and buying behavior and visit those stores in which they see the same as they are buyers. A network store creates a sense of belonging to a certain class of people. In terms of trading formats, network retailers are divided into supermarkets, hypermarkets, discount stores, etc. Depending on this, their category according to the system proposed by the authors can be stretched from 2 to 4 categories. The network supermarket can be closer to the convenience store or to a specialized store. When evaluating stores from this category, you can not compare stores of different formats, for example, supermarkets with hypermarkets.

4. Specialized store - "Esteem".

In these stores specialization is carried out by assortment. An example are specialized stores that sell only fruits, vegetables, smoked fish or trade only with meat from Argentina. Wine stores are also classified as specialized. This category of stores is clearly different from the others, as in the specialized stores there is a clear division of the product line. In this regard, such stores are difficult to compete with large networks due to the lack of choice of other categories of goods. Despite this, a deep and wide assortment of a special product is a significant advantage of specialized stores. High quality of service and qualification of personnel is one of the decisive factors when choosing this store for purchase. Buyers come to specialized stores when they want to purchase the best quality and taste, comparing it to other products from the same category. In this way, they realize the need for esteem.

5. Boutique - "Self-actualization".

A boutique is a store in which all aspects of its activities are at a high level, including the price. It is important to understand that the buyer is willing to pay a lot of money for the appearance of the store, the quality of the goods, the high level of service, etc. The assortment in such stores is rather deep (even unique), than wide. The buyer compares himself with this place, that is, for him it is no longer just a store, but a manifestation of self, self-actualization through the store. Also, the buyer can go to a boutique, because he refers himself to a social group, in which a visit, for example, of the "Azbuka Vkusa" ("The ABC of Taste") store in St. Petersburg is a manifestation of good taste and style. In most cases, these stores are preferred by high-income buyers who are willing to pay for a high class of service, beautifully decorated store interior, unique products and presentation of the best.

We can identify the five most expensive stores in St. Petersburg at the cost of some "boutique" consumer basket. The price of the basket includes the maximum price for 1 kg. cheese, the price for 1 pc. oysters and the price for the most expensive bread in the store. Below is a list of food boutiques in St. Petersburg in descending order of the value of such a consumer basket:

1. "Eliseevsky". 7740 rubles / 120 euros (http://www.kupetzeliseevs.ru);
2. The Globe of Gourmet. 5450 rubles / 83 euros (https://globusgurme.ru);
3. The ABC of Taste. 5406 rubles / 83 euros (http://spb.av.ru);
4. "Super Babylon". 5187 rubles. / 80 euros (http://www.superbabylon.ru);
5. "Le Land 24" 4851 rubles. / 75 euros (http://supermarket-land.ru).

2 Metrics

The authors identified 9 metrics, which are used to evaluate retailers. Metrics are divided into subjective and objective. Subjective metrics depend on the greater degree of the buyer's impressions, the objective metrics can be assessed by certain quantitative characteristics. Actually, the division into subjective and objective metrics is very arbitrary. Objective metrics are not calculated by formulas, and, like subjective ones, are estimated by the buyer's opinion. The authors have not yet reached the final decision on the classification of metrics.

Subjective metrics:

1. Perception;
2. Safety;
3. Accessibility to the product;
4. Showcase and display of goods.

Objective metrics:

5. Price range;
6. Assortment line;
7. Tastes;
8. Attraction;
9. Staff.

Metrics are estimated on a 5-point scale which corresponds to the classification of retailers: from the tent (1 point) to the boutique (5 points). In the next part, we will describe the methodology for estimating.

We briefly describe the metrics.

1. Perception. This metric includes the appearance of the store outside, parking, the first impression of the entrance to the store, the smells, the benevolence / tension of the overall situation in the store, the type of guard and his behavior, the cleanliness of the floor, the presence of a toilet, etc. Based on this metric, the buyer has a first positive or negative attitude to the store.

2. Safety. This metric includes hygiene and sanitation, lighting of shelves and store, type of store visitors, cleanliness and quality of trolleys and baskets, product presentation on shelves, convenient walkways, discernable price tags, etc.

3. Accessibility to the product. Is it possible to taste the product, is it convenient to take it, can you choose the fruit and vegetables yourself, is there enough of the goods on the shelf (there are plenty to choose from), are there different sizes of the packaging (for example, bottles of water of different volumes), what is the queue at the ticket offices, etc.

4. Showcase and display of goods. Window dressing, neatness of the shelves, beauty and accuracy of the laying out of the goods, absence of empty shelves, order on shelves, availability of accompanying goods, lack of redundant advertising materials, design of a trading hall.

5. Price range. Is there a division of goods into "very cheap", "cheap", "expensive", "very expensive". Whether the store uses loyalty programs for regular customers, whether there are sales.

6. Assortment line. Does the store always have the necessary goods (daily – "bread"), can you buy goods that are purchased less often and are expensive (weekly – "meat"). How much is represented in the store of product categories; whether there are goods "for the soul", which are not bought on the "list", but are picked up "to relieve stress". Whether it is possible to buy the goods for a holiday (flowers, gifts, etc.).

7. Tastes. Is there a variety of tastes in the range, for example, different varieties of tomatoes. Whether there is at shop "some special trait" in assortment ("only in this shop"). Does the store offer fresh products (freshly baked bread or fresh salads), etc.

8. Attraction (Involvement. Enthrallment) Innovation, not boredom, management of the buyer's behavior: can the buyer "come for milk, and leave shop with a whole cart of what he was not going to buy."

9. Staff. Goodwill, presence in the workplace, awareness, sociability, unobtrusiveness, ability to give the necessary advice.

3 Assessment methodology

As we have already said, the score for each metric is 5-point scale. You can put not only the whole points (1, 2, 3, 4, 5), but also fractional, for example, 2,4. Points correspond to categories of shops. The evaluation is made by experts-buyers who should have the impression of shops of each category. If an network retail store is estimated, the expert understands that for such a store a normal score is "3" and if he wants to put, for example, 3.6 points in the "Tastes" metric, then he should have for this justification that such the choice of tastes is more characteristic of a specialized store, and not of a supermarket. This is the difficulty of rating points. To obtain stable and reasonable estimates it is necessary to have a staff of experts-buyers who have the following characteristics.

1. The expert visited shops of different categories and has an idea about each of them.

DOI: 10.1051/shsconf/20173901011

2. The expert has experience of shopping in shops, because you cannot evaluate the "Attraction" metric, if you do not really try to buy something in the store and experience the involvement.

3. The expert visited many shops of the same category, for example, different retail stores and can compare the shopping experience in each of them.

The work of the buyer expert is somewhat like that of the inspector "Le Guide Rouge" of Michelin [6]. It is necessary to have a lot of experience and a formed coordinate system to evaluate the metrics. Training of experts-buyers begins with a joint visit to stores of different categories and an explanation of what character is the metric, how and for what we can raise or lower the score for a certain metric.

4 The results of the study

For the study, the network hypermarkets FMCG of St. Petersburg category 3 were chosen. This choice was because among the stores of this category there is a fierce competition and it would be desirable to assess their positions on the metrics in the competitive struggle. Another reason for choosing the third category was that corporate networks are the easiest to study, the most common compared to other categories of stores. Also, representatives of various corporate networks are easy to compare with each other, because all network stores are organized on the same principles, which you cannot say, for example, about specialized stores

The study took place in February-March 2016, it was attended by participants of the Foresight Club of St. Petersburg, students and teachers of the Peter the Great St. Petersburg Polytechnic University, in total there were 42 shops in 16 corporate networks.

The results of the metering of six stores from the trade networks "Auchan", "OKEY" (2 shops), "Lenta" and "Carousel" (2 shops), for which the greatest amount of data was obtained are given in Tables 2-4. The achieved category is calculated as a simple average of nine metrics. The indicator of the competitive position is calculated by dividing the reached category to the declared category (in our case 3) and is expressed as a percentage of the declared category. For example, if the achieved category is 3.25 points, then the competitive position for the declared category 3 will be 108.3%.

Table 2. The average value of metrics from the first to the fifth

Shop	Perception	Safety	Accessibility to the product	Showcase and display of goods	Price range
"Auchan"	2.6	2.4	2.8	2.4	3.8
"Lenta"	2.4	2.2	2.8	2.4	2.9
"OKEY" 1	4	3.7	3.8	3.8	4.1
"OKEY" 2	1.7	2.4	2.5	3	3.4
"Carousel" 1	1.8	1.8	2.4	1.8	2.8
"Carousel" 2	3.3	2.7	3	2.8	2.8

Table 3. The average value of the metrics from the sixth to the ninth and the achieved category

Shop	Assortment line	Tastes	Attraction	Staff	Achieved category
"Auchan"	3.6	3	2.6	2.8	2.9
"Lenta"	3.3	3.2	2.4	3.3	2.8
"OKEY" 1	4	3.8	3.3	3.3	3.8
"OKEY" 2	3.9	3.3	2.8	3.1	2.9
"Carousel" 1	2.9	1.8	2.3	1.5	2.1
"Carousel" 2	3.1	2.1	2.9	3.2	2.9

Table 4. Summary rresults

Shop	Achieved category	Declared category	Competitive position
"Auchan"	2.9	3	96.7%
"Lenta"	2.8	3	93.3%
"OKEY" 1	3.8	3	126.7%
"OKEY" 2	2.9	3	96.7%
"Carousel" 1	2.1	3	70%
"Carousel" 2	2.9	3	96.7%

The calculated competitive positions should be compared with the performance indicators of the stores: revenue per square meter of sales area, average purchase size [7].

5 Discussions and Conclusions

The level of competitiveness among stores that are part of the same network is not the same. There are stores "good" and "bad", but the overall level of the network is traced. The quality of the network of stores can also be characterized by the level of deviation of the achieved category from the declared category for all stores of network. Based on the results of the research, the first place was the network of hypermarkets "OKEY".

The values of the points awarded depend very much on the experts-buyers who put the valuations. Ideally, you need to have an unchanging homogeneous group of professional experts-buyers who visit the same shops. In our study, experts were people of different ages, social status, customer experience. This does not allow us to say that research gives us adequate and uniform estimates of stores.

Metering should be done several times a year for investigating changes in the values of different metrics. The most significant changes occur after repairs and reconstruction of trading floors, with the change of management of the company, mergers and acquisitions of retail networks.

 DOI: 10.1051/shsconf/20173901011

The most difficult problem is the selection and preparation of experts-buyers. Before conducting a research with experts, it is necessary to conduct theoretical and practical installation exercises on which the theory of metrics is declared. The most important part of the classes is the explanation of the categories of shops on the Maslow pyramid. Many experts in the beginning cannot perceive the scale of assessments from 1 to 5 points correctly, believe that if the store is "good", then it should get 5 points. However, a third category store can receive 5 points only in some cases, for example, for an assortment, if it corresponds to a store of the fifth category. For this, it is necessary to conduct trainings with experts in real stores of different format categories for all metrics.

The conception of the store categories and their evaluation by metrics will probably be very different for different cities and other regions. Even the same retailers in different countries offer different levels of service, assortment, are differently designed, use different trade equipment. The difference in quality and in the concepts of specialized stores and boutiques will be even more significant. In the opinion of the authors, according to the proposed methodology, one cannot evaluate the quality of stores based on data on the number of goods presented in stores, the size of stores, and big data on people's behavior [8]. The presented methodology refers to the direction of subjective, psychological instruments of perception of stores by specific buyers in a particular place.

References

1. V. K. Yurtaeva, Development of a modern retail format "hard discounter" as a tool for effective positioning in difficult economic conditions. *MMR*, **3**(129), 186-200 (2017)
2. M. G. Clifford, Classifications of retail stores and shopping centres: some methodological issues. *GJ*, **45**, 255-264 (1998)
3. A. I. Klimin, Current trends in stimulating consumers at points of sale. *STL of SPbPU ES*, **6**(209), 200-207 (2014)
4. A. Kovalenko, A. Polevoy, Competitive Strategies for Retail Food Networks. *Classification and Empirical Analysis* (2017)
5. A. H. Maslow, A Theory of Human Motivation. *PR*, **50**, 370-396 (1943)
6. L. Karpik, Michelin's red guidebook. *ST*, **42**, 369-389 (2000)
7. A. I. Klimin, A. A. Zakharova, Analysis of the effectiveness of incentive measures in retail. *AFA*, **6**, 305-331 (2009)
8. T. Khvatova. Big Data is watching YOU: opportunities and challenges from the perspective of young adult consumers in Russia, *JMM*, **33**, 719-741 (2017)

Global Banking: Transformation, Innovation & Competition

Yury Korobov[1,*]

[1]Saratov Socio-Economic Institute (branch of Plekhanov Russian Economic University), Department of Banking, Money & Credit, 410017 Saratov, Russia

Abstract. The article discusses the main transformation processes in banking sector: globalization, concentration of capital, formation of a new model of banking and new banking culture. It is shown that the driving forces of these processes are competition and innovation. The interpretation of the bank of the future as a sum of four concepts is offered: "Bank = technology", "Bank = client", "Bank = product", "Bank = life". It is hypothesized that a person in the digital economy can be described as Homo financial (a person with financial literacy and financial thinking).
Key words: globalization, banking sector, transformation processes, innovation, competition

1 Introduction

We live in a constantly changing world. Everything is in a change – social relations, technology, economy, – and the banking sector is not an exception. Moreover, the changes which are currently taking place in the banking sector are often not evolutionary, but revolutionary. The depth of these changes is such that they should be considered not as modernization, but as transformation.

One of the main driving forces of these transformations is the development of competition, which encourages banks to evolve and search for the optimal business model, as well as actively introduce *innovations* – to expand the range of banking products and services, improve the quality of service, and maximize the use of opportunities of new information technologies.

The most important **transformation processes** in banking sector, in my opinion, are:
- Globalization;
- Growth of capital concentration;
- Formation of a new model of banking;
- Formation of a new banking culture.

The purpose of the article is to disclose the content of the listed transformation processes and to show their influence on the essence of the bank of the future.

* Corresponding author: yurikor1965@gmail.com

2 Background

Various problems of globalization and its relationship to competition and innovations have been discussed in many publications, including, [1], [2], [3].

Issues of financial globalization and its impact on competition and capital concentration in banking sector have been revealed [4], [5], [6], [7].

Among studies concerning formation of the new model of banking should be mentioned publications [8], [9], as well as the author of this article [10, 11].

However, in general, it should be noted that the issues of the essence of the bank of the future, as well as the formation of a new banking culture, are not sufficiently studied in the scientific literature. This determines the relevance of the topic of this article.

3 Globalization in banking sector

One of the most important forms of globalization processes in the end of 20th – beginning of 21st century is the financial globalization. The main reasons for this are differences in the level of profitability and risks in individual countries, increased mobility of financial resources, as well as the development of the digital economy, which allows to easily overcome the boundaries between countries and regions.

The process of globalization forces to reconsider the established idea of the essence of the banking system. Traditionally, in the economic literature, the banking system used to be defined as the set of credit institutions within a specific country [12, 13, 14]. But this interpretation is correct only for the national banking system. Globalization leads to increased interdependence of banking systems of different countries, up to the formation of an international banking system. And this is a new, higher level of banking systems' development.

In the context of globalization, there is a need to improve the institutional structure of the national banking system by developing a modern banking infrastructure (including upgrading of its information subsystem, increasing the transparency of the banking market) and modernizing the regulation of the banking sector (creation a uniform standard for the regulation of banks and non-bank credit institutions, adoption of measures to combat unfair competition and curb monopolistic tendencies, etc.).

4 Concentration of capital in global banking

The current period of time is characterized by the transition from the extensive model of the banking network development to the intensive one. While in the 20th century in many countries (including Russia) there was a significant increase in the number of commercial banks and a rapid expansion of their branch network, the main trends of the 21st century are the increase in the bank capital concentration and the rationalization of the branch network.

The concentration of capital and assets in Russian banking system is particularly high. The overwhelming share of assets (more than 98%) is concentrated in the 200 largest banks (i.e., about 30% of operating banks). Even more impressive is the degree of concentration calculated for the 5 largest banks (among which Sberbank occupies a special place): they account for more than 50% of assets. Taking into account the dynamics of recent years, we should not expect these indicators to decrease.

5 A new model of banking

For many hundreds of years, banks have undergone a profound evolution. Initially, they acted as infrastructure organizations performing the functions of redistributing financial resources, managing issue and circulation of money. This model of banking could be called the *traditional model*. It is based on the concept "Bank = institution".

The development of bank competition in the 20th century led to the transformation of a commercial bank into a producer that meets the financial needs of customers. Such a banking model could be called the *marketing model*. It is based on the concept "Bank = enterprise".

What difference does it make to consider a bank an institution or an enterprise? The answer to this question is not only of theoretical issue, but also of big practical importance. The way how the bank managers and employees perceive the essence of a bank largely affects the policy of the bank, its relationship with customers [11].

When a bank is perceived as an institution, then bank employee is a clerk, and customer is a petitioner. This corresponds to the administrative type of relationship between banks and clients.

When a bank is perceived as an enterprise, then bank employee is a producer, and customer is a buyer. This is a completely different, partner type of relationship between the banks and clients.

Now, under the influence of competition and innovation, a new model of banking is being formed [8, 11]. What will be the concept behind this model? A simple and unambiguous answer is hardly possible here. In my opinion, the model of the bank of the future will be characterized by the set of several concepts.

5.1 Introduction of new technologies (the concept "Bank = technology")

In many publications, the bank of the future is considered as a technological phenomenon. And, indeed, the list of new devices and technologies implemented in banks is impressive.

First of all, the bank of the future will be characterized by even more extensive use of automation. Already today, banks are using video registration and biometrics systems, video consultations and self-service kiosks, interactive displays and touch-screen tables, electronic cashiers and automatic depositories, devices for contactless payments and electronic signatures, etc.

An increasing number of banking products is sold on the basis of remote technologies. Today, bank customers can gain access to basic banking services using Internet banking and mobile banking, however to open an account, you still have to come to the bank in person. But the introduction of digital identity identification can relieve customers from this.

Another way of technological development is the introduction of banking robots. Some of them are material devices representing the next evolution stage of ATMs, and some are purely virtual – like chat bots capable to communicate with customers via special applications.

All this forms a new image of the bank and a new quality of service. But the model of the bank of the future is not limited only to technological changes.

5.2 Increasing customer self-sufficiency (the concept "Bank = customer")

The introduction of remote banking services and increasing the financial literacy of population lead to the increase in the number of transactions that a client performs independently, without the involvement of bank employee.

The development of Internet information services makes it possible to provide a high degree of independence and freedom for bank customers who have the opportunity to compare the products of different banks and – to some extent – to design their own banking products. A financially literate client in the digital economy can become a "bank for oneself".

5.3 Transfer of customer's attention from banks to banking products (the concept "Bank = product")

For an independent and competent customer, the product is in the foreground, not the provider. Therefore, attracting the attention of customers is of particular importance for banks.

The development of non-banking payment systems, the creation of social networks in the field of finance and credit, the outsourcing of banking services lead to the increasing blurring of the boundaries between banks and non-bank credit organizations.

Also, we witness the expansion of the *financial supermarket* model in which a customer can get the widest possible range of financial products (banking, investment, insurance, etc.) "under one roof".

5.4 Integration of banks into people's everyday life (the concept "Bank = life")

Even today, a bank is not an ordinary enterprise, akin to a manufacturing or trade company, but a socially-oriented institution whose objective is not only to obtain profit.

The contemporary banks are engaged in charitable activities, give support for social initiatives. All this creates a positive image of the bank and a favorable social atmosphere for its activities, a "climate of trust".

In addition, the social role of banks is evident in the fact that, in order to attract customers, banks are making efforts to increase the financial literacy of population. Means of this are promotional publications (leaflets, brochures) which explain the content of certain banking products, information and analytical publications in mass media, as well as participation in educational events aimed to increase the financial literacy.

A separate area of banking that affects the level of financial literacy in society is the development of banking for children. Most banks today offer products (cards, deposits, loans), the target audience of which are children. Some banks go further by funding educational projects, development and publication of educational materials.

The logical conclusion of the integration of banking services into people's everyday life in the digital economy is the model of lifestyle banking, when the bank becomes an integral element of human life.

One of the manifestations of this is banking design. Banks endeavor to make the appearance of their buildings attractive, as well as the interior of banking premises.

Banks tend to make visiting their offices for customers as convenient and enjoyable as possible by offering them something more than just banking services.

The most famous example of such a strategy is the activity of *Umpqua Bank* (Portland, Oregon, USA). To change the attitude of customers to the bank, its CEO *Ray Davis* decided to use a model of retail company, turning banking into shopping, and bank branches into financial shops [15].

Moreover, today banks are no longer limited to financial services, evolving into diversified service companies. For example, *Sberbank* plans to provide public services in its offices, up to issuing civil passports.

Thus, the bank of the future combines such concepts as "Bank = technology", "Bank = customer", "Bank = product", "Bank = life". Banks become different both externally, and internally. At the same time, the society is changing – a new banking culture is being formed.

6 A new banking culture

Banking culture is one of new and insufficiently researched phenomena. Often it is treated as a kind of corporate culture, which is a system of intra-bank values, a set of principles for the bank's activities or rules of conduct for its employees. Carriers (subjects) of banking culture in this case are banks and bank employees. From this point of view, the banking culture looks like a phenomenon that has only local significance for the banking community.

This interpretation seems to be too narrow. It would be appropriate to the term "culture of a bank", but the term "banking culture" has a richer content. In my opinion, the carriers (subjects) of the banking culture are not only banks, but also other subjects involved in relations with banks – namely banking customers, and society represented by the central bank and other social institutions (legislative bodies, state regulation institutions, infrastructure organizations, etc.).

Thus, banking culture in the broadest sense can be defined as a set of principles of behavior of all subjects of banking activity – banks, banking customers and social institutions that ensure the functioning of the banking system.

According to this, there are three levels of banking culture [16]: corporate banking culture, consumer banking culture and social banking culture.

The elements of the corporate banking culture are the internal banking rules (appearance, rules of conduct in the workplace, correspondence, telephone conversations), a set of professional competencies of bank employees, the culture of communication with customers, corporate identity (branding), management culture (leadership style), banking strategy, culture of competition, principles of relations with the state and society.

Elements of consumer banking culture are the financial literacy of bank customers and their willingness to use banking services.

It is important to note that only the awareness of potential customers about the range of banking services is not enough. Banking services don't meet the primary production and personal needs, but the financial needs derived from them. Therefore, the motivation to acquire banking services from potential customers is much lower than to purchase material goods; the needs for banking services are often latent in nature, they need to be awakened. Creating incentives for more active and efficient consumption of banking products is an important task, both from the point of view of the banks themselves and from the point of view of society.

Elements of social banking culture are banking legislation, the development strategy of the banking system, regulation of bank competition (antitrust measures, combating unfair competition), banking supervision, banking culture of the mass-media.

The high risks inherent to the banking system call for the regulation of its activity by the state, and a calm and balanced attitude from mass-media.

The levels of banking culture are not separate from each other, but are interrelated. The content of intra-bank rules is influenced by the level of culture of banking customers and the framework conditions set by regulatory bodies. The task of increasing the financial literacy of clients is important for both banks and the state. The lack of the social banking culture (for example, inaction in the face of a banking crisis) can have devastating impact on both banks and their customers.

Thus, the formation of a new model of banking culture is not a local process, which is important and interesting only for bankers. Everyone who is interested in the successful and

DOI: 10.1051/shsconf/20173901012

stable functioning of the banking system, as well as improving their financial well-being, should understand the banking culture.

The formation of a new banking culture, along with the transformation of the bank into an element of lifestyle, make it possible to state the hypothesis that the person of the digital economy is no longer just Homo sapiens (an intelligent person who appeared hundreds of thousands years ago) or even Homo economicus (economically-minded person formed in the conditions of development of the market economy), but Homo financial (a person with financial literacy and financial thinking).

7 Conclusion

A special feature of the development of the banking sector in modern conditions are transformation processes – globalization, capital concentration, formation of a new model of banking and a new banking culture. The driving forces of these transformations are competition and innovation.

The result of all the transformations should be the emergence of the bank of the future, the essence of which can be represented as the sum of four concepts: "Bank = technology", "Bank = client", "Bank = product", "Bank = life". It is hypothesized that a person in the digital economy can be described as Homo financial (a person with financial literacy and financial thinking).

References

1. Y. Gorodnichenko, J. Svejnar, K. Terrell, Globalization and innovation in emerging markets. *IZA Discussion Paper*, **3299** (2008)
2. D. D. Dorn, G. H. Hanson, G. Pisano, P. Shu, Foreign competition and domestic innovation: evidence from U.S. patents. *NBER Working Paper*, **22879** (2016)
3. M. Vokoun, Globalization and Innovation: Innovation activities of multinationals in the Czech Republic 1998-2010. *Globalization and Its Socio-Economic Consequences. 16th International Scientific Conference Proceedings*, **5** (2016)
4. J. Tobin, Financial Globalization. *World Development,* 28(**6**) (2000)
5. A. N. Berger, R. DeYoung, H. Genay, G. F. Udell, Globalization of financial institutions: evidence from cross-border banking performance. *Brookings-Wharton Papers on Financial Services,* **3** (2000)
6. E. S. Prasad, K. Rogoff, S.-J. Wei, M. A. Kose, Effects of financial globalization on developing countries: some empirical evidence. *IMF Occasional Paper*; **220** (2003)
7. R. M. Stulz, Presidential address: the limits of financial globalization. *The Journal of Finance*, **60**(4) (2005)
8. B. King, Bank 3.0: Why banking is no longer somewhere you go, but something you do. *Marshall Cavendish,* (2012)
9. G. Wehinger, Banking in a challenging environment: Business models, ethics and approaches towards risks. *OECD Journal*, **2** (2012)
10. Y. I. Korobov, Развитие банковской системы: модернизация или трансформация [Development of the banking system: modernization or transformation]. *Наука и общество [Science and Society]*, **3** (2015)
11. Y. I. Korobov, Сущность банка и ее эволюция в современных условиях [The essence of the bank and its evolution in modern conditions]. *Банковские услуги [Banking Services]*, **7-8** (2005)

12. O. I. Lavrushin, *Деньги, кредит, банки [Money, credit, banks]*, 15th ed., KnoRus, (2016)
13. Ch. Pass, B. Lowes, L. Davies, *Collins Dictionary of Economics*, 4th ed., Glasgow: HarperCollings (2005)
14. BusinessDictionary, *Banking system* [online]. Available at: http://www.businessdictionary.com/definition/banking-system.html (2017)
15. R. P. Davis, A. Shrader, Leading for Growth: How Umpqua Bank Got Cool and Created a Culture of Greatness. *Jossey-Bass,* (2007)
16. Y. I. Korobov, Новое понимание банковской культуры [New understanding of banking culture], *Финансы и кредит [Finance and Credit]*, **15** (2012)

Modelling bankruptcy prediction models in Slovak companies

Maria Kovacova[1,*] and *Jana* Kliestikova[1]

[1]Department of Economics, Faculty of Operation and Economics of Transport and Communications, University of Zilina, Univerzitna 1, 010 26 Zilina, Slovakia

Abstract. An intensive research from academics and practitioners has been provided regarding models for bankruptcy prediction and credit risk management. In spite of numerous researches focusing on forecasting bankruptcy using traditional statistics techniques (e.g. discriminant analysis and logistic regression) and early artificial intelligence models (e.g. artificial neural networks), there is a trend for transition to machine learning models (support vector machines, bagging, boosting, and random forest) to predict bankruptcy one year prior to the event. Comparing the performance of this with unconventional approach with results obtained by discriminant analysis, logistic regression, and neural networks application, it has been found that bagging, boosting, and random forest models outperform the others techniques, and that all prediction accuracy in the testing sample improves when the additional variables are included. On the other side the prediction accuracy of old and well known bankruptcy prediction models is quiet high. Therefore, we aim to analyse these in some way old models on the dataset of Slovak companies to validate their prediction ability in specific conditions. Furthermore, these models will be modelled according to new trends by calculating the influence of elimination of selected variables on the overall prediction ability of these models.
Key words: bankruptcy, prediction models, company

1 Introduction

Prediction of financial distress is very critical in enterprise risk management, especially for financial institutions. In particular, financial institutions have to develop various risk management models, such as bankruptcy prediction and credit scoring models. For bankruptcy prediction, financial institutions need effective prediction models in order to make appropriate lending decisions. On the other hand, credit scoring models are used for the management of large loan portfolios and/or credit admission evaluation.

Also the 2008 global financial crisis and its aftermath have shown the ability to predict bankruptcy to be a vital management skill. So methodologies used for that purpose should

* Corresponding author: maria.kovacova@fpedas.uniza.sk

be as close to reality as possible as bankruptcy prediction is one of the most important business decision-making process [1].

Default of companies as well as default of other subjects represents occasion which happens not only in the market economy. Even with the maximum effort is not possible to avoid bankruptcy. Consequences of these defaults are extensive and have impact not only on the company itself, but it influences all stakeholders and if we look on the company as on the part of market economy it effects all its participants.

The problem of bankruptcy forecasting is one of the most actively studied nowadays among practical and theoretical issues of company management. Assessment of the current financial status and determination of bankruptcy probability are of interest to shareholders, suppliers, creditors and others aiming is to deal with perspective and reliable business partners.

By present, many models of bankruptcy forecasting have been developed, but this area remains still a field of research activity. On the other side little is known about the practical application of existing models mainly because the use of existing models is limited by the conditions in which they are developed. Another question concerns the factors that can be significant for forecasting [2].

This is given by the fact that since different economic environments have various properties that do not allow reusing models and related sets of factors in other conditions. This fact has been also confirmed by comparative studies of models for different countries [3, 4].

For developing countries, with the economic structure significantly different from the developed countries this fact is particularly significant [5].

Consequently, respecting the prerequisite that efficient methods can only be applied to companies of the same group in the same economic situation in the period for which the models were originally designed, we aim to analyse, these in some way old models on the dataset of Slovak companies, to validate their prediction ability in specific conditions.

Furthermore, these models will be modelled according to new trends by calculating the influence of elimination of selected variables on the overall prediction ability of these models.

2 Literature Review

Currently a huge number of studies are devoted to the issue of bankruptcy prediction. All researches can be divided into two groups. First group of researchers focus their attention on forecasting methodology while second group concentrate on the choice of proper variables providing the highest prediction ability.

Based on the used forecasting methodology can be bankruptcy prediction models divided into three main categories: statistical models, artificial intelligence and theoretical models [6]. Until the end of twenty century the basic prediction methodologies were based on statistical methods, such as the most common discriminant analysis [7, 8, 9], logit [10] and probit [11, 12]. These statistical methods are quiet popular among researchers and practitioners even nowadays as their prediction ability is relatively high and the way of processing not so difficult.

On the other side in the beginning of 21st century artificial intelligence expert systems including machine-learning techniques became the primary method for bankruptcy prediction. Mainly artificial neural networks (ANN) were used [13]. Although the prediction ability of ANNs is comparatively higher there are reasonable limitations such as the need to have a great experience in order to select the control parameters properly and difficulties with building the model itself [14, 15].

In addition, some authors have investigated the applicability of other artificial intelligence methods to bankruptcy prediction, for example, the principal component analysis [16, 17], support vector machines [18, 19], decision trees [20, 21], rough sets, [12, 22], data envelopment analysis [23, 24], and others.

Focusing on the second group of researchers concentrating on the choice of proper variables providing the highest prediction ability can be stated that financial indicators are the most popular [25, 21].

In addition to these indicators are added other, such as corporate management factors [26, 27], environmental factors [28, 29], level of legislation development [30] and so on.

Finally, we can summarize that forecasting models usually include indicators from all of these factors [31]. Additionally can be noted that in spite of the average number of indicators is ten, a large number of indicators doesn´t increase the predictive ability of the model [32]. It is important also to note that even though there are more than 500 prediction models worldwide, many of which have shown high predictive ability, they are poorly used in practice and the focus is mainly on developing new ones reflecting specific individual environment.

Therefore, we aim to analyse these in some way old models on the dataset of Slovak companies to validate their prediction ability in specific conditions. Furthermore, these models will be modelled according to new trends by calculating the influence of elimination of selected variables on the overall prediction ability of these models.

3 Data and Methodology

For the presented study we have used dataset of annual financial reports of Slovak companies, which were gained through the Register of financial statements, Ministry of Finance of the Slovak Republic. Calculations were provided on the basis of the year 2015.

To validate the prediction ability of selected old models in specific conditions of Slovak environment we had to set criterions to classify company as bankrupt or non-bankrupt. These criterions were set according to Slovak legal system in combination with other significant characteristics respecting specifics of Slovak environment [33, 34].

Based on the given criterions we consider company as bankrupt if:

- negative value of earnings after taxes,
- the value of financial independence indicator is less than 0.04,
- the value of current ratio indicator is less than 1,
- company has at least two liabilities 30 days after due date from different creditors,
- the total amount of payable and not payable liabilities is higher that the value of company´s assets.

According to these specifics we have randomly chosen 768 companies (50% bankrupt and 50% non-bankrupt) for which we calculated selected bankruptcy prediction models to validate their prediction ability in specific conditions.

Additionally, these models will be modelled according to new trends by calculating the influence of elimination of selected variables on the overall prediction ability of these models.

To fulfil the aim of the presented study and further calculations we selected bankruptcy prediction models shown in table 1, applied them on the dataset of Slovak companies to validate prediction ability of these models.

DOI: 10.1051/shsconf/20173901013

Table 1. Description of applied bankruptcy prediction models. (Source: self-processed based on [11,35,36, 37, 9])

Author	Year	Country of origin	Formula	Indicators	Criterion for bankrupt
Zmijewski	1984	USA	$ZM = -4,336 - 4,513X_1 + 5,679X_2 + 0,004X_3$	X1 = Net Income / Total Assets X2 = Total Liabilities / Total Assets X3 = Current Assets / Current Liabilities ZM = Overall index	$P(b) = \frac{1}{\left(1 + \exp^{(-ZM)}\right)}$ P(b) > 0,5
Taffler	1983	UK	$TM = 0,53X_1 + 0,13X_2 + 0,18X_3 + 0,16X_4$	X1 = Earnings Before Taxes / Short Term Liabilities X2 = Current Assets / Total Liabilities X3 = Short Term Liabilities / Total Assets X4 = Sales / Total Assets TM = Overall index	TM > 0,3 Enterprise creates value 0,2 ≤ TM ≤ 0,3 Grey zone TM < 0,2 Enterprise is going to bankruptcy
Fulmer	1984	USA	$H\ Factor = 5.528X_1 + 0.212X_2 + 0.73X_3 + 1.27X_4 - 0.12X_5 + 2.335X_6 + 0.575X_7 + 1.083X_8 + 0.894X_9 - 6.075$	X1 = Avg Retained Earnings / Average Total Assets X2 = Revenues / Average Total Assets X3 = EBIT / Total Equity X4 = Cash Flows from Operations / Avg Total Debt X5 = Avg Total Debt / Total Equity X6 = Total Current Liabilities / Avg Total Assets X7 = log (Avg Tangible Assets) X8 = Avg Working Capital / Avg Total Debt X9 = log (EBIT) / Interest Expense	FM > 0 Good financial health of the company FM < 0 Bad financial health of the company
Altman	2017 (originally 1968)	USA	$Z = 0,035 - 0,495X_1 - 0,862X_2 - 1,721X_3 - 0,017X_4$	X1 = Working Capital / Total Assets X2 = Retained Earnings / Total Assets X3 = EBIT / Total Assets X4 = Book Value of Equity / Book Value of Total Liabilities Z = Overall index	Z > 2,6 "Safe" Zone 1,1 < Z < 2,6 "Grey" Zone Z < 1,1 "Distress" Zone
Springate	1978	Canada	$SM = 1,03X_1 + 3,07X_2 + 0,66X_3 + 0,4X_4$	X1 = Net Working Capital / Total Assets X2 = EBIT / Total Assets X3 = Earnings Before Taxes / Short Term Liabilities X4 = Sales / Total Assets SM = Overall index	0,862 < S Good financial health of the company S < 0,862 Possible financial problems of the company

DOI: 10.1051/shsconf/20173901013

Confusion matrix and receiver operating characteristic curve (ROC) are considered as the most relevant and appropriate tools for the validation of predictive ability of tested models [38].

Table 2. Confusion matrix (Source: self-processed)

		Predictive value		
		0 (non-bankrupt)	1 (bankrupt)	
Actual value	0 (non-bankrupt)	TP	FN	TP+FN
	1 (bankrupt)	FP	TN	FP+TN
		TP+FP	FN+TN	Total

Confusion matrix (see table 2) can be stated as contingency table linking the number of proper and improper company′ classification based on the actual and predicted values and indicating following relevant characteristics:
- Type I. error (false positive) – percentage of bankrupt companies predicted by the model as non-bankrupt,
- Type II. error (false negative) – percentage of non-bankrupt companies predicted by the model as bankrupt,
- sensitivity (positive predictive value) – percentage of correct classification of non-bankrupt companies,
- specificity (negative predictive value) – percentage of correct classification of bankrupt companies,
- model accuracy – overall prediction ability of the model according to data set of companies also known as accuracy of prediction model.

4 Results and Discussion

To fulfil the given aim of the presented study we have provided calculation and application of five famous bankruptcy prediction models presented in table 1 on the dataset of Slovak companies. Results are shown in following table 3 and table 4.

Based on these results can be assumed that all calculated prediction models have reached about 50% prediction accuracy on the dataset of Slovak companies. This is given mainly because of the high type I. error in models of Fulmer (62%), Altman (99%) and Springate (78%).

On the other side in models of Zmijewski and Taffler the type I. error and type II. error were similar about 50%. Accoriding to provided calculation we can assume that we haven′t found important differences between tested model′s ability to predict future bankrupt of the company.

Table 3. Calculated confusion matrix of chosen models (Source: self-processed)

Zmijewski model		*Predictive value*		
		0 (non-bankrupt)	1 (bankrupt)	
Actual value	0 (non-bankrupt)	201	183	384
	1 (bankrupt)	186	198	384
		387	381	768
Taffler model		*Predictive value*		
		0 (non-bankrupt)	1 (bankrupt)	
Actual value	0 (non-bankrupt)	231	153	384
	1 (bankrupt)	205	179	384
		436	332	768
Fulmer model		*Predictive value*		
		0 (non-bankrupt)	1 (bankrupt)	
Actual value	0 (non-bankrupt)	273	111	384
	1 (bankrupt)	237	147	384
		510	258	768
Altman model		*Predictive value*		
		0 (non-bankrupt)	1 (bankrupt)	
Actual value	0 (non-bankrupt)	375	9	384
	1 (bankrupt)	380	4	384
		755	13	768
Springate model		*Predictive value*		
		0 (non-bankrupt)	1 (bankrupt)	
Actual value	0 (non-bankrupt)	322	62	384
	1 (bankrupt)	299	85	384
		621	147	768

Calculations of relevant characteristics regarding assessing tested models are provided in table 4.

Table 4. Calculated characteristics of chosen models (Source: self-processed)

Model	*Zmijewski model*	*Taffler model*	*Fulmer model*	*Altman model*	*Springate model*
Type I. error	0.4844	0.5339	0.6172	0.9896	0.7786
Type II. error	0.4766	0.3984	0.2891	0.0234	0.1615
Sensitivity	0.5234	0.6016	0.7109	0.9766	0.8385
Specificity	0.5156	0.4661	0.3828	0.0104	0.2214
Model accuracy	51.95%	53.39 %	54.69 %	49.35 %	52.99 %

Furthermore, we have modelled these models according to new trends by calculating the influence of removal of individual variables on the overall prediction ability of these models. Therefore, we provided new calculations for each model while removing each indicator and testing the new overall prediction ability. The aim was to assess the impact of these removal on the model accuracy, which will serve as a basis for future research.

Table 5. Calculated new model's accuracies after removal of individual indicators (Source: self-processed)

Model	Model accuracy	Indicator removed	New model accuracy
Zmijewski model	51,95 %	X1 = Net Income / Total Assets	51.43%
		X2 = Total Liabilities / Total Assets	50.65%
		X3 = Current Assets / Current Liabilities	52.21%
Taffler model	53,39 %	X1 = Earnings Before Taxes / Short Term Liabilities	53.65%
		X2 = Current Assets / Total Liabilities	51.95%
		X3 = Short Term Liabilities / Total Assets	52.73%
		X4 = Sales / Total Assets	53.91%
Fulmer model	54,69 %	X1 = Avg Retained Earnings / Average Total Assets	54.95%
		X2 = Revenues / Average Total Assets	54.95%
		X3 = EBIT / Total Equity	54.95%
		X4 = Cash Flows from Operations / Avg Total Debt	54.17%
		X5 = Avg Total Debt / Total Equity	54.43%
		X6 = Total Current Liabilities / Avg Total Assets	52.99%
		X7 = log(Avg Tangible Assets)	53.65%
		X8 = Avg Working Capital / Avg Total Debt	52.99%
		X9 = log(EBIT) / Interest Expense	54.04%
Altman model	49,35 %	X1 = Working Capital / Total Assets	49.22%
		X2 = Retained Earnings / Total Assets	49.22%
		X3 = EBIT / Total Assets	50.00%
		X4 = Book Value of Equity / Book Value of Total Liabilities	49.35%
Springate model	52,99 %	X1 = Net Working Capital / Total Assets	52.21%
		X2 = EBIT / Total Assets	51.30%
		X3 = Earnings Before Taxes / Short Term	53.52%

		Liabilities	
		X4 = Sales / Total Assets	53.26%

Results of modelling provided in table 5 have shown us that removal of individual indicators of models doesn't lead to significant improvement of their overall prediction accuracy. There are only small differences gained by provided calculations and mainly the removal has led to decrease of ability of the model to predict future bankrupt of the company.

These findings confirmed that there is a need to adjust bankruptcy prediction models to different economic environments. These contrasts cause the limited options to reuse models and related sets of indicators in other conditions. This is highly important also in the case of Slovak Republic. Our calculations have shown that the use of foreign models has led only to 50% prediction accuracy. Although these models are well known and were constructed in developed countries their applicability in developing countries, with the economic structure significantly different from the developed countries this fact is particularly substantial [5].

Our findings are in strong correlation with findings of other research studies confirming the need to develop specific models for different countries [3]. Additionally also new indicators, not only financial, should be included for forecasting in such surroundings [31].

This limitation was in some way eliminated by designation of a predictive model that can be easily adapted to any new situation, both in terms of model architecture and the indicators used [2].

5 Conclusion

In spite of numerous researches focusing on forecasting bankruptcy using not only traditional statistics techniques, but also artificial intelligence models, they are rarely used in practise. This is given mainly by the fact which was confirmed also in the presented study that the prediction ability of the model developed in specific conditions of individual country can't be successfully applied in the conditions of other country or environment.

Therefore, we aim to analyse in some way old models on the dataset of Slovak companies to validate their prediction ability in specific conditions. Results have shown that the prediction accuracy of these models was about 50%, which is quiet low.

Furthermore, these models were modelled according to new trends by calculating the influence of elimination of selected variables on the overall prediction ability of these models. This modelling hadn't led to significant improvement of the overall prediction accuracy of tested models. There were only small differences gained by provided calculations and mainly the removal had led to decrease of ability of the model to predict future bankrupt of the company.

Results obtained in the presented study have confirmed in some way conclusions of previously published researches. However, the issue of bankruptcy prediction is widely spread worldwide for many years and there have been developed numerous models, since nowadays there hasn't been developed any generally accepted bankruptcy prediction model which can be generally applied regardless specific environment of individual countries.

Acknowledgements

This work was supported by the Slovak Research and Development Agency under Grant number APVV-14-0841: Comprehensive Prediction Model of the Financial Health of Slovak Companies.

References

1. E. Kirkos, Assesing methodologies for intelligent bankruptcy prediction. *Art. Int. Rev.*, **43**, 83-123 (2015)
2. Y. Zelenkov, E. Fedorova, D. Chekrizov, Two-step classification method based on genetic algorithm for bankruptcy forecasting. *Exp. Sys. App.*, **88**, 393-401 (2017)
3. D. Liang, Ch. Tsai, H. Wu, The effect of feature selection on financial distress prediction. *Know. Bas. Sys.*, **73**, 289-297 (2015)
4. Y. Peng, G. Wang, G. Kou, Y. Shi, An empirical study of classification algorithm evaluation for financial risk prediction. *App. S. Comp*, **11**(2), 2906-2915 (2011)
5. E. Fedorova, E. Gilenko, S. Dovzhenko, Bankruptcy prediction for Russian companies: Application of combined classifiers. *Exp. Sys. App.*, **40**(18), 7285-7293 (2013)
6. M. A. Aziz, H. A. Dar, Predicting corporate bankruptcy: Where we stand? *Corp. Gov. Int. J Bus. Soc.*, **6**, 18-33 (2006)
7. E. I. Altman, Financial ratios, discriminant analysis and the prediction of corporate bankruptcy. *J Fin.*, **23**(4), 589-609 (1968)
8. R.O. Edmister, Financial ratios as discriminant predictors of small business failure. *Journal of Finance*, **27**(1), 139-140 (1972)
9. G.L.V. Springate, *Predicting the possibility of failure in a Canadian firm* (Unpublished master´s thesis), Simon Fraser University, Canada 42 (1978)
10. J. A. Ohlson, Financial ratios and the probabilistic prediction of bankruptcy. *J Acc. Res.*, **18**(1), 109-131 (1980)
11. M.E. Zmijewski, Methodological issues related to the estimation of financial distress prediction models. *J. Acc. Res.*, **22**, 59-82 (1984)
12. A.A. Kasgari, M. Divsalar, M.R. Javid, S.J. Ebrahimian, Prediction of bankruptcy Iranian corporations through artificial neural network and Probit-based analyses. *Neur. Com. App.*, **23**(3-4), 927-936 (2013)
13. J.L. Bellovary, D.E. Giacomino, M.D. Akers, A review of bankruptcy prediction studies: 1930 to present. *J Fin. Edu.*, **33**, 1-42 (2007)
14. N. Gordini, A genetic algorithm approach for SMEs bankruptcy prediction: empirical evidence from Italy. *Exp Sys. App.*, **41**(14), 6433-6445 (2014)
15. D. Santos, E. M. Sabourin, P. Maupin, Overfitting cautious selection of classifier ensembles with genetic algorithms. *Inf. Fus.*, **10**(2), 150-162 (2009)
16. M. V. Achim, C. Mare, S. N. Borlea, A statistical model of financial risk bankruptcy applied for Romanian manufacturing industry. *International Conference on Emerging Markets Queries in Finance and Business*, **3**, 132-137 (2012)
17. M. Onofrei, D. Lupu, The modelling of forecasting the bankruptcy risk in Romania. *Ec. Com. Ec. Cyb. St. Res.*, **48**(3), 197-215 (2014)
18. D. Alaminos, A. DelCastillo, M. A. Fernandez, A global model for bankruptcy prediction. *Pl. One*, **11**(11) (2016)

19. D. Zhao, C. Y. Huang, Y. Wei, F. H. Yu, M. J. Wang, H. L. Chen, An effective computational model for bankruptcy prediction using kernel extreme learning machine approach. *Com. Eco.,* **49**(2), 325-341 (2017)
20. C.F. Tsai, Y.F. Hsu, D.C. Yen, A comparative study of classifier ensembles for bankruptcy prediction. *App. S. Com.,* **24**, 977-984 (2014)
21. N, Zieba, S.K. Tomczak, J.M. Tomczak, Ensemble boosted trees with synthetic features generation in application to bankruptcy prediction. *Exp. Sys. App.,* **58**, 93-101 (2016)
22. M. Virag, T. Nyitrai. Is there a trade-off between the predictive power and the interpretability of bankruptcy models? The case of the first Hungarian bankruptcy prediction model. *Ac. Oec.,* **64**(4), 419-440 (2014)
23. I. M. Premachandra, Y. Chen, J. Watson, DEA as a tool for predicting corporate failure and success: a case of bankruptcy assessment. *Om. Int. J Man. Sc.*, **39**(6), 620-626 (2011)
24. M. M. Mousavi, J. Ouenniche, B. Xu, Performance evaluation of bankruptcy prediction models: an orientation-free super-efficiency DEA-based framework. *Int. Rev. Fin. An.,* **42**, 64-75 (2015)
25. P. DuJardin, A two-stage classification technique for bankruptcy prediction. *Eur. J Op. Res.* **254**(1), 236-252 (2016)
26. C. Salloum, N. Azoury, Corporate governance and firms in financial distress: evidence from a Middle Eastern country. *Int. J Bus. Gov. Eth.*, **7**(1), 1-17 (2012)
27. X. Bredart, Financial distress and corporate governance: the impact of board configuration. *Int. Bus. Res.*, **7**(3), 72 (2014)
28. M. Karas, M. Reznakova, To what degree is the accuracy of a bankruptcy prediction model affected by the environment? The case of the Baltic States and the Czech Republic. *Pr. Soc. Beh. Sc.,* **156**, 564-568 (2014)
29. V. Delas, E. Nosova, O. Yafinovych, Financial security of enterprises. *Pr. Ec. Fin.,* **27**, 248-266 (2015)
30. M. Rowoldt, D. Starke, The role of governments in hostile takeovers – evidence from regulation, anti-takeover provisions and government interventions. *Int. Rev. Law Ec.,* **47**, 1-15 (2016)
31. M. H. Tinoco, N. Wilson, Financial distress and bankruptcy prediction among listed companies using accounting, market and macroeconomic variables. *Int. Rev. Fin. An.,* **30**, 394-419 (2013)
32. T. Kliestik, J. Majerova, Selected issues of selection of significant variables in the prediction models. *Financial management of firms and financial institutions: 10th international scientific conference*, 537-543 (2015)
33. V. Bartosova, P. Kral, A methodological framework of financial analysis results objectification in the Slovak Republic. *European Proceedings of Social & Behavioural Sciences*, 17, 189-197 (2016)
34. M. Durica, P. Adamko, Verification of MDA bankruptcy prediction models for enterprises in Slovak Republic. *10th international scientific conference International Days of Statistics and Economics*, 400-407 (2016)
35. R. J. Taffler, M. Tseung, The audit going-concern in practice. *Pr. Acc. Mag.,* **88**, 263-269 (1984)
36. J. G. Fulmer, J. E. Moon, T. A. Gavin, M. J. Erwin, A bankruptcy classification model for small firms. *J Com. B. Len.*, 25-37 (1984)

37. E. I. Altman, M. Iwanicz-Drozdowska, E.K. Laitinen, Financial distress prediction in an international context: a review and empirical analysis of Altman's Z-score model. *J Int. Fin. Man. Acc.*, **28**(2), 131-171 (2017)
38. P. Adamko, L. Svabova, Prediction of the risk of bankruptcy of Slovak companies. *International Scientific Conference Managing and Modelling of Financial Risks*, 15-20 (2016)

Ukrainian-Chinese collaboration: Prospects of development

Oleh Kratt[1*], *Kateryna* Pryakhina[2], and *Maryna* Bilyk[3]

[1]Doctor of Economics, professor, Kremenchuk Mykhailo Ostrohradskyi National University, Ukraine
[2]PhD candidate, Kremenchuk Mykhailo Ostrohradskyi National University, Ukraine
[3]PhD student, Kremenchuk Mykhailo Ostrohradskyi National University, Ukraine

Abstract. In article the current state of the Chinese-Ukrainian relations is analysed, the priority directions of strategic cooperation are considered. The analysis of economic development of Ukraine and China with definition of positive and negative tendencies is carried out. It is certain that there is a mutual aspiration of the parties before increase in volumes of bilateral trade, diversification of its structure, development of long-term forms of economic cooperation, deepening of investment interaction between Ukraine and China. The analysis of the study is conducted during 2010-2016 years. We have identified prospective investment sectors within the framework of cooperation between Ukraine and China - Transport Infrastructure, Agriculture, Renewable energy, Telecommunications, IT, educational sphere. The current dynamics of the trade balance between Ukraine and the People's Republic of China, taking into account the considerable predominance of imports, requires a thorough review of trade policy measures: to improve the structure of trade, introducing new forms of trade relations, To develop the institutional framework for cooperation, To search for ways to reduce the energy intensity of export industries, To stimulate the means of tariff and non-tariff regulation of imports into Ukraine of mainly investment high-tech goods. The forecast of possible risks according to each perspective direction of development is made.
Key words: collaboration, development, analysis, economic relations, prospects

1 Introduction

For this time People's Republic of China is one of the world economic leaders whose value in high gear grows in many spheres of the international life. The certificate of it is election of national monetary unit of China - yuan - world reserve currency on the basis of which the cost of special drawing right of the IMF is calculated. Ukraine, in turn, has rather high economic potential, world achievements in some fields of science and technology, an advantageous geopolitical position which does it by a zone of certain economic and political interests, in particular for People's Republic of China. The mutual understanding of these realities also became the main prerequisite of formation and development of the Ukrainian-Chinese relations. The analysis of the study is conducted during 2010-

[*] Corresponding author: katerinapryakhina@gmail.com

2016 years, import and export indicators and the balance of foreign trade balance used. The forecast of possible risks according to each perspective direction of development is made.

The purpose of article is the research of character and the prospects of development of the modern Ukrainian-Chinese economic relations.

2 Data and Methods

The scientific article uses general scientific methods: statistical analysis for studying, grouping, comparison of indices of total turnover of countries, graphic method - for visual representation of tables and schemes. Promising directions for the development of relations between China and Ukraine are analyzed and the risk forecast for investors is synthesized.

3 Results

To date, a significant number of Ukrainian scholars view the Chinese vector of Ukraine's foreign policy as one of the highest priority. Works by Pron [1] devoted to the history of the formation and development of Ukrainian-Chinese relations. The author believes that countries have deep and strong historical traditions that cover multidectoral spheres of politics, economics, and culture. Among the latest publications, it is worth highlighting articles by Holod [2], which explores China's investment expansion in terms of prospects for developing relations with Ukraine. Vysotskaya [3] in her studies carried out a thorough analysis of the main trends and prospects for the development of Ukrainian-Chinese economic relations. In the scientific work of Rogovuy [4] discusses the prospects and risks of developing Ukrainian-Chinese trade and economic relations. Ukrainian-Chinese economic cooperation in the context of foreign economic security it took place in scientific works by Levkivskyi [5]. The mechanism of international economic relations between the two countries is considered and the priority promising areas for further development of foreign economic relations between China and Ukraine are singled out. Grodsky [6] researches modern trade and investment cooperation between Ukraine and China. The strategic guidelines for intensify cooperation between Ukraine and China in the context of globalization are outlined. On the special role of bilateral cooperation to strengthen economic presence of Ukraine in the Asian region is focused [6]. Position of Ukraine in the PRC’s system of foreign economic relations it took place to be in scientific works by Vlasenko L. In the article foundations of Chinese theory of international relations have been considered, differences between Chinese and classical Western concepts of international interactions were highlighted. Existing hierarchy of PRC’s strategic partnership was analyzed [7]. In scientific work from Cheng [8], the choice and strategic opportunities of China in the context of «Ukrainian crisis» was analyzed the determining factors that influenced the choice of China in the context of «Ukrainian Crisis», taking into account the geo-economic and geo-political interests of China. It is also proved that contrary to the European Union and the United States of America, who often use political tools to resolve conflicts (which often leads to disputes and conflicts escalation), China offers economic settlement mechanisms that are acceptable to all stakeholders. Halperina [9] is studying a question about System adaptation of social and economic model of China to conditions of global instability. The features of the system adapt social and economic model of China to the conditions of global instability. Thesis determined new conditions of socio-economic model of China.

In the years of Ukraine’s independence, these relations have generally followed an upward trend and reached a peak in December 2013 when the Treaty of Friendship and

Cooperation was signed.China's interest in cooperation with Ukraine is also linked to the decision of the Chinese leadership to create "foreign food bases". In particular, it includes the use of Ukraine's agricultural opportunities combined with China's investment and technological capacities. The current Ukraine-China relations are largely influenced by China's growing political and economic interests regarding the countries of the EU and New Eastern Europe (Belarus, Ukraine, Moldova, Latvia, Lithuania, and Estonia). China plans to implement its global strategy to increase exports and investment in potential markets located between Russia and the EU. Ukraine could become an important place to promote Chinese products and brands, gain access to new markets, and acquire strategic assets. Despite China's interest towards Ukraine, these relations were put on hold since the beginning of the Revolution of Dignity. Likely, this is because rapprochement with Ukraine can go only as far as China will be able to maintain its safe balance in the geopolitical US-Russia-China triangle and in the world [10].

It costs to mark, that world economic relations represent a combination of different forms of business interaction that occurs between countries of the world, and primarily include trade and economic cooperation, the flow of investment flows, scientific and technical ties, the exchange of technologies and intellectual property.

Therefore, it is advisable to look more closely at each of the forms of interaction between Ukraine and the People's Republic of China at the present time.

According to the General Customs Administration of the People's Republic of China, in 2016, the trade turnover between Ukraine and the PRC amounted to $ 6520.2 million. At the same time, Chinese exports to Ukraine - 4687.7 million. The US (increase by 32.2%), Chinese imports from Ukraine amounted to $ 1,832.5 million (decrease by 24.1%). Balance of bilateral trade in favor of PRC amounted to $ 2855.2 million [11]. The average growth rate $\bar{R}$ was used to substantiate the evaluation of the change in indicators.

Table 1. Dynamics of foreign trade turnover of Ukraine with China for 2010-2016, million USD

Year	2010	2011	2012	2013	2014	2015	2016	$\bar{R}$, %
Export	1316,6	2180	1777,2	2726,7	2674,1	2399,1	1832,5	5,6
Imports	4700,4	6268,3	7899,6	7903,2	5411,0	3771,0	4687,7	-1
Balance	-3383,8	-4088,3	-6122,4	-5176,6	-2736,8	-1371,9	-2855,2	-3
Total turnover	6017	8448,3	9676,8	10629,9	8085,1	6170,1	6520,2	1

The indicators of Table 1 show that the average growth rate of export is 5.6%, which indicates an increase in exports every year by 5.6%. In this case, the import rate decreases by 1% and Balance in 3%. The total turnover is increased every year by 1%. The highest indicator of Ukrainian export operations is for 2013 ($ 2726.7 million, Fig. 1). For the period from 2014 to 2016 there is a significant reduction in the volumes of exported goods, which is explained by the political situation in the south and east of Ukraine, under the influence of Russian aggression. This also applies to imported Chinese goods to Ukraine (a significant reduction in 2015 to $ 3771 million). The presence of a negative balance in bilateral trade with China puts Ukraine ahead of the task of finding ways to increase the volume of Ukrainian exports.

DOI: 10.1051/shsconf/20173901014

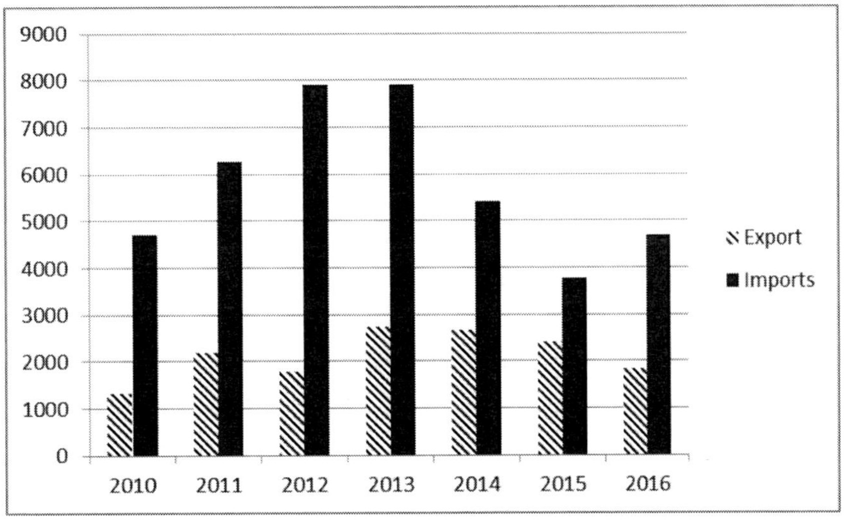

Fig. 1. Indicators of export-import operations of Ukraine with China for the period 2010-2016

Most specific gravity in the structure of export for today is occupied by supplying with mineral products, grain-crops, fats and oils of animal or vegetable origin. It is possible to mark, that the Ukrainian export to Republic of China has raw material character mostly. The main reason for this is to orient the Chinese market to self-sustaining and stimulate the domestic producer, correspondingly reducing the number of industries in which, given the lack of Chinese counterparts, significant imports would be available (Fig. 2).

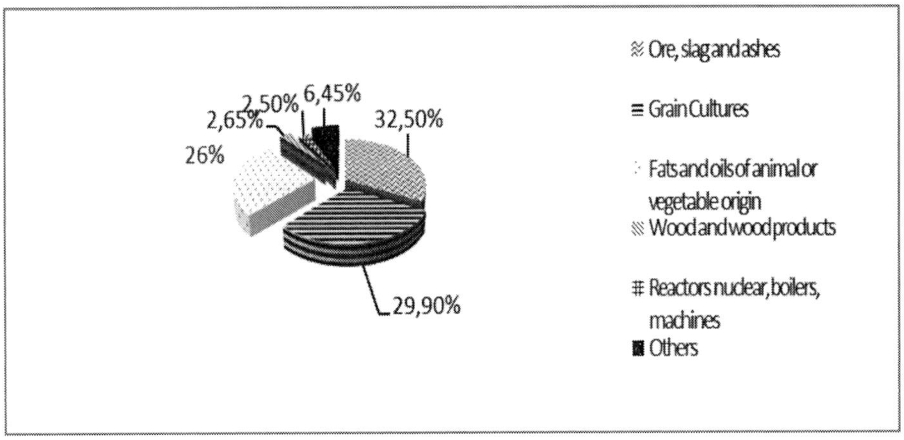

Fig. 2. Indicators of exports of Ukrainian products to China in 2016

The basis for imports from China in 2016 was: electric machines – 22.5%; nuclear reactors, boilers and machines – 17,1%; plastics, polymeric materials – 6.1%; ferrous metals – 4.4%; organic chemical compounds – 3.7%; various chemical products – 3.7%; footwear – 3.5%; articles of ferrous metals – 3% [12].

Summarizing the above-mentioned, it is possible to state that the export-import operations and development cooperation directions of the last years have shown that the increase in commodity turnover will occur due to increase in import to the Ukraine of Chinese products and services. Thus the volumes of home export in PRC have saltatory character (Fig. 1) and hardly will exceed a monthly index in the 250 million дол USA in a prospect.

In 2015 Ukraine joined initiative "One belt, signing one way" from PRC "Protocol about a collaboration within the framework of the Silk way", and now, actually, the searches of deepening and activations of trade and economic relation are conducted between countries within the framework of this protocol.

A bilateral investment collaboration does not answer possibilities of China and necessities of Ukraine as yet. The increase of volumes of import from PRC in 2016 is not accompanied by activation of investment collaboration (0,48% in the general volume of the attracted lines of foreign investments), the Ukrainian investments in PRC are practically absent also [11].

Priority for development of mutually beneficial investment collaboration are such spheres, as transport, agricultural, creation of infrastructural objects, space, innovative, aviation, scientific, technical and other spheres.

Basic investment projects are within the framework of collaboration between Ukraine and China.

1. Mechanical engineering. The Beijing Skyrizon Aviation Industry Investment Co Ltd company created with Ukrainian "Motor Sich" joint venture which is based already in China. Construction of a factory for the production and maintenance of aircraft engines was started in Chongqing City. At the AirshowChina 2016 exhibition the Beijing Skyrizon company declared a possibility of license production by joint venture of the D-136 helicopter engines, MC-500B and TV3-117VMA-SBM1V and engines for planes of general purpose A-450C, turbojet D436-148FM and D-18T, etc. Beijing Skyrizon finances production start, and the Ukrainian side provides technologiesA certain risk for our state is the lack of the strategic directions from use and loan of the Chinese technologies and innovations that could become a push to improvement of the technologies which are available in Ukraine and attraction of foreign experience.

2. Renewable energy. The CNBM International Corporation in Ukraine owns the 10 largest solar power plants located in the Mykolayiv and Odessa regions. Their capacity is 267 MW, while the power of all Ukrainian solar power plants reaches 500 MW. According to CNBM, the investment amounted to about $ 1 billion.

3. Agro-industrial complex The Chinese corporation COFCO Agri launched in 2016 in the Mykolaiv Sea Commercial Port a transshipment complex of grain and oilseeds with an annual capacity of 2.5 million tons. The investment project cost $ 75 million, it is already working properly, and from the Ukrainian water area there are steamships with grain for the People's Republic of China. Several elevators and an oil extraction plant are also controlled by COFCO Agri in Ukraine.

Beginning in 2013 around the village of Naumovka, located in the Chernihiv region in the north of Ukraine is a farm company, which is founded by the large Chinese Agricultural State Corporation Huangfantsyu. Today, In Venture Investment Group is leading a major player in the soybean market and soybean oil production in China, which is interested in purchasing or building refineries in Ukraine.

4. Sphere of telecommunications. The office of the Chinese company Xinwei Group, one of the leading telecommunication companies in the world, is opened in Kyiv in 2014. The Ukrainian telecommunications operator Prostate builds a 4-generation mobile broadband multimedia communications network based on the McWiLL technology of the Xinwei Group. 150 base stations have already been installed, which covered Kyiv and significant territory from Kharkiv to Mariupol, and the "strip" coverage runs through 4 eastern regions of Ukraine. Further informatization is an important tool for the development of all social spheres of Ukraine. It is in the common interest of Ukraine and China to continue to work on this issue [13].

 DOI: 10.1051/shsconf/20173901014

A number of investment projects which the Chinese investors in Ukraine intend to realize is perspective. From the last: waste recycling plant in the Lviv region, processing of waste in Ukraine, thermal power plant in the Zhytomyr region, the belt road near Kiev, the bridge in Kremenchuk, the airport in Zhytomyr and other infrastructure facilities.

Ukraine and China actively develop cooperation in the cultural and humanitarian sphere. The development of cooperation between Ukrainian universities, educational and scientific organizations of the Republic of China has positive dynamics and significant prospects for further growth. One of the priority areas of cooperation is the training of Chinese citizens in Ukrainian higher education institutions. The practice of creating joint institutes will test new teaching methods. Preparation and conducting internships for students, graduate students and young scientists is one of the priority areas of bilateral cooperation.

It should also be noted the intensification of the Ukrainian-Chinese dialogue at the expert level. The Ukrainian and Chinese experts often meet, communicate and exchange opinions at different conferences, meetings, forums A significant amount of expert meetings shows the mutual aspiration to find the solution for the existing problems. Expert discussions help to find compromises, to overcome differences, to strengthen bilateral dialogue and to develop a further strategy for the development of relations. We hope that current expert work will be implemented at the political and economic level in concrete intergovernmental agreements. Therefore, in our opinion, directions of cooperation between Ukraine and China in the following areas are promising (Table 2).

Table 2. Directions of cooperation between Ukraine and China

Sphere of cooperation	Development prospects	Risk forecast to investors
Transport Infrastructure	- development of a network of multimodal transport corridors using water (maritime) and rail links in the North-South and West-East directions. (One of the promising projects in the context of the development of trade routes is a deep sea port project that could actually double the capacity of the Ukrainian port infrastructure for the needs of the New Silk Road)	- legislation of Ukraine is imperfect in the field of customs, tax, investment regulation, - the renewal of the fleet of rolling stock of Ukrainian carriers is slow, - there is a low activity in the implementation of projects in Ukraine for the construction of high-speed roads (railways, motor-car), including paid, introduction and development of multimodal transport and logistics.
Agriculture	- export of meat and dairy products, seeds of oilseeds and nuts, - cooperation in the development of infrastructure and logistics, more active and close cooperation in the field of veterinary and phytosanitary, - cooperation in the field of geodatabase and geospatial data generation, - development of smart farming, - the construction of plants on the production of means of plant and fertilizer protection on the territory of Ukraine, the use of new technologies for raising the productivity and productivity of cattle breeding and poultry farming.	- the technical condition of agricultural enterprises does not meet the needs of production, - monopolization of the corporate agro-sector of the markets of resources, especially financial, channels of product sales, - restriction of the circulation of agricultural land.
Renewable energy	- energy efficiency and rational use of nature: (the branch of alternative energy - solar, river, wind, etc.).	- the risks of changing the legislative framework for a green tariff and the lack of an updated strategy for renewable energy in Ukraine by 2035, - the complexity of the tax system.
Telecommu-nications, IT	- new prospects opens for Ukraine a project for the creation of a new all-Ukrainian 4G network with the help of the Chinese side, - cooperation in the field of information technologies on the basis of Ukrainian IT Start-Up.	- the existence of legal and organizational barriers, - the complexity of the tax system, bureaucracy insufficient reliability of information infrastructure.
Educational sphere	- signing an agreement on the mutual increase of student quotas for exchange, - concluding agreements on cooperation and implementation of the China-Ukrainian double diploma program, - preparation and internship of graduate students and young scientists on the basis of mutual exchange, - conducting international educational forums, including The Forum of Rectors of Ukrainian and Chinese Universities, in order to expand the direct cooperation of universities.	- language barriers - deepening mutual learning of Chinese and Ukrainian languages, - the beginning of reforms in the educational sphere.

4 Discussion and Conclusion

Considering the conducted research it is possible to draw the following conclusions.

Firstly, Ukraine has an important geopolitical and geo-economic situation, making it an important transport and logistics hub within the framework of the Silk Road Infrastructure Project, which aims to connect Europe and Asia more closely. The Silk Road initiative gives Ukraine the technical ability to bypass the restrictions imposed by Russia on the transit of its products to the countries of Central AsiaMoreover, our country is interesting to China as a production site for further export to the EU. This aspect has a positive impact on the development of Ukrainian-Chinese economic relations.

Secondly, China is one However, the trade balance for Ukraine is negative. Imports of goods from China to Ukraine in 2016, according to state statistics, grew by 24.3% (to $ 4687.7 million), exports decreased by 24.1% compared to 2015 (before $ 1832.5 million). However there is a mutual aspiration of the parties before increase in volumes of bilateral trade, diversification of its structure, development of long-term forms of economic cooperation, deepening of investment interaction between Ukraine and China. The deepening of trade relations with the PRC positively affects Ukraine's economic growth, however, the current dynamics of the trade balance between Ukraine and the PRC, given the significant predominance of imports, requires careful review of trade policy measures. Therefore, our proposals are as follows:

1. To improve the structure of trade, introducing new forms of trade relations, in particular industrial cooperation, cooperation on a compensation basis and the organization of joint productions through the purchase of licenses.
2. To search for ways to reduce the energy intensity of export industries as it significantly affects the competitiveness of Ukrainian exports.
3. To develop the institutional framework for cooperation, which should include not only interstate macroeconomic, but also a number of sectoral bodies for establishing effective cooperation in specific areas.
4. To stimulate the means of tariff and non-tariff regulation of imports into Ukraine of mainly investment high-tech goods.
5. Creation of a common investment fund on parity terms.

Thirdly, China is interested to invest the investments in economy of Ukraine, considering its branched transport system, existence of a powerful scientific and industrial complex, access to the markets of the European Union. It is confirmed by a number of already realized investment projects (the creation of Beijing Skyrizon Aviation Industry Investment Co Ltd and Motor Sich joint venture, tart in 2016 in the Nikolaev sea trade port transshipment facility of grain and oil-bearing crops, etc). And today the complementary nature of economic systems, the presence of a significant potential for cooperation in the trade, economic, scientific and technical, and humanitarian sectors create the necessary prerequisites for the progressive development of mutually beneficial Ukrainian-Chinese relations.

References

1. S. Pron, China: Foreign Policy and Diplomacy in the Second Half of the 20th Century. *Monograph Nikolaev,* 147, (2012)
2. V. Holod, China Investment Expansion. *Prospects for Developing Relations with Ukraine, Chinese research studies*, 38-44 (2012)
3. M. Vysotskaya, Analysis of the main trends of development between Ukraine and China Strategy of development of Ukraine. *Economics, sociology, law,* 13-18 (2013)

4. V. Rogovuy, Prospects and risks of Ukrainian-Chinese trade and economic relations development. *Bulletin of Zhytomyr State Technological University*, 68 (2016)
5. V. Levkivskyi. Ukrainian-Chinese economic cooperation in the context of foreign economic security. Scientific herald of the Chernihiv State Institute of Economics and Management, *Series 1: Economics*, 95-103 (2013)
6. S. Grodsky, Modern trade and investment cooperation between Ukraine and China, *Economic analysis,* 202-206 (2013)
7. L. Vlasenko, Position of Ukraine in the PRC's system of foreign economic relations. Scientific herald of Uzhgorod National University. *Series: International Economic Relations and World Economy*, 15-19 (2017).
8. H. Cheng, The choice and strategic opportunities of China in the context of Ukrainian crisis. Foreign Trade: *Economics, Finance, Law,* 117-128 (2015).
9. L. Halperina, System adaptation of social and economic model of China to conditions of global instability. [online], Available at: http://ir.kneu.edu.ua/ (2016)
10. Euromaidan Press [online], Available at: http://euromaidanpress.com/ (2017)
11. Embassy of Ukraine in the People's Republic of China and Mongolia (in combination) [online], Available at: http://china.mfa.gov.ua/ua/ukraine-cn/trade (2017)
12. Official site of the State Statistics Service of Ukraine [online], Available at: http://www.ukrstat.gov.ua (2017)
13. Silk Road of Ukraine [online], Available at: http://silkroad.in.ua/ (2017)

 DOI: 10.1051/shsconf/20173901015

Market Opportunity and Policy Support for Chinese Old Aging Industry: An Application of Text Mining

Chenguang Li[1, 2,*]

[1]Business Management Department, North China University of Technology, 100144 Shijingshan Distinct, Beijing, China
[2]Enterprise Management Intelligent Decision Research Centre, North China University of Technology, 100144 Shijingshan Distinct, Beijing, China

Abstract. Text mining has become a vital method of natural and social science research. A policy documents text mining framework with three phases has been introduced. Using 52 policies of aging industry development, this paper explores five business opportunity and three areas policy support for Chinese old aging industry. These findings contribute both foreign and local the elder service enterprise.
Key words: text mining, market, opportunity, policy, age

1 Introduction

A growing consensus recognizes that the population aging has substantially impacted on economic growth. Classical works in population aging, focusing on the government consumption budget [1], the labour market [2], work model [3], etc. Albeit most scholars believe that social problems (such as financial strains comes from social security, Medicare and Medicaid and so on) is attributed to people are getting older faster, ageing population has been creating business opportunities for old aging industry.

China has the largest population in the world and represented about 33% of the Asian population. For the past few years, the population size was expending dramatically due to improvement in medicine and sanitation that was benefiting from rapid economic growth. And the population as a whole was aging rapidly. The investigations from National Bureau of Statistics of the People's Republic of China shows population aged over 60 has reached 231 million in 2016, and it will increase to 480 million by 2050.

When 1980s Chinese government pushed a "later, longer, fewer" approach that encouraged people to marry later, wait longer between children, and have fewer babies [4]. In 30 years, most families in cities have only one child. A few years later, the children from single-child parents will face what is known as the "4-2-1" phenomenon (when the child grows up to working age, he or she must have to care for two parents and four grandparents in retirement [5]. These middle-aged men have been bearing a tremendous burden.

* Corresponding author: dr.lcg@qq.com

So China's government encourages the development of aging industry. And most polices were set for getting more local and foreign companies involved in that industry development process. Scholars advise finding implications for public policy and identifying a profitable business opportunity is important to a company, especially in developing countries [6]. Chinese old ageing industry brings a great business opportunity. For that reason, it is necessary to analyse policies for mining market opportunity and government support.

Text mining applications have been developed for policy analysis [7], thus this study will use this method for exploring the market opportunity and government support. By reading over the official aging industry documents, the study explores the presence of meaningful clusters of terms that appear across documents in the collected corpus [8]. Then it uses semantics to make sense of the opportunity and support [9].

This article presents a framework for text mining. The remainder of this paper is organized as follows: Section 2 presents a framework of policy documents mining. Section 3 describes the text mining process. Section 4 provides the analysing results. Finally, we briefly give the suggestions in Section 5.

2 A Framework of Text Mining

Text mining, also known as text data mining or knowledge discovery, has become a vital research tool with the rising popularity of data mining [10]. This method, referring multidisciplinary theory, is the complex task with unstructured and fuzzy data processing. And the process includes information index, text analysis, feature extraction, clustering, categorization, networking, semantic analysis, machine learning, and data mining. In recent years, the commercial potential of text mining has been emphasized on account of most market information is contained in text documents.

This article proposes three phases of a policy mining process (as Fig. 1 illustrating). First, pre-treatment that transforms policy documents into a text mining set could normalize the policy text format. The policy intermediate form is semi-structured, which contains word segmentation, feature extraction, clustering, etc. by segmenting long text. There is no connected concept at this stage. Next, text mining, as a significant task, discover the relationship across features and concepts based on frequency and network analysis. In this phase, some appropriate filter and model can be useful in keywords' extraction. Then, analyzing the association of high frequency and centrality words in postprocessing of the concept intermediate form, the path semantics and topic could be explored. At last, the results are summarized from path meaning of path correlation.

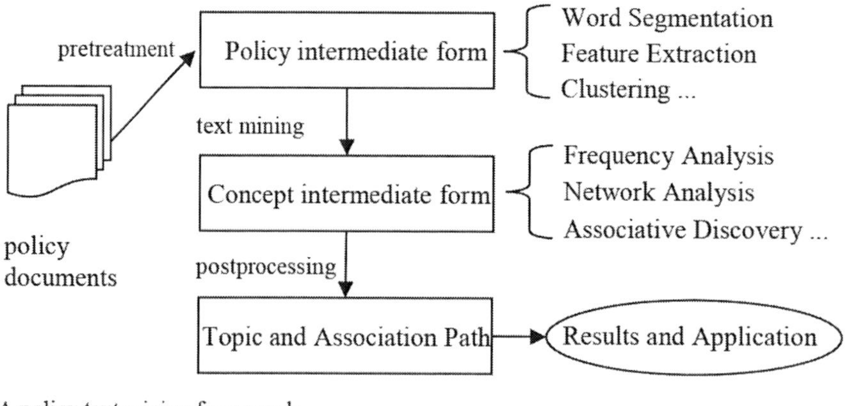

Fig. 1. A policy text mining framework

 DOI: 10.1051/shsconf/20173901015

Policy documents will be put into a set with unified format. Policy intermediate form should split statement, checking and removing interference and nonsense words. The concept form is prepared for frequency and association analysis based on word segmentation. Topic and association path mining depends on network analysis with the use of semantics.

3 Text Mining Process

3.1 Sampling

The study used 52 policies of aging industry development have been issued by Chinese Ministry of Civil Affairs, Commerce and 32 provinces' Department of Civil Affairs from 2013 to 2017 (as Table 1 shown). The sample is involving in planning, opinion, inform and law, which are the important policy forms in China.

Table 1. Aging Industry Development Policy sample

Type	Number
planning	32
Opinion	6
Inform	10
Law	4

3.2 Text Mining

Firstly, the policy documents are unified Unicode form and stored in text (.txt) document files, which are made up of the text library *V*(*d*). And $w_n(d)$ represents semantic words in each policy p_n, so we have

$$V(d) = (p_1, w_1(d); \ldots; p_i, w_i(d); \ldots; p_n, w_n(d)). \tag{1}$$

This study extracts 5251 segments after browsing through the policy text. By filtering nonsense words, there are 1741 words constitute policy intermediate form. Then, 302 words have been screened in feature extraction.

In the next place, word frequency statistics have been done for searching key words. However, to avoid topic highly correlated non-significant influence, it must remove some frequently used words such as "old", "aging", "policy", etc. Redundant words, should be added to filter list, need researchers to distinguish from synonyms.

By applying maximus matching method, the word segmentation weighting coefficient α_i can calculated with

$$a_i = \frac{TF_i * \log(\frac{N}{DF_i})}{\sqrt{\sum_k \left[TF_k * \log(\frac{N}{DF_k})\right]^2}} \tag{2}$$

where TF_i represents the frequency of words, DF_i describes the text's, *i* is occurrence, and *N*=52.

Thus, 40 highest frequency words have been found, which illustrates in Table 2.

 DOI: 10.1051/shsconf/20173901015

Table 2. Highest frequency words

Words	**Freq.**	**Words**	**Freq.**	**Words**	**Freq.**	**Words**	**Freq.**
care service	3643	culture	333	talent	202	quality	137
community	1013	service industry	290	project	200	security	134
medical treatment	831	standard	280	technology innovation	196	assistance	133
facility	660	rehabilitation	242	retirement homes	188	hospital	128
safeguard	602	resource	238	disease	186	fund	127
insurance	559	demand	232	training	181	civil rights	125
health	430	rescue	222	fitness	145	psychology	125
education	414	cooperation	221	subsidization	143	travel	120
nursing	412	sanitation	213	study	141	investment	119
family	368	market	207	charity	138	Internet & informatization	118

To clear the meaning of the words, co-word matrix has to be compiled based on the successive appearance of one word to another. Next, cluster analysis of whole high-frequency words, as Fig. 2 shown, demonstrates the association among semantics. This networking process can help to observe the distribution of internal relationship, which is described by the core words and others associated with them.

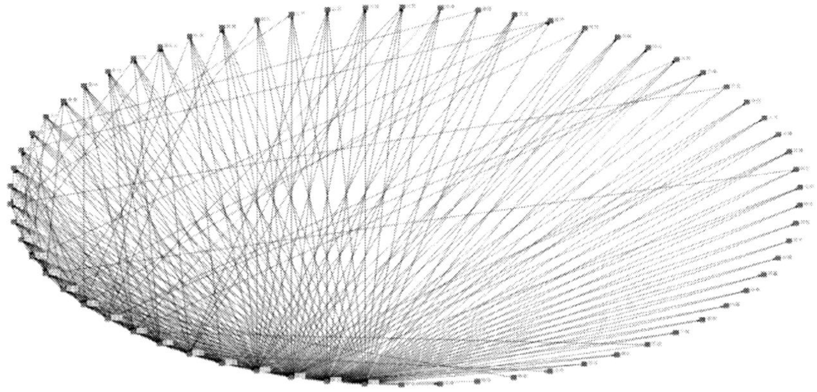

Fig. 2. Semantic network. The highest frequency words focus on care service, medical treatment, facility, safeguard, education, etc.

Then, the network analysis shows the centrality of the semantics, and measures the dispersion degree of the words. This article uses Ucinet 6.0 to calculate that parameter. As

Table 3 represents, the degree centrality of "care service" is largest while meaning it is the vital work the government encouraged.

Table 3. Centrality of semantic words

Words	Centrality	Words	Centrality	Words	Centrality	Words	Centrality
care service	47.724	culture	24.884	talent	14.544	quality	11.216
community	46.277	service industry	23.547	project	14.501	security	11.169
medical treatment	42.445	standard	23.445	technology innovation	14.385	assistance	10.801
facility	36.995	rehabilitation	22.696	retirement homes	13.767	hospital	10.762
safeguard	31.011	resource	20.439	disease	12.849	fund	9.718
insurance	29.589	demand	18.947	training	12.496	civil rights	9.143
health	29.150	rescue	18.635	fitness	12.114	psychology	9.096
education	27.409	cooperation	16.902	subsidization	11.375	travel	8.301
nursing	26.742	sanitation	15.354	study	11.375	investment	7.946
family	25.906	market	14.665	charity	11.253	Internet & informatization	6.119

To cluster around the concept network, it is necessary to clique analysis. With the centrality of semantic words, we can draw the graph for associative discovery (As Fig. 3 illustrated).

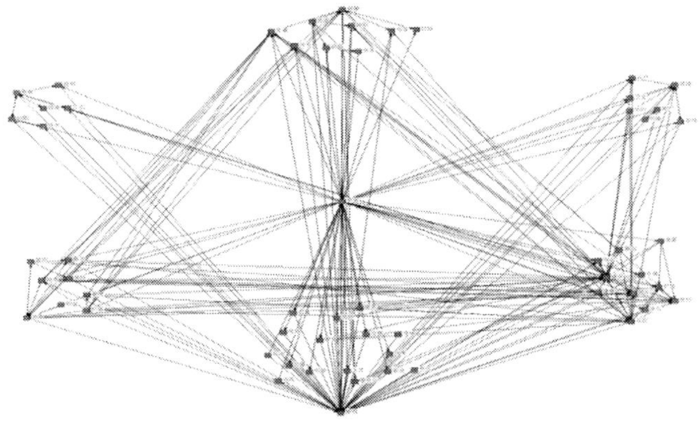

Fig. 3. Clustering semantic network. There are six cliques in the network.

Finally, we can categorize the related words from the semantic network cluster. Meanwhile, the interrelationships of the semantics component of the association path has provided foundations of decision analysis (as Table 4 shown).

Table 4. Association path.

Cliques	Semantic words
care service	community, family, nursing, standard, cooperation, market, project, retirement homes, quality, travel
medical treatment	health, rehabilitation, disease, psychology
safeguard	demand, insurance, rescue, civil rights, fitness, charity, security
facility	sanitation, hospital, Internet &informatization
education	culture, talent, technology innovation, training, study
policy support	resource, fund, subsidization, service industry, investment

4 Opportunity and Support

We can find that Chinese old aging industry, from the association discovery, focuses on care service, medical treatment, safeguard, facility and education. In these areas, elderly care service with a long path indicates huge market opportunity. For instance, community is a basic unit of social organism. And it is a large group of families. The common place of aged in China is community, where they have great fun, chatting, singing, dancing and exercising. Therefore, develop community care services is potential business prospects. Standard of service promotion is significant work to get profit, when our old aging market just emerging several years. The parents of adult who is the only child need home care for sharing care pressure. Cooperation with foreign companies, has been encouraged by the local government such as Jiangsu province, is also a wise choice of old aging industry development. Furthermore, both ORPEA from France and RIEI from Japan have built dozens of retirement homes. Each elder man charges 10,000 RMB (about 1,544 US dollars). And they can get two thousands to five thousands RMB ($309~$772) subsidies from the local government. Due to the foreign capital involved, the state enterprise has to improve the service quality for competition. Moreover, the senior tour project is another profitable business.

In medical treatment area, it is urgent for providing information to planners of health care for the aged. The importation of new medical equipment and the effective medicine can also help to develop old aging industry. In recent years, elder people's psychological health has been paid great attention, but the absence of the elderly psychiatrists is bad for their health.

The insurance for the aged is so important for developing the industry. We need more types of the insurance project for ensuring safety. The charity organization will be welcome for aiding the poor and offering social welfare projects to cope with the demand of security and fitness. Besides, elderly rights protection is an emerging market. And emergency project for rescuing is also a desired service.

With the rapid development of the Internet and mobile devices, the frequency of Chinese elder person using mobile phones is increasing gradually. The idea of Internet doctor and Mobile doctor will bring a new business in old aging industry.

Geriatric university gets a quickly develop in China. And study brings a lot of fans to the aged and voluntary activities make their lives more meaningful. In addition, training for elderly service and research new techniques for caring can create jobs for job seekers.

China's government encourages foreign cooperation and give subsidies, land resources and investment guarantee. In terms of tax, the investors will enjoy all the preferential policies for business invitation. And local governments are making a big push to streamline the permitting process.

5 Conclusion

This paper focuses on the market opportunity and policy support for Chinese old aging industry. We introduce a policy documents text mining framework with three phases, to analyze 52 reference policies. Applying association path discovery, the study finds out five business opportunity for foreign and local companies. These findings contribute to development of Chinese old aging industry. And the policy support from China's government is descripted briefly in this article. It is an useful reference for foreign enterprise.

Acknowledgements

This paper is funded by the National Science Foundation of China for Youth (No.71503011), Beijing Municipal Social Science Foundation (No.16GLC041), Fundamental Research Funds for the Advantage Discipline of North China University of Technology (No.XN081) and Science Foundation of North China University of Technology for Youth 2017 (No.1743012), and NCUT Foundations (No. XN019019, XN018036). Partially supported by the National Science Foundation of China for Youth (No.71603005).

References

1. R. Mao, J. Xu, Population aging, consumption budget allocation and sectoral growth. *China Economic Review,* **30**(44), (2014)
2. K. Lisenkova, M. Mérette, R. Wright, Population ageing and the labour market: Modelling size and age-specific effects. *Economic Modelling,* **35**, 981-989 (2013)
3. B. Beach, Ageing populations and changing worlds of work. *Maturitas*, **78**(4), 241-242 (2014)
4. M. Hvistendahl, Has China outgrown the one-child policy? *Science*, **329**(5998), 1458 (2010)
5. J. K. Miller, X. Fang, Marriage and family therapy in the people's Republic of China: Current issues and challenges. *Journal of Family Psychotherapy*, **23**(3), 173-183 (2012)
6. J. C. W. Smith, R. L. Watts, The investment opportunity set corporate financing, dividend and compensation policies. *Journal of financial Economics*, **32**(3), 263-292 (1992)
7. E. Ngai, P. Lee, A review of the literature on applications of text mining in policy making. *Proceeding of the 20th Pacific Asia Conference on Information Systems*, 343-354 (2016)
8. C. Boussalis, T. G. Coan, Text-mining the signals of climate change doubt. *Global Environmental Change*, **36**, 89-100 (2016)

9. E. Tobback, H. Naudts, W. Daelemans, E. Junqué De Fortuny, D. Martens, Belgian economic policy uncertainty index: Improvement through text mining. *International Journal of Forecasting, In Press* (2016)
10. A. H. Tan, Text mining: The state of the art and the challenges. *Proceedings of the PAKDD 1999 Workshop on Knowledge Disocovery from Advanced Databases*, (1999)

Research on the Relationship between Sino-EU Trade and Economic Growth

Yaqing Liu[1]*

[1]School of Economics and Management, North China University of Technology, China

Abstract. The dependence on foreign trade of China and European Union are both at a relatively high level, and the space for Sino-EU trade is still enormous. So, this paper adopts cointegration analysis and error correction model to study the relationship between Sino-EU trade and economic growth from a long-term and short-term perspective respectively. Cointegration analysis shows that: whether it is Sino-EU trade and China's economic growth or Sino-EU trade and EU's economic growth, there exists long-term cointegration relationship between them. Sino-EU trade plays a more important role in economic growth of China than its role in stimulating economic growth of European Union. Error correction model demonstrates that Sino-EU trade may deviate from its long-run equilibrium with economic growth in the short run.

Key words: research, relationship, trade, economic growth

1 Introduction

In recent years, the all-round cooperation between China and European Union has been steadily promoted, the economic and trade cooperation has been fruitful. For a long time, the EU has been China's largest trading partner, the largest source of technology and an important source of foreign investment. According to the data from Eurostat, the proportion of trade between the EU and China has increased by three times since 2000, from 5.5% to 14.9% in 2016, and total trade has reached 515 billion euro. In 2016, the EU's major import source was China, which was up to 20% of EU's imports from the non-EU, and China ranked second in EU export destination, which was up to 10% of EU's exports to non-EU countries. In June 2017, Li Keqiang, Premier of China's State, pointed out that there's broad prospects for economic and trade cooperation between China and EU in the twelfth EU-China Business Summit. Also, China is at an important stage of consumption upgrading and industrial upgrading, it's economy remains at around 6.5% growth at medium and high levels. So, the demand for European high-quality consumer goods, advanced technology equipment and outstanding services is increasing. At the same time, the upturn of Europe economic, steady growth of its' household consumption, the increasing of infrastructure investment will also provide new opportunities for Chinese enterprises. Therefore, with the further development of Sino-EU trade, it is of great significance to explore the relationship between Sino-EU trade and economic growth.

* Corresponding author: anny375@126.com

However, in the process of developing the trade between China and European Union, trade frictions have been a fact that cannot be ignored. It has affected the trade relations between China and European Union seriously. Therefore, with the further development of Sino-EU trade, understanding the relationship of Sino-EU trade and economic growth correctly is of great significance for further strengthen bilateral economic and trade exchanges and achieving win-win situation in economic development for both sides.

The relationship between economic growth and foreign trade has always been the focus of scholar's attention. In 1930s, economist Robertson proposed the view that foreign trade is the engine for growth for the first time. From then on, many scholars' research supported this opinion. Balassa [1] investigates the relationship between exports and economic growth in a group of eleven developing countries that have already established an industrial base, and results show that export can promote economic growth. Edwards [2] develops a simple endogenous growth model that emphasizes the process of technological absorption in small developing countries, and holds the view that countries that liberalize their international trade and become more open tend to grow faster. Liu and Sinclair [3] find that economic development, exports and FDI appear to be mutually reinforcing under the open-door policy. Makki and Somwaru [4] analyze the role of FDI and trade in promoting economic growth across selected developing countries and the interaction among FDI, trade, and economic growth, their results suggest that FDI, trade, human capital, and domestic investment are important sources of economic growth for developing countries. Evidence supports the view that foreign direct investment has a positive effect on economic growth in the host country. Saggi [5] and Schneider [6] find that foreign technology has a stronger impact on per capita GDP growth than domestic technology.

Also, many scholars study the influence of international trade on environment, technology and so on. Frankel and Rose [7] find that trade tends to reduce three measures of air pollution and there is little evidence that trade has a determinantal effect on the effect. Galiani and Sanguinetti [8] study the impact of trade liberalization on wage inequality. Their research suggests that trade deepening can only explain a relatively small proportion of the observed rise in wage inequality.

However, contrary to the conventional view on the growth effects of trade barriers, Dhawan and Biswal [9] find that trade barriers are positively and, in most specifications, significantly associated with growth. Lee [10] uses Vector Autoregressive (VAR) model to examine the relationship between real GDP, real exports and terms of trade for India during the period 1961-93. Their results reveal that there is one long-run equilibrium relationship among the three variables, and the causal relationship flows from the growth in GDP and terms of trade to the growth in exports, and the causality from exports to GDP appears to be a short run phenomenon. Lu [11] finds that tariff rates and black-market premia, interacting with the estimated share of free trade imports, have significant negative effects on the growth rate of per capita income across countries. So, there is a dispute over the relationship between economic growth and foreign trade.

And many researchers from China hold view that foreign trade will promote China's economic growth. Bao [12] analyzes the relationship between Sino-EU trade volume and economic growth of China, results showed that there is positive correlation between them. Jiang [13] studies the interdependence between China and Europe in trade and economic growth, and the governance of trade protectionism, he finds that there's close relationship and interdependence for China and Europe's trade and economic growth, the development of Sino-EU trade promotes the economic growth of China and Europe. Spulber [14] point out that the overall pulling effect of Sino-EU trade on China's added value and employment is far greater than that of the EU's added value and employment, while its efficiency of boosting added value is less than that of the European Union.

This paper studies the relationship between Sino-EU trade and economic growth from a long-term and short-term perspective respectively. The following structure of this paper is: a brief description of the characteristics of Sino-EU trade and economic growth will be presented in section 2. Section 3 will explore the short-term and long-term relationship between Sino-EU trade and economic growth through cointegration analysis and error correction model. Section 4 is conclusion.

2 Statistical analysis of Sino-EU trade and economic growth

To examine the relationship between Sino-EU trade and economic growth, this paper uses data of economic growth, foreign trade and bilateral trade of China and European Union from 2000 to 2015. The data of Sino-EU trade and total foreign trade comes from UN Comtrade Database. And we get the data of economic growth from China's National Bureau of Statistics and Eurostat respectively.

The degree of dependence on foreign trade is the proportion of total foreign trade in GDP, which is used to measure the dependence of a country's economy on international trade, or the dependence of a country's production and consumption on the international market or the world economy. Figure1 and Figure 2 show the dependence on foreign trade of China and EU. As is shown in the Figure1, the dependence on foreign trade of China is at a relatively high level, which reflects that there's a large dependence on foreign trade of China's economic growth. But the dependence on foreign showed a trend of declining to a certain extent since 2007. This is closely related to the faltering of world economy and the reducing external demands of goods since the subprime mortgage crisis in 2007. As can be seen from Figure 2, the EU's dependence on foreign trade is also at a relatively high level, but lower than China, and there has been a certain decline in recent years.

Fig. 1. Dependence on Foreign Trade of China

DOI: 10.1051/shsconf/20173901016

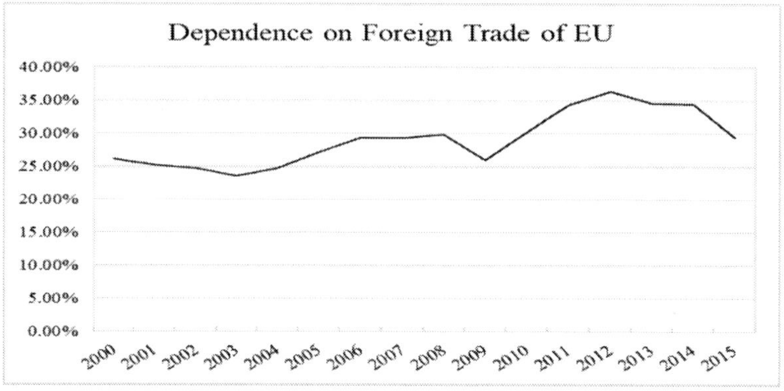

Fig. 2. Dependence on Foreign Trade of EU

Figures 3 and 4 show the proportion of Sino-EU trade in China and EU's total foreign trade respectively. It is not difficult to find that Sino-EU trade accounts for about 15% of China's total foreign trade in recent years, and remains at a relatively stable state. Also, Sino-EU trade accounts for a steady rise in the proportion of EU's total foreign trade. So, there's a close trade relation between China and EU, and the space for further trade is still enormous.

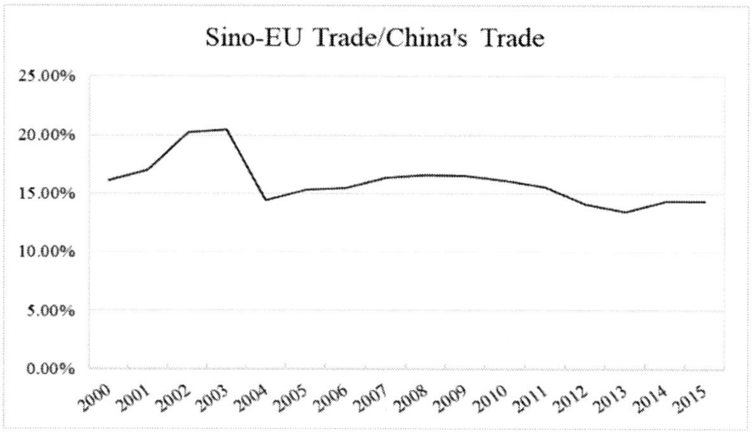

Fig.3. Ratio of Sino-EU Trade to China's Trade

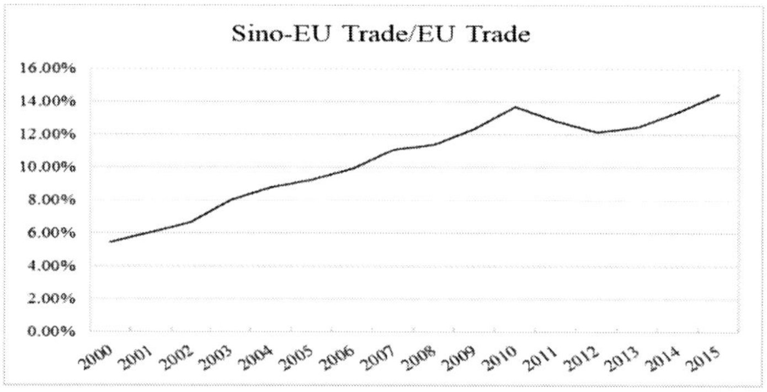

Fig. 4. Ratio of Sino-EU Trade to EU's Trade

3 Empirical analysis of Sino-EU trade and economic growth

The cointegration theory proposed by Engle and Granger provides an effective tool for the analysis of non-stationary time series and for the long-term equilibrium relationship of non-stationary time series data. And error correction model (ECM) proposed by Davidson, Hendry, Srba and Yeo is an extension of cointegration analysis. If there is a short-term deviation for some reason, it is bound to return the equilibrium state by correcting the error. The error correction model combines short-term volatility with long-term equilibrium in one model. So, this paper adopts cointegration theory and ECM to research the relationship between Sino-EU trade and economic growth. That is, we construct equation of Sino-EU trade (X) and China's economic growth (Y1) and Sino-EU trade (X) and EU's economic growth (Y2) respectively. In order to eliminate the possible heteroscedasticity of variables, this paper deals the economic variables with natural logarithm.

3.1 Unit root test

Before cointegration analysis, the unit root test is needed to test whether the time series is stationary, and the result is shown in Table1.

Table 1. Unit Root Test

Variables	Testing value of ADF	P-values	Conclusions
lnX	0.8577	0.9989	Non-stationary
lnY1	-0.6368	0.8340	Non-stationary
lnY2	-0.9393	0.9226	Non-stationary
ΔlnX	-3.584	0.0805	Stationary
ΔlnY1	-5.426	0.0056	Stationary
ΔlnY2	-4.669	0.0139	Stationary

As is shown in table 1, results of ADF unit root test indicate that time series lnX, $\ln Y_1$ and $\ln Y_2$ are non-stationary, while its' first order difference ΔlnX, $\Delta\ln Y_1$ and $\Delta\ln Y_2$ are stationary. That is, X, $\ln Y_1$ and $\ln Y_2$ are all integrated of order one I(1). So, cointegration analysis can be carried out.

3.2 Cointegration analysis

Using the E-G two step methods, we construct cointegration regression between Sino-EU trade and China's economic growth, Sino-EU trade and EU's economic growth to test their cointegration and long-term equilibrium relationship.

The cointegration equation between Sino-EU trade and China's economic growth is:

$$\ln Y_1 = 0.3578 + 0.7838 \ln X \quad (1)$$

Among which, each coefficient is tested by the significant level of 5%, $R^2 = 0.9260$. And the ADF unit root test of residual series (testing value of ADF is -2.204, P-values is 0.05) indicate that the series is stationary.

The cointegration equation between Sino-EU trade and EU's economic growth is:

$$\ln Y_2 = 19.65 + 0.3913 \ln X \quad (2)$$

Among which, each coefficient is tested by the significant level of 5%, $R^2 = 0.9570$. And the ADF unit root test of residual series (testing value of ADF is -1.818, P-values is 0.06) indicate that the series is stationary.

The significance of cointegration is that it reveals whether there is a long-term stable equilibrium between economic growth and Sino-EU trade. The economic variables that satisfy the cointegration cannot be deviated from each other too far, and a shock can only make them deviate from the equilibrium position in a short time and automatically revert to equilibrium in the long run. Therefore, whether it is Sino-EU trade and China's economic growth or Sino-EU trade and EU's economic growth, there exists long-term cointegration relationship between them. Also, Sino-EU trade plays a more important role in economic growth of China than its role in stimulating economic growth of European Union.

3.3 Error correction model

Cointegration analysis reflects the long-term equilibrium relationship between economic growth and foreign trade. The relationship between economic growth and foreign trade can be unbalanced, and this dynamic structure of the short-term disequilibrium relation can be described by the error correction model.

The error correction model of Sino-EU trade and China's economic growth is:

$$\Delta \ln Y_1 = 0.1191 + 0.1765\, \Delta \ln X - 0.2359 e_{t-1} \quad (3)$$

$$t = (10.03) \quad (3.074) \qquad (-5.117)$$

Among which $R^2 = 0.7370$.

The error correction model of Sino-EU trade and EU's economic growth is:

$$\Delta \ln Y_2 = -0.0101 + 0.4875\, \Delta \ln X - 0.402 e_{t-1} \quad (4)$$

$$t = (-2.6679) \quad (5.821) \qquad (-1.908)$$

Among which $R^2 = 0.7462$.

The parameters of equation (3) and (4) are all tested by significant test. What's more, error correction term e_{t-1} is also tested by significant test, which means that Sino-EU trade may deviate from its long-run equilibrium with economic growth in the short run, and its coefficient is negative, which shows that it conforms to the error correction mechanism.

4 Conclusions

Sino-EU trade plays an important role in the economic growth of China and European Union just as demonstrated by the empirical research in this paper. Also, Sino-EU trade may deviate from its long-run equilibrium with economic growth in the short term, but we cannot deny its driving force for economic growth.

Therefore, in order to further promote Sino-EU trade, the following suggestions are put forward. Firstly, practical measures must be taken to reduce trade protectionism in some European countries and to create a favourable market and institutional environment for Sino-EU trade and economic growth. Secondly, China and European Union are at different stages of economic development, and differences in the industrial structure and factor endowments between them are huge. So, we should make full use of China and EU's comparative advantages and give full play to their complementary trade. Also, technology trade improves the quality of innovation by increasing the pool of R&D experiments from which the best technology is chosen (Spulber, 2008). So, some measures should be taken to promote the development of technology trade. In a word, extensive cooperation should be carried out in eras such as environmental protection, technology, economic and so on. Both sides should create opportunity to promote mutual trade and economic growth with respect to foreign investment growth, household consumption and the demand for qualified labour.

References

1. B. Balassa, Exports and economic growth: further evidence. *Journal of development Economics*, **5**(2), 181-189 (1978)
2. S. Edwards, Trade orientation, distortions and growth in developing countries. *Journal of development economics*, **39**(1), 31-57 (1992)
3. X. Liu, P. Burridge, Sinclair, Relationships between economic growth, foreign direct investment and trade: evidence from China. *Applied economics*, **34**(11), 1433-1440 (2002)
4. S. Makki, A. Somwaru, Impact of foreign direct investment and trade on economic growth: Evidence from developing countries. *American Journal of Agricultural Economics*, **86**(3), 795-801 (2004)
5. K. Saggi, Trade, Foreign Direct Investment, and International Technology Transfer: A Survey. *World Bank Research Observer*, **17**(2), 191-235 (2002)
6. P. Schneider, International trade, economic growth and intellectual property rights: A panel data study of developed and developing countries. *Journal of Development Economics*, **78**(2), 529-547 (2005).
7. J. A. Frankel, A. K. Rose, Is Trade Good or Bad for the Environment? Sorting Out the Causality. *Review of Economics & Statistics*, **87**(1), 85-91 (2005)
8. S. Galiani, P. Sanguinetti, The impact of trade liberalization on wage inequality: evidence from Argentina. *Journal of Development Economics*, **72**(2), 497-513 (2003)
9. U. Dhawan, B. Biswal, Re-examining export-led growth hypothesis: a multivariate cointegration analysis for India. *Applied economics*, **31**(4), 525-530 (1999)
10. J. W. Lee, International Trade, Distortions, and Long-Run Economic Growth. *Staff Papers*, **40**(2), 299-328 (1993).
11. F. Lu, Correlation Analysis between Sino-European Union Trade and China's Economic Growth. *Journal of international trade*, 03 47-50 (2005)
12. J. Bao J., On interdependence of trade and economic growth between China and Europe and policy choice for new protectionism, **5**, 29-37 (2010)
13. X. Y. Duan, Jiang, The impact of Sino-EU trade on bilateral economy and employment. *Journal of international trade*, **8**, 29-39 (2012)
14. D. F. Spulber, Innovation and international trade in technology. *Journal of Economic Theory*, **138**(1), 1-20 (2008)

The importance of ROE for calculating EVA Equity: the case of Motor Jikov Strojírenská, a.s.

Jan Mareček[1], *Zuzana* Rowland[2,*]

[1]Institute of Technology and Business, School of Expertness and Valuation, Okruzni 517/10, 37001 Ceske Budejovice, Czech Republic
[2]University of Žilina, Faculty of Operation and Economics of Transport and Communications, Univerzitná 8215/1, 01026 Žilina, Slovakia

Abstract. A remarkable success depends on a variety of factors, one of which is the ability to motivate the employees, increasing motivation of managers, measuring business processes and measuring and observing values of the company. The tangible results are also shown by the indicator of return on equity (ROE). The article aims at reflecting the importance of ROE for the EVA Equity (economic value added for shareholders) calculation of Motor Jikov Strojírenská, a.s. The data come from Albertina database. These are details of financial statements from 2000-2015. As a matter of fact, weighted average costs of capital, alternative cost value of equity capital and EVA Equity are calculated. Software Statistica and its sophisticated tool data mining - automated neural networks was used for finding a correlation between EVA Equity indicator and ROE. In addition, 10,000 neural networks were generated, five of which with the best results have been stored. The results show that EVA Equity is not dependent on the ROE rate.
Key words: return on equity (ROE), economic value added for shareholders (EVA Equity), correlation, company management, artificial neural networks

1 Introduction

The ability to motivate the employees to the greatest extent is one of the most important factors. For that purpose, many coherent motivational theories were developed in time. These theories help the company get the maximum out of its employees [1].

Increasing motivation of managers is a most significant factor for companies [2]. developed a theory which generates motivation of managers, i.e. 'Motivational Factor Inventory' (MFI). MFI consists of 47 items grouped into 6 motivational aspects (human interaction, task, general working conditions, strengthening personal development and compensation). The results showed that MFI constitutes a sophisticated tool for motivating managers. Hawthorne Effect may be also considered as a coherent theory. This theory

* Corresponding author: rowland@mail.vstecb.cz

suggests that employees are more productive upon knowing of their work being measured and analysed. In order to motivate the employees for higher performance, they require appropriate recognition for the good job and assurance that their opinion will be taken into account [3]. Maslow's hierarchy of needs [4], theory X and Y [5] and other theories also rank among motivational theories. All these methods help to the overall growth in the value of the business.

Apart from motivation of managers and employees, measuring the performance of business processes also comprises an important factor. This activity became a central topic not only in academic, but also in business sector, since organizations are required to produce concrete and tangible results [6]. According to the modern scientific literature, the most cited models come from measuring performance from Balanced Scorecard Company, model EFQ etc. However, models focused on the whole company and models focused on a single business process must be discriminated [7].

In addition, shareholders must constantly measure and observe the company value. Above all, creating a capital budget belongs among ingenious devices which measure the business performance [8]. This applies to selecting investment projects which maximize the value of the shareholders [9]. A benchmark which enables a direct comparison of the company performance with another similar, prosperous company [10], key indicators of the productivity for clarifying business strategies [11]. Balanced Scorecard Method focused on strategic organization management [12]. Six Sigma Method focused on permanently improving all organization processes and eliminating procedural errors [13].

In addition, an indicator of return on equity (ROE) is also essential for the company management. It shows how much of the net profit falls on one Czech Crown of the invested capital [14]. The ROE indicator is calculated as a ration of the net profit and equity capital [15]. The ROE indicator is essential mainly for shareholders, co-partners and other investors [16]. Kyriazis and Anastassis [17] investigated the relative explanatory power of the EVA model with respect to stock returns and firms′ market value, compared to established accounting variables – net income or operating income. Their results showed that EVA, even though useful as a performance evaluation tool, need not necessarily be more correlated with shareholder′s value than established accounting variables. The results showed that firms using EVA exhibit a higher percentage of institutional ownership and a lower percentage of insider ownership than non-adopters [18]. Prospector companies tend to use EVA less than defender firms and providing appropriate incentives may be more complex than the developers of EVA imply. Rowland [19] decomposed EVA Equity, which is viewed as the strategical goal of a company, into partial indicators that can be incorporated into a company′s tactical plan in the form of operative plan indicators.

Therefore, the article focuses on appreciating the importance of ROE for calculating EVA Equity of the particular company.

The hypothesis has been formulated as follows: EVA Equity value is dependent on ROE value at least from 70%.

2 Data and methods

We have the financial statements of Motor Jikov Strojírenská, a.s. at our disposal. The data have been derived from Albertina database applying to financial statements from 2000 to 2015. The data will be compiled to a statistical table. Each line will contain information about one particular year. The information about financial statements will be drawn up in columns. In addition, Economic Value Added for shareholders (EVA Equity) in each analysed year will be added in the next step.

The sources of the particular data about financial statements are suggested in the table 1.

Tab. 1. The sources of the particular data from financial statements

Name	Statement	Item
Capital	Balance	(C.I+C.III+C.IV+D.I.2) - (B.III+B.IV.2+B.IV.3+C.I.1)
Activity total	Assets balance	A + B + C + D
Stock	Assets balance	C.I.
Debt	Assets balance	C.III.
Short-term financial assets	Assets balance	C.IV.
Current assets	Assets balance	C
Equity capital	Liability balance	A.
Bonds and bills of exchange	Liability balance	B.II.6 a B.II.7
Common and financial help	Liability balance	B.IV.2 a B.IV.3
Long-term help	Liability balance	B.IV.1
Short-term commitments	Liability balance	B.III
Retained earnings	Liability balance	A.III.2+A.IV+A.V
Debt	Liability balance	B
Short-term debt	Liability balance	B.III + B.IV
Money sources	Liability balance	A + B.IV. + B.II.6 + B.III.9
Revenues and sold goods	Profit and loss	I.
Cost of sold goods	Profit and loss	A.
Productivity	Profit and loss	II.
Productivity consumption	Profit and loss	B.
Personal expenses	Profit and loss	C.
Write-offs	Profit and loss	E.
Interest expenses	Profit and loss	N.
Gross yields for the accounting period	Profit and loss	***
Total yields	Profit and loss	I. + II. + … + XIII.
Total costs	Profit and loss	A + B + … + T
Profit before interest and tax	Profit and loss	*** + S+ Q + N
Revenues	Profit and loss	I. + II.1.

It requires an exact calculation of weighted average costs of the capital [20]:

$$WACC = r_f + r_{LA} + r_{entreprenarship} + r_{FinStab} \tag{1}$$

Where:

r_f risk free yield derived from the interested rate of ten-year government bond,

r_{LA} function of indicators defining the liquidity,

$r_{entrepreneurship}$ function of indicators defining the creation of, production power,

$r_{FinStab}$ function of indicators defining the relations between, company assets and sources of their covering.

Subsequently, a value of the rate of equity will be set [20]:

$$r_e = \frac{WACC * \frac{UZ}{A} - (1-d) * \frac{U}{BL+B} * \left(\frac{UZ}{A} * \frac{E}{A} \right)}{\frac{E}{A}} \quad (2)$$

Where:

r_e	rate of equity,
WACC	weighted average cost of capital,
UZ	money sources (equity capital, interest-bearing debt),
A	assets,
E	equity capital,
BL	bank loans,
B	bonds,
$U / BL + B$	interest rate – *i* (interest),
d	the rate of income tax – *t* (tax).

Economic value added for shareholders will be derived from the relation [20]:

$$EVA\ Equity = (ROE - r_e) * E \quad (3)$$

Where:

ROE	return on equity.

Subsequently, the table will be imported to Statistica Software of DELL Company in version 12. We will be looking for the extent of correlation of the indicator of EVA Equity and ROE.

Firstly, data mining, automated neural networks (ANNs) and regression will be dealt with. EVA Equity will be used as a dependent variable, whereas ROE will be employed as an independent variable. The data set will be divided into three aggregates – training, testing and validating. The first group will contain 70% of input figures. Neural structures will be generated from the training data set. The two remaining groups will include 15% of input figures. Groups may provide the reliability verification of the neural structures i.e. the found model [21]. 10,000 neural networks will be generated, the best five of which will be preserved[1]. The hidden layer of the multi-layer perceptron network (MLP) will contain from 2 to 10 neurons. In case of the neural network of radial basic function, the hidden layer will include only 5 to 15 neurons. Multiple perceptron networks will be considered for these distribution functions in the hidden layer and the output layer:

- Linear,
- Logistic,
- Atanh,
- Exponential,
- Sinus.

The picture 1 illustrates a pyramid disintegration of ROE indicator in INFA Model the Neumaiers.

In case the correlation of EVA Equity and ROE is high, the correlation of constituent variables entering ROE calculation should be systematically analysed. In case of success

[1]The method of least squares will be used. Generation of networks will be finished when there is no improvement, i.e. when there is no reduction in the sum of the squares. We thus retain the neural structures whose sum of squares of residues to actual development of gold will be as low as possible (zero, ideally).

DOI: 10.1051/shsconf/20173901017

and the high correlation between EVA Equity and the selected indicator entering EVA Equity calculation, a future development of economic value added for owners of the indicator may be predicted. In this way, a simple tool for indicating the state of the company will be created. The picture 1 shows variables entering ROE calculation which apply only to the net return on equity, EBIT (i.e. profit without interests and taxing), added value, personal expenses etc.

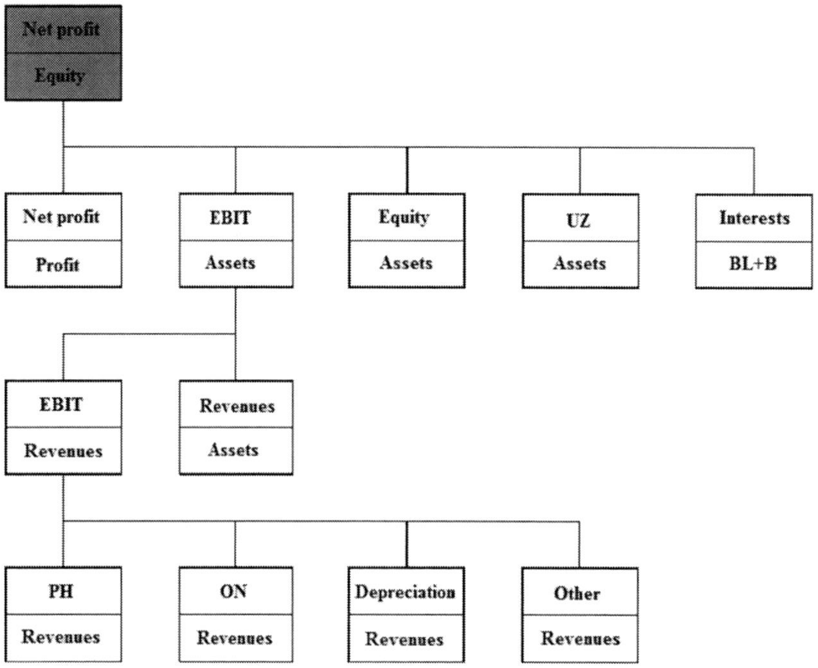

Picture 1. Disintegration of ROE indicator in INFA Model

3 Results

The table 3 suggests basic description of the data sets. It analyses the minimum, maximum and standard deviation of input data. The data are analysed in regard to the training, testing and validating data aggregate.

Table 2. Statistic characteristics entering the calculation

Characteristics	ROE (%)	EVA (thousands CC)
	Input var.	**Output**
Minimum (Training)	-19.69	9.36
Maximum (Training)	13.25	34811.48
Average (Training)	0.47	10214.67
Standard deviation (Training)	10.1829	11162.02
Minimum (Testing)	3.11	2100.99
Maximum (Testing)	10.52	3561.82
Average (Testing)	6.815	2831.4
Standard deviation (Testing)	5.2397	1032.96
Minimum (Validating)	-80.48	10.8
Maximum (Validating)	2.66	20991.66
Average (Validating)	-38.91	10501.23
Standard deviation (Validating)	51.21	14810.28
Minimum (Overall)	-80.48	9.36
Maximum (Overall)	13.25	34811.48
Average (Overall)	-3.6594	9327.58
Standard deviation (Overall)	22,4105	10609

Deviations of the individual characteristics for three different data sets suggest three markedly different data sets. However, the result will be rather negative. For example, take the minimum ROE of each data set. The minimum of the training data set was spotted at the value of 19.69, testing data set at 3.11 and validating data set at -80.48. The differences are big, especially the data interval, of which the maximum value in each data set is (13.25; 10.52 or 2.66).

Acquired and at the same time preserved neural structures are suggested in table 3.

Table 3. Generated and preserved neural networks

Index	Network	Train. perf.	Test. perf.	Valid. perf.	Training error	Testing error	Valid. error	Training algorithm	Error function	Activation hidden layer
1	MLP 1-2-1	0.355191	1	1	50000528	16465105	32538328	BFGS 0	Total sq.	Logistic
2	RBF 1-5-1	0.718692	-1	1	27608817	9556342	2586642	RBFT	Total sq.	Gauss
3	MLP 1-3-1	0.310595	1	1	51609109	12888885	44741428	BFGS 0	Total sq.	Logistic
4	RBF 1-5-1	0.584094	-1	1	40170595	2036325	64055783	RBFT	Total sq.	Gauss
5	RBF 1-5-1	0.583614	-1	1	37836882	3172625	53546896	RBFT	Total sq.	Gauss

Two MLP neural layers were preserved, i.e. the first and the third network. Both of them activate inner neural layers by Logistic Function and outer neural layers by Sinus Function. Algorithm Quasi – Newton was used for constructing both neural networks. Furthermore, three RBF networks were preserved. All networks activate the hidden neural layer by Gauss Curve and outer neural network by Identity.

The value of individual networks is dependent mainly on their performance. This is measured by the correlation coefficient. Values of correlation coefficients of individual networks and data sets are depicted in Table 4.

Table 4. Correlation coefficients of neural networks according to the input data sets

Neural network	Training set	Testing set	Validating set
1.MLP 1-2-1	0.355191	1	1
2.RBF 1-5-1	0.718692	-1	1
3.MLP 1-3-1	0.310595	1	1
4.RBF 1-5-1	0.584094	-1	1
5.RBF 1-5-1	0.583614	-1	1

As a matter of fact, we are looking for a neural structure with an optimal value of the correlation coefficient in all data sets while considering their identical amount. In regard to the formulated hypothesis, we are looking for a correlation coefficient above 0.7, which applies only to network No. 2. However, a value of correlation coefficient -1 occurs in the testing data set, the fact of which means that the indirect proportion is to be dealt with. On the other hand, correlation coefficient 1 occurs in the validating data set.

The whole situation is illustrated in Table 5.

Table 5. Statistic characteristics of predictions

Characteristic	1.MLP 1-2-1	2.RBF 1-5-1	3.MLP 1-3-1	4.RBF 1-5-1	5.RBF 1-5-1
Minimal prediction (Training)	4476.7	304.5	5615.3	2800.3	728.1
Maximal prediction (Training)	16157.5	28353	16811.9	31470.2	21751.2
Minimal prediction (Testing)	6320.4	-1320.3	6576.3	3889.3	2779.8
Maximal prediction (Testing)	10494.1	5894.5	9176.8	4936.1	5576.5
Minimal prediction (Validating)	10705.4	-538	9333.8	4501.9	4122.7
Maximal prediction (Validating)	17019.4	17822.2	30585.8	5627.6	6946
Minimal residua (Training)	-13898.4	-12211.6	-12372.1	-10909.4	-17995.7
Maximal residua (Training)	21356.6	13415.3	22979	20218.9	13060.2
Minimal residua (Testing)	-6932.3	-3793.5	-5614.9	-2835.2	-3475.5
Maximal residua (Testing)	-42194	4882.1	-4475.3	-327.4	782.1
Minimal residua (Validating)	-10694.6	548.8	-9594.1	-4491.1	-4111.9
Maximal residua (Validating)	3972.2	3169.4	-9323	15364	14045.6
Minimal standard residua (Training)	-2	-2.3	-1.7	-1.7	-2.9
Maximal standard residua (Training)	3	2.6	3.2	3.2	2.1
Minimal standard residua (Testing)	-1.7	-1.2	-1.6	-2	-2
Maximal standard residua (Testing)	-1	1.6	-1.2	-0.2	0.4
Minimal standard residua (Validating)	-1.9	0.3	-1.4	-0.6	-0.6
Maximal standard residua (Validating)	0.7	2	-1.4	1.9	1.9

The table illustrates statistic characteristics of predictions. The table suggests big differences between the predictions of individual neural structures. This is still acceptable since not all the networks are applicable in practice. Nevertheless, dramatic differences between identical characteristics of the same network according to the individual data sets may be seen. These are again very big, although residua, of which the sum is close to zero and there are no big differences between preserved networks, may be excluded. Table 6 shows results of the sensitivity analysis.

Table 6. Sensitivity analysis

	ROE (%)
1.MLP 1-2-1	1.271928
2.RBF 1-5-1	2.665624
3.MLP 1-3-1	1.15558
4.RBF 1-5-1	1.456952
5.RBF 1-5-1	1.684762
Average	1.646969

Except for No. 3, the analysis indicates positive results. Changes in ROE correlate with a dramatic shift in EVA Equity.

The hypothesis in the introduction has been formulated as follows: EVA Equity value depends on ROE value at least from 70%; thus, the hypothesis has been rejected. Broadly speaking, EVA Equity does not depend on ROE value in its calculation.

4 Conclusion

The aim of this article was to appreciate the importance of ROE for the calculation of EVA Equity of the specific company.

Based on the calculation, we conclude that EVA Equity does not depend on ROE (at least in case of Motor Jikov Strojírenská, a.s.). In this case, it is not possible to carry on in the analysis of variables entering ROE calculation and simplify EVA Equity calculation, or alternatively, find the motion indicator of EVA Equity.

All the same, as long as we analyse the way EVA Equity was calculated, the conclusion is not very logical and cannot be agreed with. More likely, a distinct lack of data may be blamed for, or alternatively, data errors which entered the calculation. ROE is useless predictor of EVA in particular companies. In our case there was no relationship in other the relationship is to be found. There is most likely a third variable which influences the relationship between ROE and EVA. Still, the situation may be successfully dealt with. The data sets of companies might be used for further calculations; and in this way, on the one hand they may develop the data base for the calculation and on the other, they may generalize the result at a certain level.

References

1. P. Rong, Analysis and Application of Modern Motivation Theory in Public Relations Management. *Proceedings of the 11th International Conference on Innovation and Management*, Vaasa, Finland, 1010-1014 (2014)

2. S. Seiler, B. Lent, M.Pinkowska, M. Pinazza, An integrated model of factors influencing project managers' motivation – Findings from a Swiss Survey. *International Journal of Project Management,* **30**(1), 60-72 (2012)
3. H. Lück, E. Helmut, The Hawthorne effect – an effect fo many occasions? [Der Hawthorne-Effekt – ein Effekt für viele Gelegenheiten?]. *Group dynamics and organizational consulting [Gruppendynamik und Organisationsberatung],* **40**(1), 102-114 (2019)
4. V. Pereira, M. G. C. Campos and G. Camarini, Maslow's Theory of Motivation - Hierarchy of Needs: Results from Construction Operational Workers in São Paulo City, Brazil. *Applied Mechanics and Materials,* **174-177**, 2339-2342 (2012)
5. P. M. Medeiros, N.M.S. Galvão, J.G.N. Araújo, U.C.T. Lagioia, A.P.S. Ferreira, Perception of Accounting Sciences Students Regarding Behavior in the Workplace: A Study Based on Theory X and Y. [Percepção dos Discentes de Ciências Contábeis Quanto ao Comportamento no Ambiente de Trabalho: Um Estudo Baseado na Teoria X e Y]. *Journal of Management, Finance and Accounting [Revista de Gestão, Finanças e Contabilidade]*, **6**(1), 152-172 (2016)
6. A. Van Looy, A. Shafagatova, Business process performance measurement: a structured literature review of indicators, measures and metrics. *SpringerPlus,* **5**(1), (2016)
7. V. Burja, R. Mărginean, The Study of Factors that may Influence the Performance by the Dupont Analysis in the Furniture Industry. *Procedia Economics and Finance,* **16**, 213-223 (2014)
8. A. Kasych, M. Vochozka, Theoretical and methodical principles of managing enterprise sustainable development. *Marketing and Management of Innovations*, (2), 298-304 (2017)
9. G. Andor, S.K. Mohanty, T. Toth, Capital budgeting practices: A survey of Central and Eastern European firms. *Emerging Markets Review,* **23**, 148-172 (2015)
10. M. Tvaronavičienė, K. Razminienė, L. Piccinetti, Cluster efficiency study through benchmarking. *Entrepreneurship and Sustainability Issues,* **3**(2), 120-128 (2015)
11. J. Sofiyabadi, B.Kolahi, Ch. Valmohammadi, Key performance indicators measurement in service business: a fuzzy VIKOR approach. *Total Quality Management*, **27**(9-10), 1028-1042 (2015)
12. A. H. Aly, M. E. Mansour, Evaluating the sustainable performance of corporate boards: the balanced scorecard approach. *Managerial Auditing Journal*, **32**(2), 167-195 (2017)
13. A. Pugna, R. Negrea, S. Miclea, Using Six Sigma Methodology to Improve the Assembly Process in an Automotive Company. *Procedia - Social and Behavioral Sciences*, **221**, 308-316 (2016)
14. V. Stehel, M. Vochozka, the analysis of the economical value added in transport. *Nase More*, **63**(3), 185-188 (2016)
15. A. Kijewska, Determinants of the return on equity ratio (ROE) on the example of companies from metallurgy and mining sector in Poland. *Metalurgija – Sisak then Zagreb*, **55**(2), 285-288 (2016)
16. I. Berzkalne, E. Zelgalve, Return on equity and company characteristics: an empirical study of industries in Latvia. *Proceedings of the 8th International Days of Statistics and Economics*, Prague, Czech republic, 94-103 (2014)

17. D. Kyriazis, Ch. Anastassis, The Validity of the Economic Value Added Approach: an Empirical Application. *European Financial Management*, **13**(1), 71-100 (2007)
18. L. Lovata, M. Costigan, Empirical analysis of adopters of economic value added. *Management Accounting Research*, **13**(2), 215-228 (2002)
19. Z. Rowland, Decomposition of EVA Equity to the Sub-operational Plans of a Company. *Littera Scripta*, **9**(3), 52-65 (2016)
20. I. Neumaierová, I. Neumaier, Financial analysis of industry and construction in 2007 [Finanční analýza průmyslu a stavebnictví za rok 2007]. *Analysis of the Ministry of Industry and Trade [Analýzy MPO]*, **1**, 1-187, (2008)
21. M. Vochozka, Decomposition of a main objective of an international company set with the help of artificial neural networks into daily plans. *Proceedings of the 16th International Scientific Conference on Globalization and its Socio-economic Consequences*, Zilina, Slovakia, 2383-2390 (2016)

The influence of public debt on the performance of the economy

Jan Mareček[1], *Veronika* Machová[1*]

[1]Institute of Technology and Business in České Budějovice, School of Expertness and Valuation, Okružní 517/10, 370 01 České Budějovice

Abstract. Thanks to public debt, nations maintain a stable tax rate and at the same time raise expenses for suppressing budget crises. Public debt is considered risk-free and is issued by the national government. However, extensive public debt raises interest rates, extrudes private investments, worsens fixed balance, widens short-term fluctuation and has a negative impact on economic growth. The performance of the economy can be measured mostly by GDP, which is expressed as the monetary value of all finished products and services produced in a certain time period in one country. The aim of this paper is to explore the relationship, or rather identify the mutual correlation of public debt of one country and its performance (GDP). The research includes countries of the European Union, which have an identifiable timeline of public debt and GDP. The source of this data is the database of the World Bank. The information is from the period of years from 1995 through 2015. The correlation of the whole file of countries and then individual countries is ascertained. On the basis of the results, we can state that in the explored sample of countries, high dependence between public debt and GDP has been proven.
Key words: performance of the economy, public debt, HDP, taxes

1 Introduction

Public debt enables countries to maintain stable debt rates and at the same time raise expenses for suppresing budget crises [1]. It is therefore a kind of debt that is issued by the national government. It is theoretically considered risk-free because the government may use certain measures to ensure payment, for example raise taxes, lower expenses or print more money, and therefore erase the need for interest to be paid [2].

Public debt can be divided into internal or external debt. Internal debt is debt towards creditors inside the country of subject and external debt refers to debt from loans from outside the country of subject [3]. Public debt may also be classified in regards to the length of its maturity – short-term and long-term public debt. Short-term debt usually matures within one year, while long-term debt is usually matured after 10 years or more [4].

Current studies argue that large public debt raises interest rates, extrudes private investments, worsens fixed balance and widens short-term fluctuation. Empirical studies have discovered that government debt has a significantly negative effect on economic

* Corresponding author: machova@mail.vstecb.cz

growth [5]. On the basis of these arguments it is therefore very important to ask if the public debt in a subject country is excessively high and potentionally harmful, and eventually try to change the current state using certain resources [6].

Public debt is usually done on the basis of lending government bonds and issuing securities [7]. Countries which are less creditworthy than others borrow directly from world-wide organizations such as the World Bank and other international financial institutions [8].

This paper analyzes the influence of public debt on the performance of the economy. This performance can be measured mainly by the gross domestic product (GDP). GDP is therefore one of the most important economic indexes which represents the outlook of the national economy. It is therefore very important to determine the forecasts and estimates of GDP according to official statistic data [9].

GDP measures the financial value of all production of all goods and services in a subject country during a certain time period. It creates the economic performance of a country or regions and is important for comparing the differences between the standard of living in subject countries. In general, there are three methods for determining GDP, specifically on the basis of production, expenditure and revenues [10]. Individual methods will be described later. GDP is therefore expressed as the monetary value of all finished products and services produced in a certain time period in one subject country. GDP is usually calculated on a yearly basis, it is however possible to calculate it quarter-yearly [11]. GDP encompasses all private and public consumption, investments, government spending, private resources, paid construction expenses and foreign commercial balance. GDP thus determines the economic performance of the country [12].

GDP is therefore a measure of consumer expenditure (C) plus business investment (I) and government spending (G), as well as net exports, which are exports without imports (X-M). The dynamics of the GDP structure by revenue category reflects a real reduction in household incomes and a growing tax burden [13]. GDP can be measured by the expenditure or production method. The expenditure method is based on the ascertainment and sum of the amounts for purchases of final products and services in a single year, including consumer spending, capital expenditures, government expenditures and net exports. The production method, also referred to as a pure product method, is based on the value of the produced goods or services, usually per year. Firstly, the gross value of production is estimated in all sectors. Intermediate consumption, such as the cost of materials, supplies and services used in final output, is then deducted, and then gross output is reduced by intermediate consumption to develop net output [14, 15]. GDP is measured not only by the expenditure and production method, but also by the pension method. This method of determining GDP is complemented by all incomes of households and firms. Total expenses for all final goods and services are income earned as wages, profits, rents and interest. By combining all wages, profits, rents and interest, it is possible to determine GDP [16, 17].

The aim of the paper is to examine the relationship, or rather to identify the correlation between the amount of the country's debt and its performance.

2 Data and methods

The survey will include countries in the European Union where we will be able to identify the timeline of public debt and gross domestic product (GDP). Data will be obtained from the World Bank database (The World Bank, 2017). The data will be from a period of twenty one years, namely 1995-2015. These countries will be Belgium, Bulgaria, the Czech Republic, Denmark, Estonia, Finland, France, Ireland, Italy, Lithuania, Latvia,

DOI: 10.1051/shsconf/20173901018

Luxembourg, Cyprus, Hungary, Malta, Greece, Slovakia, Slovenia, Spain, Sweden and the United Kingdom.

All financial data is in millions of Euro. We will be basing it on the following formula:

$$\rho_{x,y=}\frac{Cov\,(X,Y)}{\sigma_X\cdot\sigma_Y} \qquad (1)$$

where $\overline{\mathbf{x}}$ **a** $\overline{\mathbf{y}}$ are the mean values of GDP and public debt.

We will examine the correlation of both the entire selection, i.e. for the whole set of countries, and for each country of the data set.

3 Results

As already indicated in the methodology, the correlation of the entire set of states will first be examined, then the correlation for each country of the data file.

3.1 Measurement of the whole set of countries

When we apply all data from the dataset, that is to say, for all observed countries, we get a correlation coefficient of 0.882. This is a very high dependence of both quantities. The result predicts interesting results on the national level. If we want to interpret the value of 0.882, we must unequivocally state that the level of GDP is very much influenced by the size of the public debt of the surveyed sample of countries. We can even conclude that the higher the value of the public debt, the higher the total GDP of the surveyed group of countries. It will certainly be very interesting how the country's correlation will develop. We will surely have to focus our attention on, let's say, the countries troubled in the past. Specifically in regards to problems with debt (ideally in relation to GDP). In particular, we should continue to focus on Greece, which is still very much confronted with the effects of the global economic crisis. A very interesting country from this point of view will also be Italy, whose debt is also extremely high in relation to GDP.

3.2. Measurement of individual countries of the examined file

Table 1 provides the country-by-country correlation coefficients and the GDP of each country of the surveyed file.

Table 1. Correlation coefficients of public debt and GDP of countries surveyed.

Country	Correlation coefficient	Country	Correlation coefficient
Poland	0,978795773	Portugal	0,847339537
Germany	0,953788009	Cyprus	0,812400833
Czech Republic	0,949450649	Slovenia	0,79948063
Hungary	0,94506547	United Kingdom	0,780284517
Belgium	0,920808469	Finland	0,759296546
Slovak Republic	0,914040138	Bulgaria	0,727560281
Austria	0,905395381	Ireland	0,726590126
Romania	0,887252931	Malta	0,675399045
Italy	0,884444598	Spain	0,616071299
France	0,878382158	Greece	0,587727189
Latvia	0,877541173	Denmark	0,538750292
Luxembourg	0,857657409	Sweden	0,53259842
Lithuania	0,857575759	Netherlands	0,398606968
Estonia	0,847928938		

The results are sorted by the correlation coefficient. For a better view, the data was inserted into the chart in Figure 1.

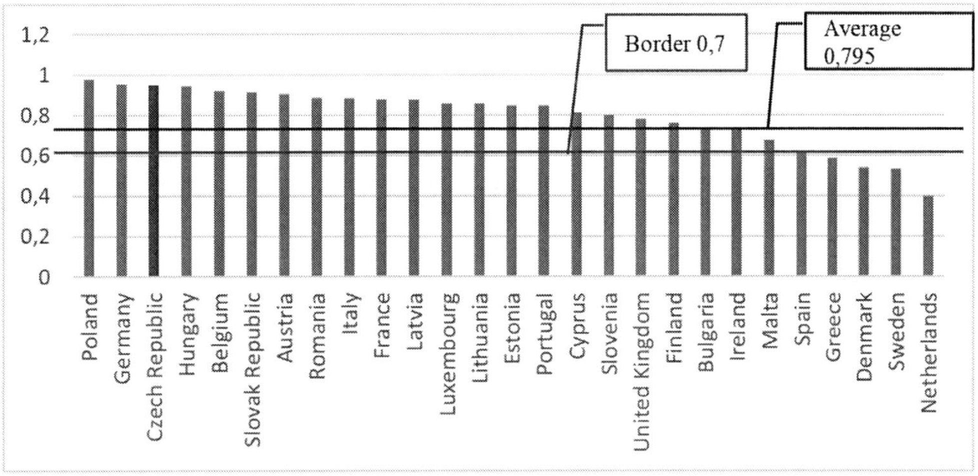

Fig. 1. Coefficients of country correlation.

The figure shows that Poland's value was the highest, almost 0.98. On the contrary, the lowest correlation value can be observed for the Netherlands, almost 0.4. We can consider the value 0.7 to be the border of large interdependence of the monitored quantities. The average correlation coefficient of the monitored set of countries is 0.795. This is a very high value. This average basically means that the correlation coefficient of all data surveyed has linked national debt and GDP not only within the country but also across the surveyed sample. However, this does not change the fact that even the average correlation coefficient is very high. Of the 27 countries surveyed, 17 countries are above the average correlation coefficient. For 22 countries, we can talk about a high degree of dependence. Only 5 countries have a lower degree of dependency. In addition, only one country, namely the Netherlands, is below the correlation coefficient of 0,5.

The Czech Republic is highlighted in red. In this case, the correlation coefficient is 0.949. In the list of countries surveyed, it occupies the third place. The interdependence of state debt and GDP is exceptionally high in the Czech Republic.

We will now focus on selected countries and analyze the evolution of public debt and GDP over the period under review.

First, let's look at Poland, which shows the highest degree of public debt and GDP in all the countries surveyed. Development is the subject of figure 2.

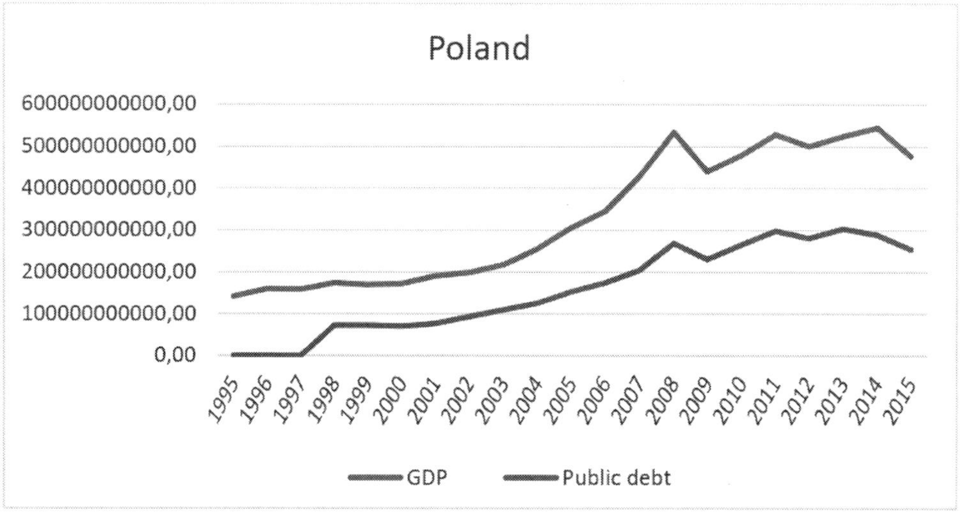

Fig. 2. Poland – Public Debt Development and GDP.

It is clear from the picture that public debt and GDP have developed over the whole period under review in exactly the same way. Any fluctuations in public debt developments are immediately reflected in the GDP. We can say that the higher the state debt, the higher the GDP. It can be seen that Poland is using its available resources very efficiently. In addition, its economy has high potential. If Poland does not reach its production caps, further increases in public debt may be recommended.

The second country to be studied is the Czech Republic. The situation is depicted in the figure in Figure 3.

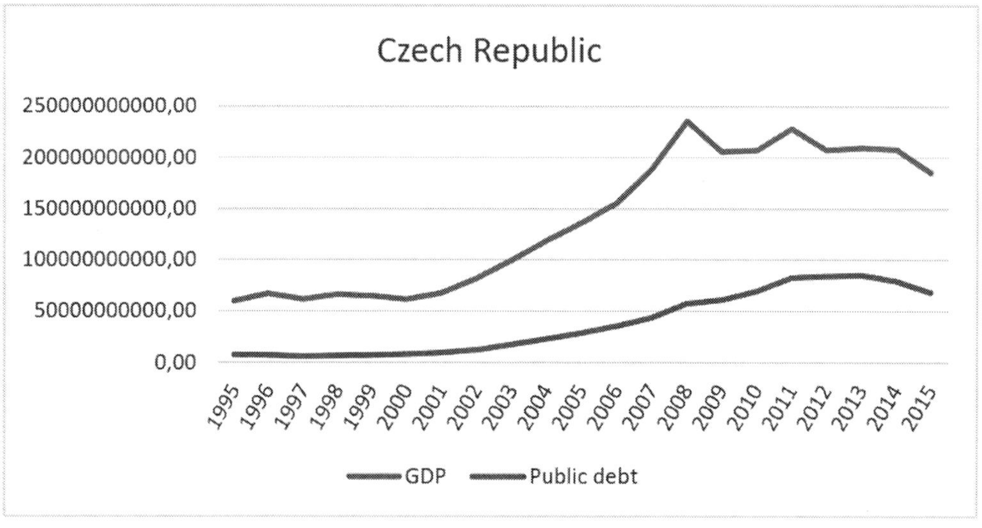

Fig. 3. Czech Republic – Public Debt and GDP Developments.

DOI: 10.1051/shsconf/20173901018

The Czech Republic also has a very high dependence on public debt and GDP. However, it is clear from the picture that the two variables do not develop as plastically as in the case of Poland. We can state that the Czech Republic also uses its resources in an efficient way. We can also say that the Czech Republic has a great potential. However, there is a certain delay between the use of foreign funds (ie public debt) and the result in the form of GDP. In addition, fluctuations are eroding, particularly in 2008. This could of course be caused by the economic crisis. However, in the case of Poland, the delay did not occur. Figure 4 shows the situation in Italy.

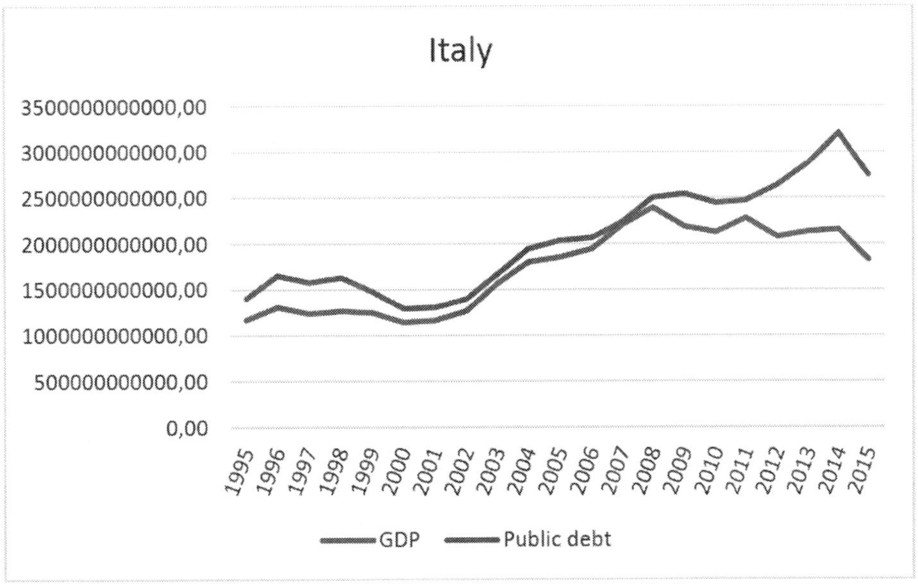

Fig. 4. Italy – Developments in public debt and GDP.

Unlike Poland and the Czech Republic, Italy has had a relatively high debt ratio throughout the period under review. The Czech Republic and Poland had their public debts at around 60% or less of GDP. By contrast, Italy's public debt is above GDP. It is clear that the interest that Italy has to pay for its debts is enormous and GDP growth slows down. In 2015, Italy's public debt is about 130% of GDP. However, despite this high debt ratio, Italy has a rate of public debt and GDP of 0.884. It can be said that with lower public debt, Italy's economy performance would be significantly lower. We can conclude that Italian national economists are at risk. At present, when Italy is able to pay interest on the use of foreign funds, where the economy is very efficient and overall has great potential, this strategy appears to be very effective, albeit rather risky. If we look from the other side to the dependence of the two variables, it is clear that the correlation is not 1. This means that the dependence is not absolute and the remaining 0.116 (some independence, some inefficiency in the use of foreign sources) should be used by Italy to reduce its public debt, at least at the time of its growth.

From the point of view of the interpretation of the state debt and GDP development, Greece is a very interesting country that has faced problems since the world economic crisis. Government debt and GDP growth in Greece is the subject of figure 5.

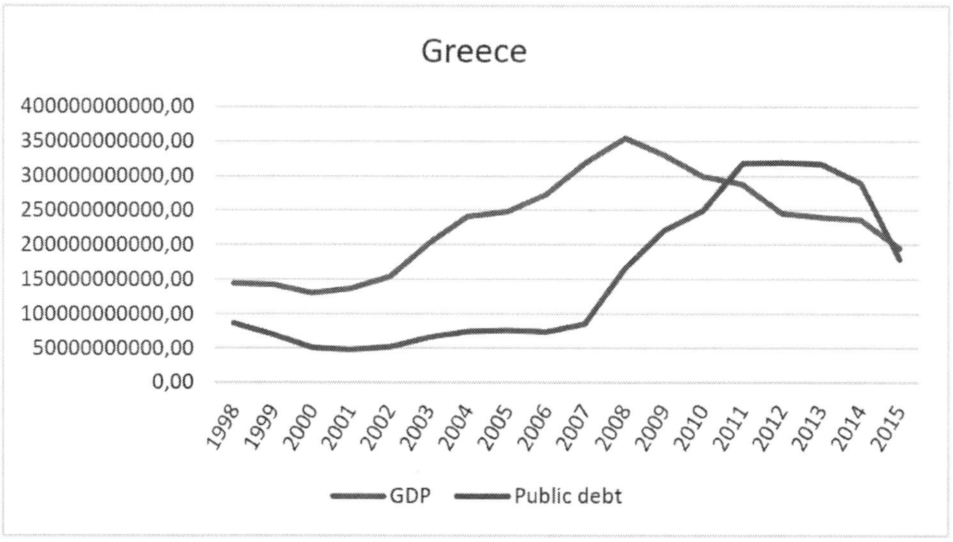

Fig. 5. Greece – public debt and GDP growth.

The correlation coefficient of public debt and Greece's GDP is around 0.588. This is the dependence of the two variables, though not extreme. Dependence can be seen in the picture by 2007. However, in the first year of the global economic crisis, the situation is changing. GDP is beginning to fall rapidly. Conversely, public debt is rising sharply. In 2011, the public debt exceeds the level of GDP. Subsequently, public debt stagnates and begins to decline. In 2015 it is again below GDP. This is the result of the Greek government's saving measures (including the overall restructuring of the Greek public debt) and the pressure of the international community. However, it is clear that Greece's public debt is a burden for its country and, unlike other countries, slows down the performance of the economy. This is due to the amount of interest paid and probably the reluctance of other countries to lend money to Greece. Sweden is also interesting for the result, as shown in Figure 6.

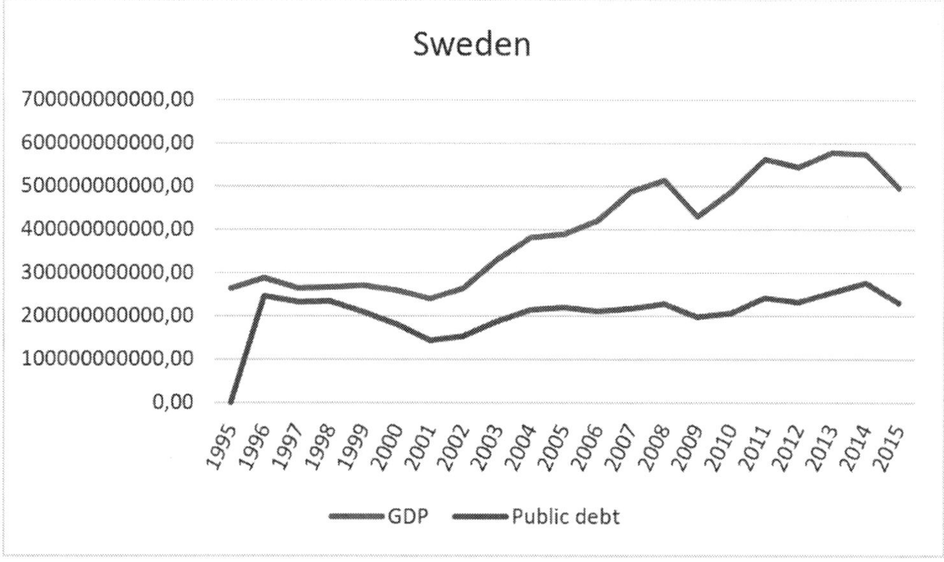

Fig. 6. Sweden – Public debt and GDP growth.

Sweden has a correlation coefficient of 0.532. This is the second lowest value in the surveyed countries. The public debt of Sweden is relatively low in relation to GDP. It is clear that Sweden does not need foreign funds to grow, and does not use these funds.

4 Conclusion

The aim of the paper was to explore the relationship, or rather to identify the correlation between the amount of the country's debt and its performance.

On the basis of the results, we can state that a large dependence of public debt and GDP has been demonstrated in the sample of countries surveyed. In the case of the 22 countries surveyed, headed by Poland and Germany, a rise in public debt may be recommended. The result will most likely increase their GDP.

In 5 countries, it is not advisable to recommend a rise in public debt. Either a country with a high state debt (Greece) or a country where debt growth will not mean a rise in the economy's performance (Sweden, the Netherlands).

The goal of the paper was met.

References

1. M. A. Allen, M. Digiuseppe, Tightening the Belt: Sovereign Debt and Alliance Formation. *International Studies Quarterly*, **57**(4), 647-659, (2013)
2. F. Fastenrath, M. Schwan, Ch. Trampusch, Where states and markets meet: the financialisation of sovereign debt management. *New Political Economy*, **22**(3), 273-293 (2016)
3. C. Bowdler, R. P. Esteves, Sovereign debt: the assessment. *Oxford Review of Economic Policy*, **29**(3), 463-477 (2013)
4. L. Agnello, V. Castro, J. T. Jalles, R. M. Sousa, Financial stress and sovereign debt composition. *Applied Economics Letters*, **23**(9), 678-683 (2015)
5. M. Caner, T. Grennes, F. Koehler-Geib, Finding the tipping point: When sovereign debt turns bad. Sovereign Debt and the Financial Crisis: *Will this Time be Different,* 63-75 (2011)
6. J. Xu, X. Zhang, China's sovereign debt: A balance-sheet perspective. *China Economic Review*, 31, 55-73 (2014)
7. P. Rousek, M. Vochozka, Public production of private and public goods in the Czech Republic in 2013. Proceedings of the 7th International conference *Economic Challenges in Enlarged Europe*, 1-9, (2015)
8. M. Ehrmann, M. Fratzscher, Euro area government bonds – Fragmentation and contagion during the sovereign debt crisis. *Journal of International Money and Finance*, 70, 26-44 (2017)
9. V. Bartoseviciene, S. Janusauskaite, S. Motuziene, Prognostication of the gross domestic product. *Proceedings of the 2nd International Scientific Conference on Globalisation and Integration Challenges to Rural Development in Eastern and Central* Europe, Kaunas City, Lithuania, 211-213, (2005)
10. J. Vrbka, Predicting Future GDP Development by Means of Artificial Intelligence. *Littera Scripta*, **9**(3), 154-167 (2016)

DOI: 10.1051/shsconf/20173901018

11. B. He, Consideration of risk in the measurement of gross domestic product and its influence on behavioral change. *2009 ISECS International Colloquium on Computing, Communication, Control, and Management*, IEEE, 268-271 (2009)
12. X. Xu, China's gross domestic product estimation. *China Economic Review*, **15**(3), 302-322 (2004)
13. I. Steblianko, Gross domestic product in the categories of production, income and end use. *Economic Annals*-XXI, **160**(7-8), 35-38 (2016)
14. K. Vltavska, J. Sixta, M. Simkova, J. Zeman, Historical time series of GDP employing standard esa 2010. *Proceedings of the 9th International Days of Statistics and Economics*, Prague, Czech Republic, 1675-1682 (2015)
15. T. Proietti, M. Marczak, G. Mazzi, Euromind- D: A Density Estimate of Monthly Gross Domestic Product for the Euro Area. *Journal of Applied Econometrics*, **32**(3), 683-703 (2017)
16. D. M. Rees, D. Lancaster, R. Finlay, A State-Space Approach to Australian Gross Domestic Product Measurement. *Australian Economic Review*, **48**(2), 133-149 (2015)
17. B. Yang, C. Li, M. Li, K. Pan A. D. Wang, Application of ARIMA model in the prediction of the gross domestic product. *Proceedings of the 6th International Conference on Mechatronics, Computer and Education Informationization*, Shenyang, China, 130, 1258-1262 (2016)
18. The World Bank, *The World Bank - USA* [online], Available from: http://www.worldbank.org/ (2017)

The offshorization of economy: the present realities

Olga Maslak[1,*], *Natalya* Grishko[1], *Olha* Hlazunova[1] and *Kateryna* Vorobiova[1]

[1]Kremenchuk Mykhailo Ostrohradskyi National University, Ukraine

Abstract. In this article the essence and main types of offshore zones have been studied. It has been analyzed the main reasons for cooperation with the tax havens, established that the main reason of the Ukrainian economy offshorization is not only the desire of economic entities to optimize taxes, but also the desire to protect their assets from the negative impact of the crisis phenomena in the national economy during 2010-2016. The foreign experience of the anti-offshore regulation has been researched. It has been defined that the general anti-offshore regulation vector should be based on the principles of financial transparency, active cooperation with the international community, an optimal combination of instruments that restrict the capital outflows and the effective instruments that stimulate the business development in the country. It has been also established that the mechanism of deoffshorization should be aimed not only at elimination of the territories with the preferential taxation, but also at receiving benefits from their use on the basis of the existing financial infrastructure of the leading world offshore centers. The experience of China in attracting additional foreign investments, gaining wider access to financial resources through the cooperation with an offshore financial centre has been researched.
Key words: offshore, economy, taxes, regulation

1 Introduction

Today the most important problem of the sustainable economic development in many countries is the high level of their offshorization, the withdrawal of capital outside the state in order to optimize taxation, the "super profits" obtaining and the maintaining the incognito status at the same time.

Offshores are traditionally considered as the countries as well as the individual districts in some countries which have the preferential tax regime or no taxes in certain cases. Some tax havens represent the regime of the full or partial confidentiality of information concerning business transactions, cash flows, and foreign companies' property. Sometimes tax havens even ensure the anonymity of the company owners [1].

* Corresponding author: oimaslak2017@gmail.com

2 Data and Methods

In the article it has been used such scientific methods as the method of analysis and synthesis, generalizations, comparison, induction and deduction, graphical methods.

Research and analytical materials, data provided by the famous information portals and agencies, articles of the scientists-economists devoted to the chosen problems were used for the determination of the offshores' essence, the modern tax havens types, and the analysis of the world practice of their use. In order to identify the main motives of Ukrainian residents for the capital withdrawal from the domestic jurisdiction, relevant statistical information concerning the socio-economic situation in the country was used. For this purpose the official figures provided by the State Statistics Service of Ukraine and the National Bank of Ukraine were used.

3 Results

Global Financial Integrity published the data concerning the capital's illegal outflow from the developing countries. Herewith, China was the leader of the capital outflow. It should be noted that annually near 139 billion dollars are illegally withdrawn from this country. In the second place was Russia ($ 104,977 billion annually) [2; 3]. Ukraine was in the 14th place in the ranking with about $ 11,7 billion of the annual amount of illegal capital outflow.

There is no single list of the offshore zones. The International Monetary Fund (IMF) and the central banks of different countries perform the control of them [4]. Due to those opportunities which a country has in the sphere of the tax optimization, we can group them as follows (Table 1).

Table 1. The offshore zones grouping based on taxation (based on [5; 6])

Group of countries	Features of taxation
United Arab Emirates, Belize, British Virgin Islands, Seychelles, Cayman Islands, Cook Islands, Dominica, Panama, Anguilla, St. Kitts and Nevis, Commonwealth of the Bahamas, Turks and Caicos, Saint Vincent	The offshore zones with zero taxation. Companies registered in these offshore zones have the right to conduct business all over the world, with the exception of the country of registration; they are completely exempted from taxation and information reporting. The simplified company registration process, confidentiality, flexible and understandable legislation
Hong Kong, Singapore, Gibraltar, Isle of Man	Companies registered in these offshore zones have the right to conduct business all over the world, with the exception of the country of registration. The largest financial centers, companies registered under their jurisdiction, are exempt from taxes, but in any case the information reporting is obligatory. If the company operates in Singapore or opens accounts in Singapore banks, the tax will be 17%
Cyprus (tax is up to 10%), Great Britain (20-26%), Malta (5-35%)	Low tax companies in the non-offshore jurisdictions. Companies registered in these jurisdictions have the right to conduct business all over the world, including the country of registration. These jurisdictions are not "offshore" and are not subject to possible restrictions on the movement of financial flows

Cyprus is the leader in the number of offshore companies with the Ukrainian owners. There are 136 enterprises on this island which theoretically could pay taxes to the budget of

DOI: 10.1051/shsconf/20173901019

Ukraine. The British Virgin Islands and the Seychelles are in the 2nd and 3rd place with 50 and 20 Ukrainian companies accordingly. The next offshores have the smaller number of Ukrainian companies. For example, Luxembourg – 5, Switzerland – 4, Jersey – 3, Malta – 3, the Federation of Saint Kitts and Nevis – 3, Hong Kong – 2, the Cayman Islands – 2, the Republic of Singapore – 2, The United States Virgin Islands – 1, the Bermuda Islands – 1, Bulgaria – 1, Gibraltar – 1, the Marshall Islands – 1, the United Arab Emirates – 1, the Isle of Man – 1. And there are also 28 companies, which are listed in declarations as registered in Russia [7].

Accordingly, the volumes of funds invested from the territory of Ukraine abroad are distributed. The prevailing part of the Ukrainian direct investment goes to Cyprus: 5925, 1 million USD in 2014, 5926, 1 million USD in 2015 and 5923, 7 million USD in 2016. These sums represent 88,4%, 91,9% and 93,8% respectively of the total amount of the invested domestic capital (Table 2).

Some funds, which have been withdrawn from the national economy in this way, are returned in the form of the direct investments through the affiliated companies of domestic entrepreneurs in the offshore areas. In the structure of the foreign direct investment directed to Ukraine Cyprus has a significant share. It was 33% in 2014, 3,4% in 2015 and 28,3% in 2016 (Table 3). In 2016, Ukraine received 6184,7 million USD from the Netherlands, which represented 17,1% of the total volume of foreign investments.

The main reason for the national economy offshorization is the economic entities' desire to optimize taxes and to obtain additional profit from this. According to the World Bank, the tax system of Ukraine is partly burdensome. This applies not only to the tax rates, but also to the processing and filing tax information. It takes 356 hours to fill all necessary documents and pay taxes in Ukraine. At the same time, these procedures take much less time in most European countries, for example, in Poland – 271, in the Czech Republic – 234, in Germany – 218, in Sweden – 122 hours [8].

Tab. 2. Direct investments from Ukraine to the economies of the world [9; 10]

Partner country	2010	2011	2012	2013	2014	2015	2016
million USD							
Total	**5760,5**	**6402,8**	**6435,4**	**6568,1**	**6702,9**	**6456,2**	**6315,2**
Netherlands	-	-	0,1	13,5	14,2	12,5	11,2
Russian Federation	166,1	190,3	236,6	292,6	377,9	196,4	122,8
Cyprus	5336,1	5900,1	5899,7	5917,6	5925,1	5926,1	5923,7
British Virgin Islands	20,8	25,8	25,8	25,8	25,8	25,8	51,3
China	0,7	0,7	0,9	0,9	1,3	1,4	1,3
in % to the total volume							
Netherlands	-	-	0,00	0,21	0,21	0,19	0,18
Russian Federation	2,88	2,97	3,68	4,45	5,64	3,04	1,94
Cyprus	92,63	92,15	91,68	90,10	88,40	91,79	93,80
British Virgin Islands	0,36	0,40	0,40	0,39	0,38	0,40	0,81
China	0,012	0,011	0,014	0,014	0,019	0,022	0,021

Tab. 3. Direct investments from the economies of the world to Ukraine, at the beginning of year [9; 10]

Partner countries	2010	2011	2012	2013	2014	2015	2016
1	2	3	4	5	6	7	8
million USD							
Total	**38992,90**	**45370,00**	**48197,60**	**51705,30**	**53704,00**	**40725,40**	**36154,50**
Austria	1674,7	1798,9	2317,5	2476,9	2314,0	1648,7	1559,8
United Kingdom	2234,1	2229,9	2536,4	2496,9	2768,2	2153,4	1790,3
Netherlands	7461,3	11389,8	9323,8	8727,6	9007,5	6986,7	6184,7
Germany	6009,6	5001,2	5329,8	4496,3	2908,4	2105,2	1598,2
Russian Federation	1900,2	2692,7	2876,1	3040,5	3525,9	2338,9	3036,9
France	1381,1	2105,4	1993,1	1510,3	1520,5	1394,6	1305,4
Switzerland	785,7	852,7	939,3	1097,6	1351,0	1391,5	1390,8
Sweden	674,2	1108,9	1141,9	1084,4	439,3	360,2	328,9
Cyprus	8603,1	9620,5	12700,8	15907,7	17725,6	12769,4	10239,5
British Virgin Islands	1283,6	1384,9	1580,2	1888,2	2275,9	1988,3	1715,0
USA	1260,0	1107,6	966,6	976,5	934,7	701,6	634,1
Hong Kong	4,6	13,0	14,4	23,0	20,6	12,4	9,3
China	11,8	12,7	14,8	18,2	25,5	21,0	18,6
in % to the total volume							
Austria	4,29	3,96	4,81	4,79	4,31	4,05	4,31
United Kingdom	5,73	4,91	5,26	4,83	5,15	5,29	4,95
Netherlands	19,14	25,10	19,34	16,88	16,77	17,16	17,11
Germany	15,41	11,02	11,06	8,70	5,42	5,17	4,42
Russian Federation	4,87	5,93	5,97	5,88	6,57	5,74	8,40
France	3,54	4,64	4,14	2,92	2,83	3,42	3,61
Switzerland	2,01	1,88	1,95	2,12	2,52	3,42	3,85
Sweden	1,73	2,44	2,37	2,10	0,82	0,88	0,91
Cyprus	22,1	21,2	26,4	30,8	33,0	31,4	28,3
British Virgin Islands	3,29	3,05	3,28	3,65	4,24	4,88	4,74
USA	3,23	2,44	2,01	1,89	1,74	1,72	1,75
Hong Kong	0,012	0,029	0,030	0,044	0,038	0,030	0,026
China	0,030	0,028	0,031	0,035	0,047	0,052	0,051

In 2016 due to the criterion of the tax payments convenience Ukraine ranked 84th. The general tax rate, including all types of fiscal payments, is 51.9%. Although there is a world tendency to reduce the tax burden on business. So that the general tax rate of the rating leaders does not exceed 20%: for example, in the United Arab Emirates it is 11,3%, in Qatar – 15,9% [11].

Nevertheless, in the modern economic conditions the desire to protect the assets from the influence of internal and external risks and from the threats to the effective entrepreneurial activity become the more important determinant of capital withdrawal abroad by residents of Ukraine. The increasing of these factors' importance is caused by the destabilization of the socio-economic situation, the reduction of the Ukrainian economy investment attractiveness and the increasing of political and legal risks impact. Thus, in recent years there has been a significant decrease in the physical volume of the gross domestic product of Ukraine: in 2014 – by 6, 6%, in 2015 – by 9,9% (Fig. 1).

DOI: 10.1051/shsconf/20173901019

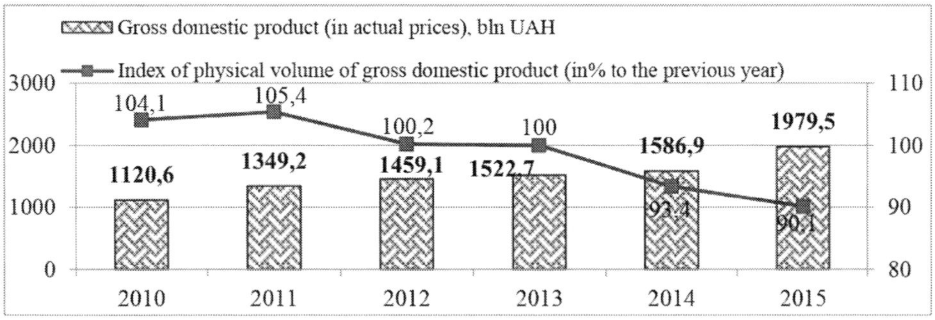

Fig. 1. The Ukrainian GDP dynamics in actual prices and prices of the previous year (% to the previous year) [9]

At the same time, the nominal GDP was characterized by the positive dynamics. Such a "fictitious" increase of the gross value added amount created on the Ukrainian territory in 2014-2015 was due to, first of all, the significant inflation increase (Fig. 2).

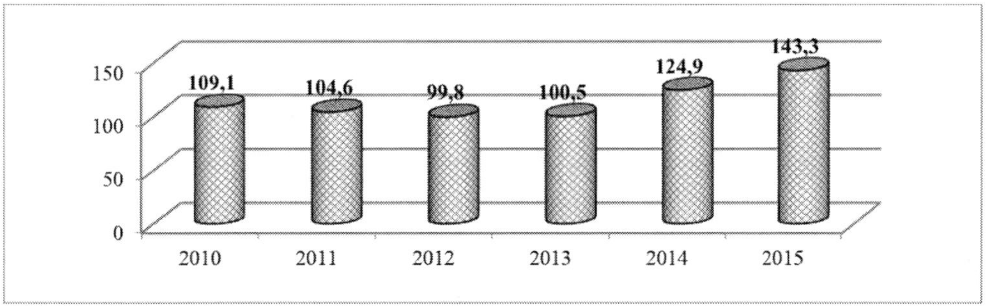

Fig. 2. The Ukrainian consumer price index dynamics during 2010-2015, % [11]

In Ukraine the consumer price index was 124,9% in 2014 and 143,3% in 2015. The depreciation of the national currency is an indicator of a significant weakening of the national economy and a threat of the further regression of the domestic business entities development.

In the period from 2013 to 2014, the gross external debt of Ukraine increased: by 6, 65% in 2013 and by 5,54% in 2014 (Fig. 3). At the same time, in 2014 the amount of the external debt exceeded the gross domestic product of the same period by 7,79%. The attention should be paid to the fact that in 2015 and 2016 the amount of the state debt of Ukraine decreased by 11,1% in 2015 and by 6% in 2016. However, in 2015 the external debt amounted to 139, 39% in relation to the GDP. It was a historic record and it was the result of a decrease in the amount of the value added created in Ukraine and further devaluation of hryvnia. At the same time, the results of domestic and foreign researches indicated that the amount of the public debt of Ukraine could be economically safe at the level of 35% of GDP [12].

 DOI: 10.1051/shsconf/20173901019

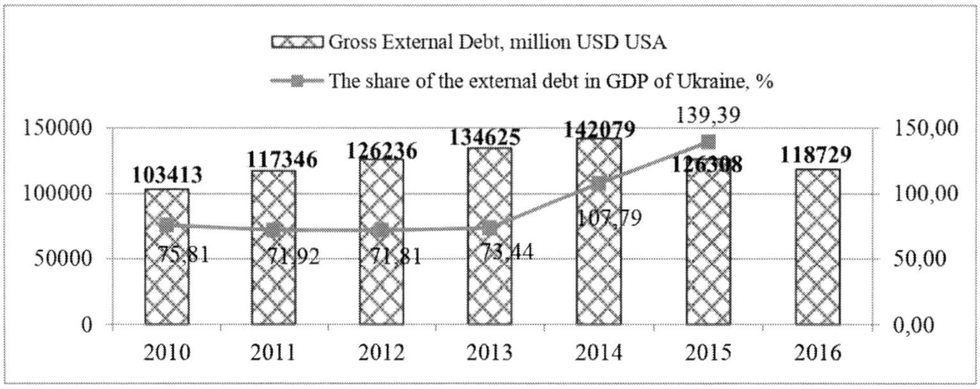

Fig. 3. The Ukrainian gross external debt dynamics during 2010-2016 [10]

The growth of capital outflows has a significant negative impact on the national currency stability, financial security level and general economic stability of the state.

That is why today the deoffshorization process is gaining momentum both at the national and international levels. The latest trend in the EU is the implementation of the European Parliament directive aimed at combating the money laundering and tax evasion, in particular through the use of the offshore schemes. New requirements for law firms, banks and other financial institutions concerning the conduction of the register of clients' final beneficiaries and providing this information to the financial monitoring services are introduced. The first financial information disclosure is announced in 2017 by such countries as the British Virgin Islands, Cyprus, Lithuania, Latvia, Luxembourg, the Netherlands and Seychelles. In 2018 Austria, Hong Kong, Monaco, Russia, Singapore, Switzerland and others countries will exchange the information for the first time. But Panama and Ukraine are not represented in this list [13].

The work of the international organizations in the sphere of the offshore business regulation is aimed, first of all, at the development of legislative regulations and rules intended to limit and regulate the offshore business, as well as to monitore their implementation completeness.

A variety of tools are used to ensure the high level of financial transparency and to prevent capital outflows through the tax havens at the national level. In this case, both loyal preventive measures and rigid (aggressive) methods, which include penal sanctions and other types of legal proceedings, are implemented (Table 4 in Discussion).

Thus the deoffshorization's tools can be nominally divided into three big groups: tools with the restrictive nature, tools that stimulate respondents to conduct business under the domestic jurisdiction and tools that help to benefit from the use of the offshore schemes.

The tools with the restrictive nature include the ensuring of exchange the information concerning the domestic taxpayers' accounts in foreign jurisdictions with the international financial organizations, the conclusion of the double taxation avoidance contracts, currency restrictions, etc. There are the following stimulating tools: the national tax regime optimization, the increase of the favorable investment climate in the state. The tools that enable to get the benefits include the creation of its own offshore zones, the use of offshores for the purpose of foreign economic expansion.

Especially interesting is the experience of China. This country uses offshore zones to attract the additional foreign investment and to gain wider access to financial resources.

Herewith, the use of the offshore zones is carried out within the framework of state programs, but the monetary control of the capital movement is also maintained. China has concluded more than 120 tax information exchange agreements, including the major offshore jurisdictions. This situation allows us to detect schemes for the tax evasion. Most of foreign investment goes to China through Hong Kong. The Hong Kong Special Administrative Region serves as an example of a "traditional" international financial center, which represents some signs of the offshore at the same time.

Hong Kong is the leader of economic freedom ranking which is published by The Heritage Foundation. In addition, Hong Kong has a high level of legal culture that attracts investors and entrepreneurs from around the world. Hong Kong refers to the countries with the preferential taxation, aimed mainly at stimulating the inflow of foreign investors in certain sectors of the economy, including the securities market. In Hong Kong the income tax is 17,5%, while dividends, interest and capital gains are not taxed. The transparency of Hong Kong's economy as a whole and the financial system in particular is so high and the taxes are so low, that this fact allows the Bank for International Settlements to classify Hong Kong as the offshore center.

In the mid-2000s, the Chinese government took a number of steps to intensify the use of foreign markets, first of all, the market of Hong Kong, to attract capital to Chinese enterprises and banks. As a result, Hong Kong financial market became a significant source of capital not only for local companies, but also for companies registered in China. This caused a very serious interest in this market from foreign investors.

4 Discussion

Tab. 4. Foreign experience in the sphere of the economy deoffshorization [14; 15; 16; 17]

Country	The main tools and methods of the economy deoffshorization
1	2
Poland	The government of the country set the course towards the creation of favorable conditions which could contribute to the flow of capital into the country. Cooperation with the offshore zones is quite legitimate in Poland; there are no obstacles for this process at the institutional level. The state has its own list of the countries with the "harmful tax competition". There are certain rules which are applied in relations with the countries from this list: – income tax on transactions, royalties, interest and certain types of intangible services is 20%; – the Polish tax authorities can decide on the determination of the income amount from transactions with a company registered in an offshore area. Such an operation may be subject to an additional 50% tax on the difference between the income declared by the taxpayer and calculated by the tax authorities; – reimbursement of VAT on the received or rendered services in or from an offshore zone is prohibited; – it is important to conduct a more detailed review of the transfer pricing, which means the need to provide tax and customs authorities with more detailed documentation on transactions with companies that are registered in the countries with so-called harmful tax competition
Estonia	Revenues, received by enterprises, which are immediately redistributed, are not taxable. Export and transit transactions are not subject to VAT. In this regard, Estonia is considered as an offshore zone in the jurisdictions of many countries, although this is not quite the case. In the absence of the income tax Estonian companies maintain accounting records. They report twice a month to the tax authorities and every "unauthorized" use of funds leads to the additional tax payment in the amount of 79,9% of the withdrawn sum. There are no restrictions in Estonia on cooperation with the offshore zones. However, the website of the tax department presents an informal "black list". Payments to the states from this list are the subject to the 21% profit tax.

USA	In 2008 the Stop *Tax Haven Abuse* Act was passed in the USA. According to this Act the fiscal and controlling authorities obtained the widest powers towards the taxpayers who resorted to offshore schemes. The amount of the fine could reach 150% of the gained profit. The adoption of the Foreign Account Tax Compliance Act in 2010 became one of the key events in the development of the anti-offshore legislation in the United States. The law obliged citizens to declare all their incomes, regardless of the country of their origin. The use of offshores is allowed only for the business development. In all other cases, the presumption of the enterprise guilt is established, it has to refute the presence of the hidden revenues in the tax havens. After the adoption of FATCA (Foreign Account Tax Compliance Act) the US Department of the Interior got the right to receive any information about the accounts and assets of its citizens abroad. Banks, that refused to comply with such requirements, were required to pay 30% of any transactions passing through the US, otherwise their accounts in financial institutions of the United States might be closed. A new Tax Reform project for the commercial sector, which was presented to the public in 2012, stipulated the obligatory payment of the tax on income received in the territory of other countries. The same document lifted a deferral from the payment of such tax on income until the moment of its repatriation (return under the domestic jurisdiction)

Continuation of Table 4

1	2
UK	First of all, the government of this country seeks to optimize the state tax system. For this purpose in 2011 the corporate tax rate was reduced from 28% to 26%, while the rates for small businesses decreased to 20%. There are the following main legislative acts regulating the residents' offshore activities: 1. The Offshore Funds (Tax) Regulations, 2009; 2. Taxation (International and Other Provisions) Act, 2010; 3. Companies Act, 2006. These legislative acts define the types of legal entities that can register non-residents on the territory of the country and regulate their taxation. In 2011 new fines for taxpayers were introduced in the UK. They are used by affiliated companies registered in the offshore jurisdictions to avoid full compliance with tax obligations
France	The use of the anti-offshore regulation's "preventive" measures is typical for France. There is no special legislation in the sphere of banking secrecy in the country. At least once a month banking institutions provide information regarding the facts of opening and closing accounts, the size of the interest or income from the capital received by customers. In turn, the tax service directs and makes the appropriate changes to resident bank accounts' centralized unified database. France belongs to the countries with an active position in the fight against the offshorization. The banks of this country took a decision to discontinue operations on the territory of the countries which were in the "gray list" of the OECD. All the financial institutions should report on transactions with jurisdictions that have not signed the tax information exchange agreement
China	The offshorization is considered as the inevitable result of the world economy's functioning mechanism transformation. China wants to benefit for the national economy from the use of the offshore jurisdictions. It uses the offshore zones as the mean to increase the industrial exports competitiveness, to get wider access to the financial resources, to increase the volumes of foreign investments into the raw material extraction and infrastructure development. The goal is also to provide access to Hong Kong's financial resources, including the direct lending, issue of yuan bonds at the Hong Kong Stock Exchanges, attraction of funds, listing on foreign exchanges of Chinese companies operating in industries closed to foreign companies.

5 Conclusion

Thus, the problem of the national economy offshorization is particularly acute for Ukraine in terms of financial, economic, political and legal instabilities; therefore, the development of the effective instruments for regulating the relations of residents with "tax havens" needs to be given increased attention. In the general sense, the mechanism of the national economy deoffshorization is a set of interacting tools and subsystems that ensure of the state tax system optimization and return to domestic capital jurisdiction through the establishment of the effective communication between economic entities, national and international institutional bodies, fiscal services and the financially-credit system of the country.

References

1. O. Volotovskaya, M., V. Zakharova. *Offshore schemes as a threat to the economic security of the Russian Federation* (2014)
2. D. Sushkevych, *The capital outflow from developing countries: causes and consequences.* Case study of the republic of Belarus (2016)
3. Global Financial Integrity [online], Available at: http://www.gfintegrity.org/ (2017)
4. Offshore zones: classification, location, methods. *Reference* [online], Available at: https://ria.ru/economy/20090403/166953816.html (2009)
5. Offshore zones – the list and comparison of offshore zones of the world [online], Available at: http://www.bcs-offshore.com/rating.html (2017)
6. L. Opekarová, V. Mulačová, H. Ezrová, R. Píchová, Tax harmonisation in free trade EU – reflection on the possibilities of tax evasion – part 1, *Littera Scripta*, **4**(1), 61-77 (2011)
7. Top 5 of the offshore owners: Liovochkin's sister headed the list – infographics [online], Available at: http://zik.ua/news/2016/11/27/top5_vlasnykiv_ofshoriv_spysok_ocholyla_sestra_lovochkina__infografika_998527 (2017)
8. Y. Gryhoreva, Theoretical and methodological approaches to assessing the consequences of offshore business for the development of modern international capital flows. *Economy and society*, **6**(25) (2016)
9. State Statistics Service of Ukraine [online], Available at: http://www.ukrstat.gov.ua/ (2017)
10. National bank of Ukraine [online], Available at: https://www.bank.gov.ua/control/uk/index (2017)
11. O. Liakhovets, Entrepreneurship as a component of modernization institutional environment. *Scientific Journal of International Humanitarian University* (2016)
12. M. Ivanchenko, K. Dobrolezha, A. Chubak, The analysis of Ukraine debt at the present conditions. *Global and National Problems of Economy*, **10**, 792-797 (2016)
13. Global deoffshorization: its consequences for Ukraine [online], Available at: http://forbes.net.ua/_ua/opinions/1414948-globalna-deofshorizaciya-yaki-yiyi-naslidki-dlya-ukrayini (2017)
14. Ukraine and the United States signed an agreement within the FATCA law on the fulfillment of requirements to the foreign accounts [online], Available at: https://economics.unian.net/finance/1764856-ukraina-i-ssha-podpisali-soglashenie-v-ramkah-zakona-fatca-o-vyipolnenii-trebovaniy-k-inostrannyim-schetam.html (2017)
15. A. Faruki, A. Tsaruev, EC Policy Towards Offshores. *Scientific student's conference, Offshores in international business practice*, **4**(23) (2012)

 DOI: 10.1051/shsconf/20173901019

16. N. Yurchenkova, The deofshorization is gaining momentum: the first results of the reform. *Corporate lawyer*, **6**, 10–14 (2016)
17. Y. Danilov, The experience of financial markets reforms in the countries – competitors of Russia in the global capital market. Research Paper Series, **2016**(171), 156 (2016)

Evaluation of selected financial aspects in the Czech Republic: a research study about the international activities of Czech business

Lucie Meixnerová[1,*]

[1]The Institute of Technology and Business in České Budějovice, Okružní 517/10, 370 01, České Budějovice

Abstract. The paper deals with the evaluation of selected financial aspects of micro, small and medium-sized enterprises in the Czech conditions in the context of the international activities. The aim of this paper is to identify and evaluate whether there are dependencies between these indicators in the context of domestic and international environment by using selected methods. In the article are evaluated the results of our own survey of the development of micro, small and medium enterprises in the years 2009–2016, with a total of 650 respondents (SMEs). The author calculates and describes traditional financial indicators and international activities represented by ROA, ROE and ROC. Research findings confirm that the companies operating in the Czech environment show higher growth at those micro-enterprises without international activities. Companies operating in an international environment shown stronger growth among medium-sized businesses. This is due to the benefits of international cooperation, the size and development of the company, and the growing influence of the international environment on business entities.
Key words: evaluation, business activities, financial aspects

1 Introduction

In a constantly changing economic environment, the irreplaceable role of small and medium-sized enterprises (SMEs) and their importance in the entire national economy is discuss more frequently. In this environment, the entrepreneurial subjects – the SMEs – also undergo changes in the management system which are part of this environment and compose of factors and forces that have a significant influence on the company and the company relations with its target market in terms of regional and international activities. The existence of competitive environment enables and international activities good functioning to only such companies which have mastered not only the entrepreneurial skills of business activity but also financial skills which play an important part. No company can thus be managed without an analysis of its financial situation, meaning the evaluation of financial soundness of the company which detects various things about both the inner and outer state of the company [1].

* Corresponding author: luciemeixnerova@seznam.cz

The objective of the paper is to analyze the selected financial aspects of the micro, small and medium-sized enterprises of the Czech Republic in the international context by using selected mathematical-statistical methods. These dependencies are assessed in relation to the international activities from the profit or loss statement, assets, and capital structure of the analyzed subjects for the period 2009-2016. For the specific enumerations of dependencies among the examined factors, the correlation and regression analysis was used. The paper follows the research implemented in the field of small and medium-sized enterprises [1]. Currently, there is no similar research about Czech business with international activities in the Czech Republic. This paper as well as the implemented research and other similar international papers have become integral parts of the information and confirmation of business entities development in the Czech Republic. The paper follows the research [1] and the authors′ own research. The information provided is valuable for the future and prospects of businesses and regions and can be used not only for own business needs but also for the public and other organizations (investors, business groups, state institutions or foreign institutions, etc.) which are closely connected with international activities, either economically or financially.

1.1 Literature Review

The role of SMEs is quite well researched and documented, as can be easily demonstrated by several studies [2], [3], [4], [5]. All these papers state the importance of SMEs in GDP or GNP and employment. These studies also research the above mentioned issues on the national level, the level of the whole country. There are studies which brings in the topic of international and are connecting SMEs and their international links, such as the ownership [6] which is focused on debt costs or [7] which is focused on behaviour and performance or [8] which is focused on performance. Empirical study has confirmed that the Czech Republic, Hungary and Slovakia has poorer results added values than Sweden, Belgium, Denmark and Finland [9]. Growth of efficiency of economic processes, added value and overall business success in small and medium enterprises depends on selection of the right sphere of business, application of research and development, adequate investments and cooperation between individual sectors or on the level of investment activity The study from authors Hu, Chang, Hsu [10] deals with the relationship between the capital structure and the business performance. It was found out that the capital structure has not sustained effective funding required for the growth and development of the business. The use of equity and foreign capital for funding also affects the tax base. The issue of the capital structure of a business in the context of the taxation by the corporate and personal income tax was analyzed by DeAngelo and Masulis [11] when the expense of foreign capital – interest – is a tax shield. Brennan and Schwartz [12] state that for funding the assets a combination of equity and foreign sources is always used, because the optimal capital structure will almost entirely consist of debt. The analysis of the factors affecting the involvement of the foreign capital in a business is further described by Lee and Hyun [13]. The choice of the capital structure affects not only the degree of debt but also the amount of profit or the operating leverage indicator. In search for optimal sources for funding the assets, it is appropriate to respect the golden balance rule of funding, i.e. the fixed assets are financed by long-term resources – equity and fixed foreign capital - and the current assets are financed by short-term sources. Dittmar [14] monitored the structure of funding in the selected businesses found out that the representation ratio of equity and foreign sources depends on the business field and the nature of activities of the analyzed entity. High-growth firms choose lower leverage ratios. Firms seem to use debt financing too conservatively, and the leverage of stable, profitable firms appears particularly low [15].

In the Czech Republic, the factors affecting the capital structure were analysed [16]. Strýčková [16] states that the decisive factors are the cost of capital and the indebtedness of the business. To fund its assets, each business should choose such a resource structure so that the average total costs of capital were the lowest. The cost of capital includes two components, the cost of debt and the cost of equity, learn more about the average total costs in Aliyev-Ada [17]. The influence of the interest rate on funding was examined [18] conclusions confirmed that larger firms tend to have higher debt ratios than smaller firms. Rodríguez-Masero [19] confirmed the fact that the phase of the economic cycle also affects the capital structure.

1.1.1 Accounting as a primary source of data for the financial analysis

It is appropriate to examine the company performance evaluation also from the perspective of the financial reporting and it provides information on the financial position and performance, which is important for economic decisions of a wide range of users [20]. Accounting provides the pivotal information on the economic phenomena of the business, sources of funding and the structure of assets as well as the cost and revenue structure especially through the financial statement accounts. In the Czech Republic conditions, this issue is enacted by the provisions of the Act No. 563/1991 Coll. on Accounting, the Decree No. 500/2002 Coll. implementing particular provisions of the Act on Accounting, and Czech accounting standards. Financial statements, the items of which are exhaustively defined by the Act on Accounting, consist of the balance sheet, profit and loss account, and annex.

It is appropriate too to compare the results of the financial performance evaluation not only in the context of its development in time, but also for example according to CZ-NACE categorization codes, as the field of activity has an impact on financial and economic performance of firms [21]. It is appropriate to examine the company performance evaluation also from the perspective of the categorization of business entities. From the perspective of business economics, we divide companies to small, medium and large companies. The term 'company' is absent from accounting, this term is replaced by the term 'accounting unit'. The categorization of an accounting unit is based on the European Union directive and its categorization (micro, small, medium and large accounting unit) is dependent on the implementation of the directive in the particular country. In accordance with the rules of the European Union (Recommendation 2003/361/EC), an enterprise is considered medium-sized if the number of employees does not exceed 250, the annual turnover EUR50 million and the annual balance sheet total EUR43 million, the classification limits for a small and micro enterprise are lower.

2 Data and methodology

The paper follows the research implemented in the field of micro, small and medium-sized enterprises [1] and the author′ own research. The primary objective of the research study was the analysis of performance of small and medium-sized enterprises in terms of international activities. Author observed a total of 650 respondents - business untities (accounting units) during the years of 2009-2016 from. The research was conducted using data from March 2013 till July 2017 and the research still continues. The basic preconditions to include a business in the research were the categorization of accounting units in the Czech Republic and company financial reports. The categorization of business entities (hereinafter referred to as the size of business) was determined by the indicators according to the Czech legislation valid from 1st January 2016. Data was collected data mostly from financial reporting (balance sheet, profit and loss statement, notes in annual reports) and also from their web pages and other official lists and databases (such us Business Register, Czech Statistical Office register etc.). The individual respondents were continuously investigated in terms of the changes which occurred in the examined period and were related to the development of their business activities, in direct continuity with a focus on accounting and overall business activity of the monitored businesses. The research results have been continuously processed and the data of 600 business entities out of the original 650 business entities before the statistical adjustment were used. The qualitative data obtained by the primary research were evaluated using the Microsoft Office Excel application and the statistical software IBM SPSS 19. In addition to standard scientific methods such as description, analysis, synthesis and comparison, authors also used selected methods of the regression and correlation analysis and ANOVA method. The dependency ratio between the selected characters x and y is evaluated by the correlation coefficient, the form of which is determined by (1),

$$r = \frac{\sum_{i=1}^{n}(x_i-\bar{x})(y_i-\bar{y})}{\sqrt{\sum_{i=1}^{n}(x_i-\bar{x})^2 \sum_{i=1}^{n}(y_i-\bar{y})^2}}, \tag{1}$$

For expressing the progress of the dependent variable y on the independent variable x, the regression analysis is used. The form of the regression function is determined by (2),

$$y = B_0 + B_1x + \mathcal{E}, \tag{2}$$

where $\mathcal{E}$ is a random error, B_0 and B_1 are the population model coefficients.

The determination index R2 (3) expresses to what extent the ascertained regression function depicts the dependence of Y on X,

$$R^2 = \frac{S_T}{S_y}, \tag{3}$$

where ST is the theoretical sum of squares and SY is the total sum of squares. The variance of Anova, which is dealt with in more detail in [22] [23] and the aim is to find out if one statistical character affects the second character, whether the factor affects the values of the second statistical set. According to the study from Hinke [24] no used a probabilistic neural network, a generalised regression neural network, a radial basis function network, a three-layer perceptron network and a four-layer perceptron network for planning financial statements because the method is limited, only for a stable company, not for studies more enterprises.

The individual conclusions listed in own surveys and in correlation and in regression and their subsequent evaluation are influenced by the respondents and, therefore, with regard to the given objective, Indicators were chosen related to the enterprises performance in international activities:

- the micro accounting unit, in the paper marked "micro",
- the small accounting unit, in the paper marked "small",
- the medium accounting unit, in the paper marked "medium",,
- return on assets, in the paper marked "ROA",
- return on equity, in the paper marked "ROE",
- return on sales, in the paper marked "ROS",
- the international business are activities on foreign market.

DOI: 10.1051/shsconf/20173901020

On the basis of present knowledge, experience and above-mentioned information, the following hypotheses were formulated:

H0: International activities have no effect on company performance at micro enterprise (a), small enterprise (b), and medium enterprise (c) level.

H1: International activities have an effect on company performance at micro enterprise (a), small enterprise (b), and medium enterprise (c) level.

3 Results

With regard to the categorization of accounting units which follows the transposition of the Directive of the European Parliament and Council 2013/34/EC (the Directive) into the Czech accounting legislation, the accounting units in the category of micro business units were represented among the interviewed respondents most often. Implementing this Directive, the national accounting has got closer to the IFRS system. Learn more about the IFRS influence e.g. in [24]. Fig. 1 demonstrates the structure of micro, small and medium-sized enterprises according to number of employees in the Czech Republic.

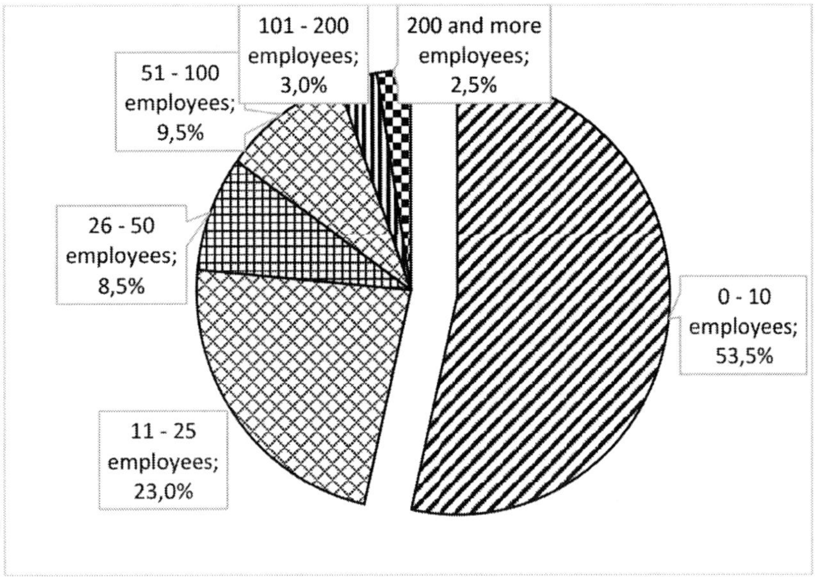

Fig. 1. Categorization of accounting units by the number of employees.

The absolute majority of micro, small and medium-sized enterprises (specifically 400 of the research sample) in the Czech Republic belongs to the first group – they have 0-10 employees and 53.5% share so they rank first. In the Czech Republic, there operate many small traders who usually cannot afford to employ more people. At the same time, building a business with more employees is very often associated with high initial expenses for which the starting entrepreneurs usually do not have enough financial resources. The second largest representation are businesses with 11-25 employees and 23% share. On the contrary, the smallest ratio of the total number of examined entities are businesses with 200 or more employees which reach 2.5%. It can be stated that, according to the number of employees, the micro and small businesses with fewer than 25 employees rather prevail and they participate in the total number of businesses with more than three-quarters. It can be therefore said that the influence of these businesses in the Czech Republic is considerable.

Micro and small enterprises are the least active in the international environment as was found out on the basis of own research of business entities in 2016 and by selected factors: business entity categorization by the number of employees and the length of operation in the domestic and international markets. The representation according to age is the largest in international markets by the medium-sized enterprises, which is confirmed in the Fig. 2.

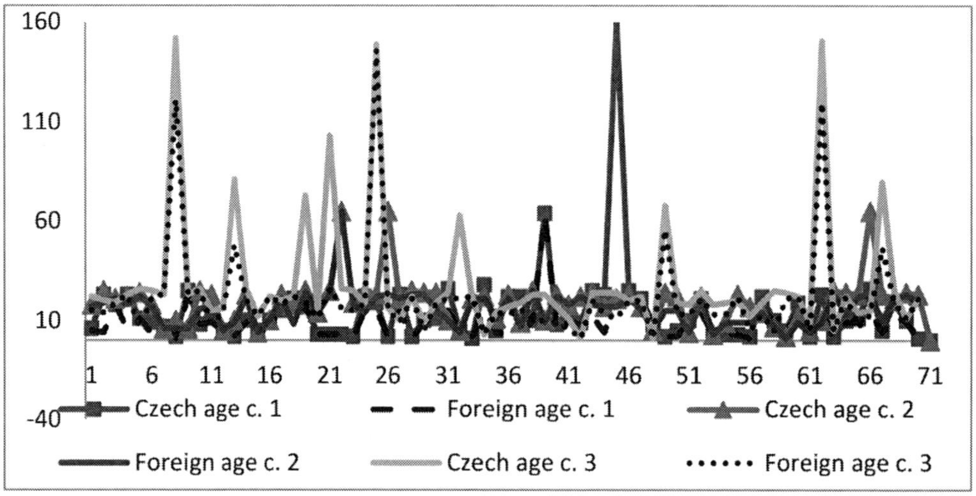

Fig. 2. The length of SMEs operating in the domestic and international environment.

In order to enter international markets, business entities operating in the Czech Republic have the following characteristic features: high industry specialization, lack of financial resources, narrow environment and above all an impact of the free movement of goods and capital, development of modern communication technologies, transport and logistics systems, and other factors that form a stimulus for the development of international business. We include language barriers, knowledge of the market including regulations and legislation, the lack of information about the environment, etc. among the reasons for the low level of activity in the international environment. Operation in international markets has a positive significance for micro, small and medium-sized enterprises.

3.1 Enterprises operating in the Czech Republic

Diagnostics of collinearity through correlation, Table 1, was carried out before the analysis of SME´s in the international environment. The selected indicators, the dependence of which was tested, include the categorization of accounting units and profitability. All the mentioned items can be found in balance sheet and profit and loss statement. We can be count (profitability) it that is why their interdependence is analyzed. The item total profit or loss was included in profitability indicators because as the subject often shows profit but has no funds and need capital for own business.

Table 1. Correlation analysis of SMEs.

Micro enterprises				Small enterprises			Medium enterprises		
	ROE	ROA	ROS	ROE	ROA	ROS	ROE	ROA	ROS
ROE	1.00			1.00			1.00		
ROA	0.47	1.00		0.30	1.00		0.76	1.00	
ROS	0.42	0.29	1.00	0.57	0.35	1.00	0.72	0.83	1.00

As shown in Table 1, the relationships between individual variables show a mostly moderate dependency for micro and small enterprises. The ROA indicator confirms the effectiveness of using the asset structure of the business. Micro and small-sized enterprises show a mostly moderate dependency on foreign capital. Medium-sized enterprises are able to receive and repay foreign resources, which confirms the moderately strong positive dependency of indicators. Enterprises with a long-term existence must be profitable and strive for a financial structure with minimal cost. Table 1 confirmed growing profitability indicators of SMEs.

3.2 Enterprises operating in the international environment

Table 2 shows the dependency between individual variables for micro, small and medium enterprises in the international environment. Individual variables such as profitability indicators such as foreign profit and sales.

Table 2. Correlation analysis of SMEs in the international environment.

Micro enterprises				Small enterprises			Medium enterprises		
	ROE	ROA	ROS	ROE	ROA	ROS	ROE	ROA	ROS
ROE	1.00			1.00			1.00		
ROA	0.02	1.00		0.10	1.00		0.88	1.00	
ROS	0.03	0.00	1.00	0.29	0.25	1.00	0.92	0.77	1.00

As shown in Table 2, the relationships between individual variables show do no a dependency for micro enterprises. Small enterprises show a mostly moderate dependency and low of foreign capital. Medium enterprises are able to receive and repay foreign capital which confirms the moderately strong positive dependency of indicators. One of the reasons is the long-term company operation in the international environment. Strong dependency of ROA, ROE, and ROC indicators points to positive capital appreciation which is one of the most important indicators of the financial performance of the company. Businesses appropriately manage the entrusted financial resources. As hardly any business can manage without foreign capital these days [24], a typical expense is for example, the loan interest. All the monitored indicators are important documents for the evaluation of business and the international environment. Subsequent analyses were conducted to further analyse these relationships and confirm the hypotheses. Companies must not be only reasonably profitable, but they must be able to reconcile their needs, i.e. bind capital in short-term parts of assets that must be paid by costs associated with their funding and also must seek financial structure with minimal costs of capital, both their own and foreign capital. Compliance with the use of foreign capital for financing the asset part should always correspond to the degree of a company's liquidity and financing rules that significantly affect profitability.

Table 3 confirms the statistically significant indicator of international activities for medium enterprises.

Table 3. Regression statistics and ANOVA for medium enterprises.

Regression statistics	
Multiple R	0.892192
R-squared value	0.796007
Set R-squared value	0.785628
Aver. value error	20.358406
Observation	75

ANOVA					
	Difference	SS	MS	F	Significance F
Regression	2	187606.095984	93803.047992	2.263234	1.663234
Residues	72	48077.904016	414.464690		
Total	75	235684.000000			

Table 3 confirms the statistically significant indicator of international activities for medium enterprises. H0 is rejected and the selected indicators affect the development of international business. For enterprises, it is important to operate in international markets, which makes them able to achieve higher performance compared to the local environment. Sales, profit or loss, own and foreign capital and for example inventories, demand, prices and competition etc., these affect the outputs of the company in the international environment.

Table 4 confirms no the statistically significant indicator of international activities for micro enterprises. H1 is rejected and selected indicators do not affect the development of international activities.

Table 4. Regression statistics and ANOVA for micro enterprises.

Regression statistics	
Multiple R	0.662212
R-squared value	0.438525
Set R-squared value	0.378469
Aver. value error	0.711808
Observation	321

ANOVA					
	Difference	SS	MS	F	Significance F
Regression	2	10.288777	5.144389	5.918733	10.153312
Residues	318	13.173445	0.506671		
Total	321	23.462222			

Table 5 confirms no the statistically significant indicator of international activities for small enterprises. H1 is rejected and selected indicators do not affect the development of international activities.

 DOI: 10.1051/shsconf/20173901020

Table 5. Regression statistics and ANOVA for small enterprises.

Regression statistics	
Multiple R	0.553746
R-squared value	0.306634
Set R-squared value	0.260868
Aver. value error	1.117797
Observation	189

ANOVA					
	Difference	SS	MS	F	Significance F
Regression	2	20.444936	10.222468	1.178	3.181444
Residues	186	46.230383	1.249470		
Total	189	66.675319			

As shown in Table 4, the relationships between individual variables show do no a dependency for micro enterprises. Small enterprises show a mostly moderate dependency and low used of foreign capital (Table 5). Medium -sized enterprises are able to receive and repay foreign resources, which confirms the moderately strong positive dependency of indicators (Table 3). Businesses appropriately manage the entrusted financial resources. As hardly any business can manage without foreign capital these days, the typical expense is for example the loan interest. All monitored indicators are important documents for the evaluation of business and the international environment.

The author concluded that the long-term company operation in the international environment and the international activities are important for balanced financial structure in all enterprises in the Czech Republic. Enterprises must not be only reasonably profitable, but they must be able to reconcile their needs, i.e. bind capital in short-term parts of assets that must be paid by costs associated with their funding and also must seek financial structure with minimal costs of capital, both their own and foreign capital. The low level of financing by the long-term resources by micro and small enterprises and subsequent high interfirm indebtedness of companies or only using owner capital regarding as serious problems of companies. Compliance with the use of foreign capital for financing the asset part should always correspond to the degree of the company's liquidity and financing rules that significantly affect profitability. Other influences that affect the asset and capital structure are: the cost of capital, inventories, accounts receivable, interest, taxes, liquidity, dividend policy, risk, etc., these are absent from this paper.

4 Discussion and conclusion

The aim of the article was to analyze selected financial aspects of micro, small and medium enterprises in the Czech Republic in the context of international activities. Evaluation of companies operating in the Czech environment showed that higher growth of micro-enterprises without international activities. Companies operating in an international environment have shown stronger growth in medium-sized businesses. This is due to the benefits of international cooperation, the size and development of the company, and the growing influence of the international environment on business entities. Unresolved issues persist in the field of accounting harmonization in Europe and related laws, which have a significant impact on financial accounting and are reported by all enterprises. These are not included in this document.

Research has found that indicators as ROA, ROE and ROC (contain the development of revenue from sales, profits or losses, own and foreign capital and its forms) are important indicators of the assessment of the company's situation. Using these and other indicators, control controls and then evaluates the activity that creates the image of independent business management. However, the indicators examined are not a measure of company success because they do not reflect business risk: the risks of using foreign capital and the related liquidity of the company and the potential threat of insolvency. They only rate the past and take no account of future benefits. Especially growing profitability Indicators, the growing sales and profits of the business entity tell us about the healthy functioning of the company operating on the international market. It is difficult to compare the performance of companies in a highly competitive and international environment. The competition is located in different countries, operating under different conditions and with different legal regulations.
Hence the author concludes, that internalization activities are important for all businesses. It has an influence on their financial structure and on profitability. Therefore the author believes that international activities have a positive impact on SMEs in the Czech Republic mostly in the area of labour employment and growing the size of companies.

References

1. L. Meixnerová, Š. Zapletalová, Z. Stefanovová, *International Business: Selected strategic, managerial and finance aspects [Mezinárodní podnikání: vybrané strategické, manažerské a finanční aspekty]*. 1st ed. Prague: C. H. Beck, 203 p. (2017)
2. C. Nicolau, Are SMEs still profitable in an economic crisis? Qualitative research on Romanian entrepreneurship and crisis management. *Bulletin of the Transilvania University of Brasov*, **8**(2), (2015)
3. H. Karadag, Financial Management Challenges In Small and Medium-Sized Enterprises: A Strategic Management Approach. *EMAJ: Emerging Markets Journal*, **5**(1), (2015)
4. A. Dixit, P. Kumar, SMEs and Economic Growth in India: Cointegration Analysis. *IUP Journal of Financial Economic*. **9**(2), (2011)
5. R. Nanu, C.M. Spulbar, R. Buziernescu, Considerations regarding growing financing capacity of SME's. Annals of The University of Oradea, *Economic Science Series*, **17**(3), (2008)
6. E. Di Maria, R., SMEs' growth in international markets: export intensity, export diversification and distribution strategies. *Economia e Politica Industriale,* **44**(3), (2017)
7. P. Newbold, W. L. Carlson, B. M. Thorne, Statistics for Business and Economics. *Harlow: Pearson*, (2013)
8. A. Majocchi, A. Zucchella, Internationalization and performance findings from a set of Italian SMEs. *International Small Business Journal,* **21**(3), (2003)
9. M. Vochozka, M. Psárska, Factors supporting growth of added value, performance and competitivness of SMEs and selected EU countries. *Innovation Management Entrepreneurship and Corporate Sustainability*, (2016)
10. Y., P., Hu, I. C. Chang, W. Y. Hsu, Mediating effects of business process for international trade industry on the relationship between information capital and company performance. *International Journal of Information Management,* **37**(5), (2017)
11. H. DeAngelo, R. W. Masulis, Optimal Capital Structure under Corporate and Personal Taxation. *Journal of Financial Economics*, **8**, (1980)
12. M. Brennan, E. Schwartz, E. Corporate Income Taxes, Valuation and the Problem of Optimal Capital Structure. *Journal of Business*, **51**(1), (1978)

13. J. Lee, Y. S. Hyun, Small and Medium-Sized Japanese Joint Venturtes in Korea. *International Small Business Journal*. **5**(1), (1986).
14. O. Dittmar, Capital Structure in Corporate Spin-Offs. *The Journal of Business*, **77**(1), (2004)
15. A. Strebulaev, Do Test of Capital Structure Theory Mean What They Say. *Journal of Finance*, **62**(4), (2007)
16. L. Strýčková, Factors Determining the Corporate Capital Structure in the Czech Republic from the Perspective of Business Entities. *E+M*, **18**(2), (2015)
17. F. Modigliani, M. Miller, M, The Cost of Capital, Corporation Finance and the Theory of Investment. *The American Economics Review*, **48**(3), (1958)
18. R. Aliyev-Ada, D. Hájková, I. Kubicová, The Impact of Monetary Policy on Financing Czech Firms. *Czech Journal of Economics and Finance*, **65**(6), (2015)
19. N. Rodríguez-Masero, Determinants of Capital Structure of Spanish Firms: The Case of Listed Companies in Time of Crisis. *Investment Management and Financial Inovation*, **13**(1), 2016)
20. B. Elliot, J. Elliot, Financial Accounting and Reporting. *Harlow: Pearson Education*, (2013)
21. P. Santis, A. Albuquerque, F. Lizarelli, Do Sustainable Companies Have a Better Financial Performance? A Study on Brazilian Public Companies. *Journal of Cleaner Production*, **133**, (2016)
22. R. Buglear, Quantitative Methods for Business and Management. *Harlow: Pearson*, (2012)
23. M. Vochozka, Using neural networks to determine the financial plan. *Innovation Management Entrepreneurship and Corporate Sustainability 2016*, (2016)
24. J. Hinke, IAS/IFRS and business performance assessment [IAS/IFRS a hodnocení výkonnosti podniku]. Prague: Alfa publishing (2013)
25. S. C. Myers, Determinants of Corporate Borrowing. *Journal of Financial Economics*, **5**(2), (1977)

Developing the algorithm for assessing the competitive abilities of functional foods in marketing

Liudmila Nilova[1], *Svetlana* Malyutenkova[1,*]

[1]Peter the Great St.Petersburg Polytechnic University, Institute of Industrial Management, Economics and Trade, Saint-Petersburg, Russia

Abstract. A thorough analysis of competitive factors of functional foods has made it possible to develop an algorithm for assessing the competitive factors of functional food products, with respect to their essential consumer features — quality, safety and functionality. Questionnaires filled in by experts and the published results of surveys of consumers from different countries were used to help select the essential consumer features in functional foods. A "desirability of consumer features" model triangle, based on functional bread and bakery products, was constructed with the use of the Harrington function.
Key words: functional foods, marketing, competition, Harrington function

1 Introduction

Production of functional food products is now one of the popular directions of expanding the assortment of food products in most countries. This segment of the food market actively develops by expanding the share of goods that play an essential role in human's diet – bread and bakery products, dairy products, beverages and other foods that are consumed daily [1-3].

At the present time, international trade of functional fermented dairy products, primarily probiotic dairy products, shows stable growth. This sector demonstrates a high level of integration between the suppliers of raw materials, manufacturers and consumers; a differentiated range of products and continuing innovations in the range of goods. Most of the market belongs to large international corporations, such as Chr. Hansen Holding A/S, Danone, Yakult Honsha Co., Ltd., Nestle, Probi AB, Danisco A/S, Lallemand Inc., and Arla Foods, Inc. [4].

Companies introduce innovative methods of interaction with customers; invest considerable funds in scientific development of more effective probiotic strains. The increase of the regional market share leads to merging and take-over of small manufacturers by bigger ones. According to the analysis, this market segment will continue to intensively develop, primarily in the Asian and Pacific regions. The growing popularity of probiotics around the world has attracted many new players to this market segment.

* Corresponding author: malutesha66@mail.ru

DOI: 10.1051/shsconf/20173901021

Along with probiotic dairy products, sales of dairy products, that combine probiotics and edible plant ingredients, also demonstrate stable growth. For example, the international company "Danon", positioning itself as a manufacturer of exceptionally healthy products, is constantly expanding the range of probiotic yoghurts enriched with edible plants – fruits, cereals [3].

In the bread and bakery market, the production of functional foods is rather one-sided. Only grain additives – bran, whole grains, and multi-grain mixtures that enrich bread and bakery products with dietary fibers and vitamins that get destroyed in the process of flour production, are now widely used. Ready-made bakery mixtures are produced by both large corporations, such as "IREKS", and small companies [5]. Other directions of developing varieties of bread and bakery products, oriented for healthy nutrition, made with the use of edible plants (crushed or powdered fruit and vegetable additives) are still in the early stage of development, although some proof of their beneficial influence on the human health by enriching the human system with biologically active substances and antioxidants has already been obtained [6-9]. For example, "Fazer" produces bread with the addition of cabbage and carrots; potatoes and onions; pumpkin and apple.

The problem of mastering the production of new functional foods is quite complex. The life cycle of food products can begin only after the market-researchers have done their work – completed some market analysis, found product analogues, assessed the competitive factors of the new foods. There are different approaches to assessing the competitiveness of goods. For food products, the most appropriate definition is the ratio of the set of characteristics of the produced products to the aggregate of the requirements of the consumer to a similar product or to a set of characteristics of the products of competitors.

The unique benefits and superior value to the customer are the number one driver of new product and indicators of competitiveness of goods [10]. But in this case, Porter's five force structures cannot be neglected especially by the bargaining power of customers [11].

When assessing competitiveness of functional bread and bakery products, one must take into account that they belong to the category of essential food products that are consumed daily. Of course, one should not deceive consumer expectations. Now most consumers believe that bread should only have a traditional composition. But consumer tastes are changing. There is already a group of people who have completely stopped using bread, considering it a high-calorie and yeast-containing product. Another group of consumers believe that bread should be not only a low-calorie product, but also promote human health. At the same time most of consumers believe that, regardless of its composition, bread should be a tasty product [12, 13]. The purpose of the work is the use of the Harrington method in the algorithm for assessing the competitiveness of functional food products for use in marketing.

2 Data and Methods

To develop a model of competitiveness factors of functional food products, a direct questionnaire survey was used; to reduce the dimension of many factors of competitiveness assessment, the method of peer review was used. Consistency of the opinions of experts and non-random nature of consent was assessed using the concordance coefficient, the importance of factors of competitiveness - the method of ranking factors. The Harrington method was used to construct a model for calculating the competitiveness of functional foods.

DOI: 10.1051/shsconf/20173901021

3 Results and discussion

Competitiveness of food products can be assessed with the use of:

– a *production panel*. It is based on the demand of specific types of foods by trade companies. Disadvantages of this method – sometimes trade companies cancel their orders due to purely technical reasons, or because of subjective factors. Besides this method does not consider consumer demand for this or that product;

– a *trading panel*. This method is based on the trade turnover of a particular product by a trading company. The main disadvantage of this method – a chaotic layout of the goods, an inflated retail price, which leads to a decrease of the consumer demand; and different marketing ways to attract buyers (by discounts, or promotions), leading to an artificial increase of the consumer demand.

– a *consumer panel*, based on direct questioning of consumers. The advantage of this method – one can easily trace dynamics of goods consumption in concrete households.

In the world practice, the quantitative assessment of a food's competitive abilities consists of three stages:

1. Market analysis and selection of the most competitive sample product as a comparison base for determining the level of competitiveness of the foods;
2. Definition of the compared parameters of two food products;
3. Calculation of the integral index of competitiveness of the foods.

Selecting the right sample food to be used as a comparison base is very important. If similar foods are produced by other manufacturers, or by the same manufacturer, but according to another recipe (for example, bread enriched with bran), this product would be used as a comparison base. However, brand new functional food products, (for example, for the first time enriched with vegetable additives), most often have no analogues. Therefore, when assessing their competitiveness by pair-wise comparison, it is rather difficult to choose a sample for the comparison base. If the analogue cannot be found, one can use for comparison a traditional food product similar to the functional one, but without additives; or a functional food product of the similar composition, but with other functional additives.

Another problem lies in selecting the criteria (parameters) for the comparison base. A number of studies [12-15] have established that sensory features, along with safety and health-promotion features play an important role in purchasing functional foods. The ratio of price and quality is also of a great importance. At the same time, it is also necessary to take into account the social importance of food products that are used daily in human nutrition. Based on the analysis of advantages and disadvantages of a new functional food product in comparison with the food product used as a comparison base, a manufacturer can compile a list of criteria for assessing the product's competitive powers, i.e. consider competitiveness as a multi-factor model, including even technological factors.

To identify the factors that determine the competitiveness of functional food products, a direct questionnaire was conducted by consumers who were asked to name factors of purchase motivation and evaluate them from 0 to 10 points. To determine the significance of the production of functional food products by industrial enterprises, the survey conducted with employees of bakery enterprises. As a result, the most significant factors identified is shown in Fig. 1.

There is no need to study the impact of each factor separately. To make a quick decision, one can use the method of reducing the dimensionality of multidimensional data, which will help to reveal the correlated factors and use them as the basis to form two or three essential features of the food product. The importance of factors was determined by the method of expert assessments. For functional food products, the following factors should be chosen as essential: functionality (health-promotion), quality (organoleptic and physicochemical characteristics), and safety (depending on the edible plant additives used) which is formed during the production process. The final decision on the competitiveness of functional foods can be carried out by pairwise comparison or the Harrington method. As a result, the algorithm for assessing the competitive abilities of functional foods is shown in Figure 2.

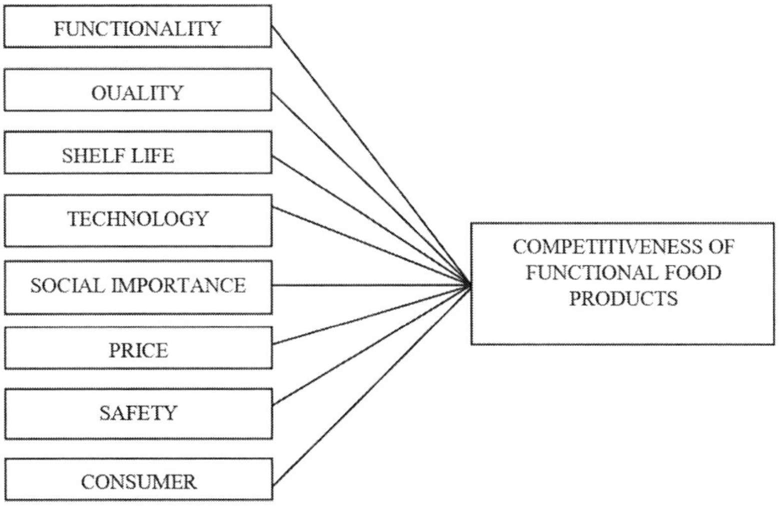

Fig. 1. Competitive factors of functional food products.

More often than not, to assess competitive powers according to the selected factors (parameters), manufacturers use the method of pair-wise comparison with a prototype (analog). But what's to be done if a functional food product can be a fundamentally new composition with new functional properties? In this case, the use of the pair-wise comparison will not result in an objective evaluation.

To obtain an objective comparative evaluation of the product's competitive powers by various factors, one can encode their natural values with the use of the *Harrington desirability function*. The construction of this function is based on the transformation of natural values of various qualities that characterize consumer features, into a scale of desirability preference. This helps to optimize consumer features of various foods by a variety of factors; to assess the selected consumer features by generalizing a set of properties and product characteristics into a single quantitative feature. The product displaying the greatest value of the generalized desirability function is considered the optimum.

The possibility of applying this method is considered on an example for assessing the level of competitiveness of functional bread and bakery products. Such products can be made of a variety of wheat and rye flours, with the use of different recipes, with various edible plant additives. Each of these products will have its own set of consumer features. This is why it is very difficult to select the best product among the whole lot. As objects of research we have taken bread and bakery products made of wheat flour of different yields, and enriched with cedar flour and with powdered sea buckthorn berries.

At the first stage, the assessment criteria were selected. It is very difficult to select the ultimate criterion from a variety of different ones, and give it a certain i-rank. A survey of the experts' opinions provided a good assistance. Studies that were conducted in the UK, Germany, Italy and Finland showed that most consumers of functional bread consider it as a food product that helps maintain health conditions, but is not expected to improve health. At the same time, they believe that functional ingredients should not change the sensory properties of bread [12]. Chinese consumers believe that safety of functional foods is their most important characteristic [14]. For bread and bakery products, safety criterion is typically associated with the use of fats containing trans fats and the formation of acrylamide during the baking process [16].

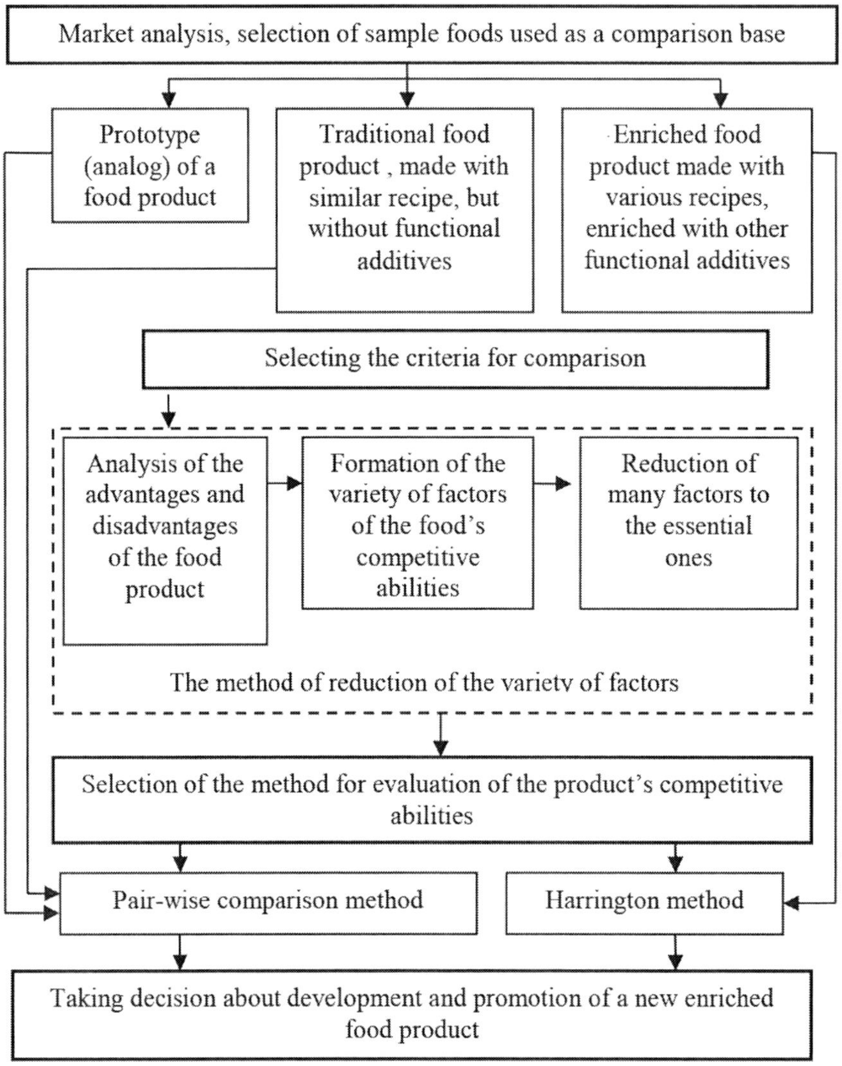

Fig. 2. Algorithm for assessing the competitive abilities of functional foods.

Based on the questionnaire filled in by experts Nilova and Malyutenkova [5] and on the consumers opinion [12-15], the authors defined the main factors for assessing the competitive powers of functional bread and bakery products – functionality, safety and organoleptic characteristics, the latter directly dependent on the physical and chemical properties of the foods. Each of the selected consumer features is complex and therefore evaluated by a range of different characteristics. The first place and – therefore – rank "1", was assigned to the most important of all properties, the last place (and rank "n") – to the least important one; the remaining characteristics were ranked from place to 2 to "n-1" (Table 1). The consistency of the experts' opinions and the non-random nature of their agreement were evaluated, respectively, by the concordance coefficient C according to Equation 1. The values came up to 0.93:

$$C = \frac{12\,S}{g^2\,(n^3 - n)} \quad (1)$$

g – the number of experts, n – the number of properties, S – the sum of squared deviations of all properties.

Functionality of the food product was evaluated by its capacity to supply the human body with physiologically important macro- and micronutrients; one portion containing not less than 15% of the physiologically needed human daily norm of energy and nutrients. Safety was assessed by the content of trans fats and acrylamide. The quality of the food (organoleptic and physicochemical characteristics) was evaluated according to its compliance with regulatory documents. These data, both group and individual, based on the experts' and consumers' surveys were then ranked (Table 1). The rank of the i-th index corresponded to its ordinal number when ranking the characteristics in order of decreasing importance. Since organoleptic and physico-chemical features characterize one consumer feature and are therefore equal, they were given the same rank – the average of their ordinal numbers.

To calculate the index of significance, the authors arranged a bakery product's features in a sequence from 1 to n in accordance with the direction of increasing the amount of all ranks assigned to it by the experts. The index of significance of each feature corresponds to the place that it occupies in the obtained row, and is calculated by Equation 2:

$$\beta_u = \frac{u}{2^{u-1}} \quad (2)$$

u –the rank assigned by the experts to this feature.

The values of β_u are given in Table 1.

Table 1. The importance of competitive powers of functional bakery products.

Consumer features	Rank, u	Importance indicator, β_u
Safety	1	1,00
Functionality (health-promotion)	2	1,00
Quality:		
Organoleptic characteristics	3	0,75
physical and chemical characteristics	3	0,75

Each of the identified consumer features of the studied bread and bakery products has its own meaning and its own dimension, so they need to be converted into dimensionless figures according to the similar dimensionless scale (the Harrington desirability function) and compared with each other.

The value of the private indicator du, converted into a dimensionless desirability scale, is called *private desirability*. The desirability scale lies in the interval from 0 (du = 0, corresponding to the unacceptable level of desirability) to 1 (du = 1, the best desirability

 DOI: 10.1051/shsconf/20173901021

value). Each food product was thus evaluated and certain values of consumer desirability, corresponding to such categories as "very good", "good", "satisfactory", "bad", and "very bad" were attributed to it. Standard grades on the desirability scale are shown in Table 2 (columns 1, 2). They correspond to certain points on the curve, given by Equation $d = (e^{-e})^y$.

Table 2. Standard marks on the desirability scale.

Scale of desirability	Marks on the scale of private desirability
Very good	1,00–0,80
Good	0,80–0,63
Satisfactorily	0,63–0,37
Bad	0,37–0,20
Very bad	0,20–0,00

Desirability values from 0 to 1 are marked on the y-axis. The values of the consumer features, rendered in a conditional scale, are indicated on the abscissa. The encoded values of consumer features of bread and bakery products are placed symmetrically to 0 on the y-axis (the encoded scale). Limits of the acceptable values for individual indicators can be either one-sided $y_n \leq y_{max}$ or $y_n \geq y_{min}$ (for bread and bakery products this stands for porosity, specific volume, vitamin C presence, etc.) or bilateral $y_{min} \leq y_n \leq y_{max}$ (moisture percentage, acidity value, vitamin E presence, etc.).

The evaluation of quality characteristics was carried out through their specific desirabilities, established by experts. The y_i properties of a particular food under study were compared with those proposed by the experts for each individual desirability function. After converting individual values into individual desirability functions for each characteristic of consumer features of bread and bakery products, a generalized desirability function D was calculated as the mean geometric value of the individual desirability functions adjusted to the significance of each feature according to Equation 3:

$$D = \sqrt[n]{\prod_{u=1}^{n} d_u^{\beta}} \tag{3}$$

n — the number of features; β_u — the index of significance of each product's feature; u — the concrete feature's number in the ranked sequence.

For each group characteristics (safety, functionality, quality), the individual desirability function d_u was calculated, as the mean geometric value of individual desirability indicators of the d_i of the i-th individual index. These results for bakery products is shown in Table 3.

The obtained results showed that bakery products enriched with cedar flour were characterized by greater functionality and a better quality of products, by both organoleptic and physicochemical factors. Their advantage lay in the fact that while they showed no differences in organoleptic characteristics compared to bread and bakery products made by traditional recipes, they displayed a better taste. Bakery products enriched with sea buckthorn powder had a characteristic orange color, an acidic taste, which could be improved only by adding a large amount of sugar. At the same time, the studied bread and bakery products showed different values of individual desirabilities, although the generalized function D, according to the desirability scale, classified them in the group of consumer features as "very good".

Table 3. Valuation of consumer properties of functional bakery products on the desirability scale.

Consumer features/ individual desirability function	Bread made of wheat flour with the addition of cedar flour	Bakery product made of wheat flour with the addition of cedar flour	Bakery product made of wheat flour with the addition of sea-buckthorn powder	Baked bread made from wheat flour with the addition of sea-buckthorn powder
Safety (d_1)	0,9219	0,9357	0,9529	0,0595
Functionality (d_2)	0,7327	0,6918	0,5943	0,6215
Organoleptic characteristics (d_3)	0,9700	0,9800	0,8900	0,9500
Physico-chemical characteristics (d_4)	0,7046	0,7719	0,6566	0,6353
Quality / (d_5 (d_3 * d_4)	0,8267	0,8698	0,7644	0,7769
The generalized value of individual desirability functions (D)	0,8732	0,8791	0,8226	0,8354

For a visual analysis of the "desirability" of the consumer properties of functional breads and bakery products, a comparison chart (Figure 3) is displayed, showing group complex private desirability characteristics – safety, functionality and quality, ranked from 0.5 to 1.0 on the desirability scale, which may be used in the marketing activities of any food production company.

Equation 4 (the weighted Euclidean distance) was used for a quantitative evaluation of the proximity of breads and bakery products B, C, D, E to the ideal hypothetical product A within the obtained values of complex individual desirabilities of the group characteristics (safety, functionality, quality). The smaller the R, the closer the product stands to the ideal food.

$$R = \sqrt{\sum_{i=1}^{n} k(A_i - P_i)} \quad (4)$$

R – represents the distance between the products A and P, and P represents the corresponding products B, C, D and E;
A_i – the complex value of the individual desirability characteristic of the group index du of the ideal product;
P_i – the complex value of the individual desirability characteristic of the group indicator du of products B, C, D, and E;
k – the weighting coefficient of the individual desirability characteristic of the group exponent d_u

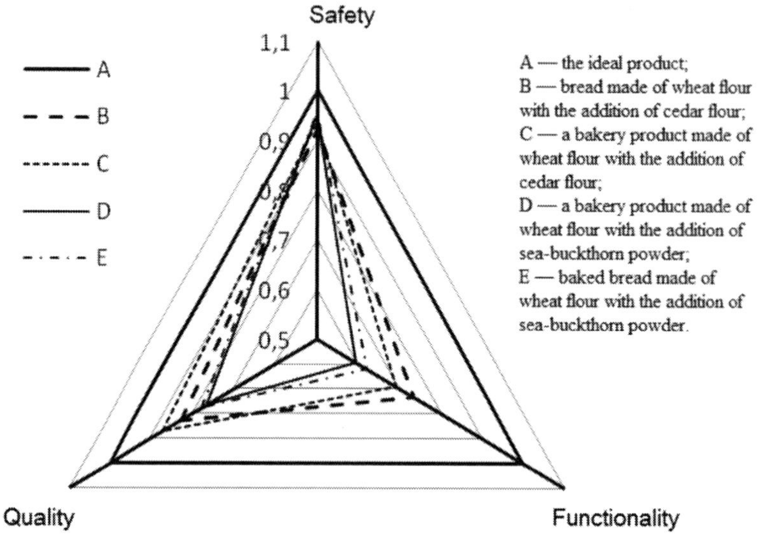

Fig. 3. Diagram of comparison "Triangle of desirability of consumer features" of functional bread and bakery products.

As a result, the following R_{AP} values were obtained:

R_{AB} = 0.3043, R_{AC} = 0.2896, R_{AD} = 0.3917, R_{AE} = 0.3664. Hence it follows that in our case the bakery product made of wheat flour with the addition of cedar flour is closest to the ideal product.

4 Conclusion

Competitive ability of functional food products is considered as a multi-factor model, taking into account both consumer features, social significance and technological factors. Based on the analysis of the experts opinions and studies of consumer preferences in different countries, basic consumer features that play an essential role in assessing the competitive abilities of functional foods have been identified.

An algorithm for assessing competitive abilities of the foods has been developed. It uses the method of encoding important consumer features with the help of Harrington's generalized desirability function. This approach allows us to compare the consumer features of functional food products of different composition. The graphical method makes it possible to visually demonstrate the optimal consumer features of functional food products using bread and bakery products as well as the "triangle of desirability of consumer features".

References

1. Z. E. Martins, O. Pinho, I.M. Ferreira, Food industry by-products used as functional ingredients of bakery products. *Trends in Food Science & Technology Journal,* 67, 106-128 (2017)
2. L.M. Beltrán-Barrientos, A. Hernández-Mendoza, M.J. Torres-Llanez, A.F. González-Córdova, B. Vallejo-Córdoba, Invited review: Fermented milk as antihypertensive functional food. *Journal of Dairy Science,* **99**, 4099-4110 (2016)

3. F. Bimbo, A. Bonanno, G. Nocella, R. Viscecchia, G. Nardone, B. De Devitiis, D. Carlucci, Consumers' acceptance and preferences for nutrition-modified and functional dairy products: A systematic review. *Appetite Journal,* **113**,141-154 (2017)
4. Probiotics Market Analysis and Reports. *Market Analysis - Probiotics 2017,* [online]: Available at: http://market-analysis.conferenceseries.com/probiotics-market-reports (2017)
5. L.P. Nilova, S.M. Malyutenkova, Innovative Foodstuff in Formation of Regional Commodity Systems. *Krasnoyarsk Science Journal,* **5**(38), 161-174 (2016)
6. L. Nilova, O. Orlova, J. Nasonova, The role of cyclic amides in the formation of antioxidant capacity of bakery products. *Agronomy Research Journal,* **13**(4), 1020-1030 (2015)
7. L.P. Nilova., T.V. Pilipenko, Otsenka antioksidantnykh svoystv obogashchennykh khlebobulochnykh izdeliy v eksperimente na laboratornykh zhivotnykh. *Voprosy Pitania.* Journal, **85**(6), 39-47 (2016)
8. D. Dziki, R. Różyło, U. Gawlik-Dziki, M. Świeca. Current trends in the enhancement of antioxidant activity of wheat bread by the addition of plant materials rich in phenolic compounds. *Trends in Food Science & Technology,* **40**(1), 48-61 (2014)
9. A.A. Basov, I.M. Bykov, *Voprosy Pitania,* **82**(3), 77-80 (2013)
10. K. B. Kahn, The PDMA handbook of new product development. 3rd ed. Hoboken, N.J.: Wiley (2013)
11. M. E. Porter, The five competitive forces that shape strategy. *Special Issue on HBS Centennial, Harvard Business Review,* **86**(1), 78-93 (2008)
12. M. Vassallo, A. Saba, A. Arvola, M. Dean, F. Messina, M. Winkelmann, E. Claupein, L. Lähteenmäki, R. Shepherd, Willingness to use functional breads. Applying the Health Belief Model across four European countries. *Appetite Journal,* **52**(2), 452-460 (2009)
13. T. Pohjanheimo, R. Paasovaara, H. Luomala, M. Sandell, The Customer is NOT Always Right? Marketing Orientations. *Appetite Journal,* **54**(1), 170-180 (2010)
14. P. Yee Lee, K. Lusk, M. Mirosa, I. Oey, Effect of information on Chinese consumers' perceptions and purchase intention for beverages processed by high pressure processing, pulsed-electric field and heat treatment. *Appetite Journal,* **73**(1), 95-104 (2014)
15. H. Rahnama, S. Rajabpour, Factors for consumer choice of dairy products in Iran, *Appetite Journal,* **111**(1), 46-55 (2017)
16. M. Przygodzka, M.K. Piskula, K. Kukurov, Z. Ciesarov, A. Bednarikova, H. Zielinski, Analysis of the antioxidative properties and Maillard reaction products in ginger cakes enriched with rutin. *Journal of Cereal Science,* **65**, 96-102 (2015)

Business relations between Ukraine and China: development options

Andrew Pochtovyuk[1, *], *Victoria* Semenikhina[2]

[1]D.of Economics, Professor, Management Department, 39600 Kremenchuk Mykhailo Ostrogradsky national University, Kremenchuk, Pershotravneva str. 20, Ukraine
[2]Senior lecturer, Management Department, 39600 Kremenchuk Mykhailo Ostrogradsky national University, Kremenchuk, Pershotravneva str. 20, Ukraine

Abstract. In the article were analyzed the present state of business relations of Ukraine and People's Republic of China (PRC). Investigations of state threats to economic security were made; options of changes in the commodity flows structure were suggested, as well as the prospects of cooperation between the two countries on the basis of maximizing the volume of domestic high technology products exports and minimizing Ukraine's dependence on the ever-increasing role of China in the global economic space were considered.
Key words: business relations, commodities, export, economic security

1 Introduction

The Chinese role in the world economy, with its stable growth trend, the prospects of the international Ukrainian-Chinese relations in general and trade in particular, have today both favorable and negative impact on the Ukrainian economic development .

In this regard, the external economic relations between Ukraine and the People's Republic of China (the PRC) are subject to a thorough studying of both Ukrainian and Chinese economists. Thus, Ukrainian scientists, Vlasenko [1], Vysotskaya [2], Levkivsky [3] in their researches noted that the current problem: substantiation of the basic directions of interaction, the impact on the State economic security , the problems of relationships strengthening and broadening require thorough analysis and studying with the subsequent development of recommendations on optimization of the Ukrainian-Chinese relations structure.

2 Data and Methods

Methodological study basis accommodates the general and special scientific conceptions, theories and methods. The methods of obtaining knowledge, such as dialectical, historical, and systematic and others are used as methodological basis of the study. Thus, the dialectical method is used for investigation of the phenomena "the technological and innovative component of the Ukrainian-Chinese trade" that gives the ability to review them

* Corresponding author: andrey.pochtovyuk@gmail.com

as a dynamic category. Historical method was used first of all in the analysis of business relations of Ukraine and People's Republic of China. Systematic method was used in evaluation of the creation business relations between Ukraine and China at this stage and selecting the options for their development.

3 Results and Discussion

The role of China in world markets is constantly growing, and the share of world trade has increased from 0.75% in 1948 to 12% in 2015, and now it's about 4 trillion US dollars [4]. According to the Ukrainian State statistics service in the structure of external trade turnover for 2001-2015 China's share amounted on an average of 4.69% (from 3.08% of total export, and 6.04% of total import) [5] (Table 1).

For China, as a state with more than a billion people (1.379 billion people – 2016), the most pressing is food problem. In the ranking of World health organization, China ranks second place in the world after India in the number of people suffering from malnutrition (about 130 million people). The increase food and agricultural products trade is the real breakthrough for Ukraine in realization of domestic natural resource potential, providing the sustainable development of agricultural sector.

However, the increase in agricultural products exports can cause fallout. Thus, in 2015, the main exports of agricultural products to China were maize (542,596 million US dollars or of 22.61% of exports to China) and sunflower oil (438 million US dollars or 18.27% of export volumes) [6]. The keen interest of China to Ukrainian sunflower oil import creates threats related to excessive expansion of sunflower cultivation that will ultimately lead to the soils depletion, and will cause irreparable harm to Ukrainian agriculture for a long period.

It is should be noted that China remains one of the main business partners of Ukraine in the military-industrial products market. In 1992 Ukraine and China signed about 200 contracts in the military sector. During 1992-2008 the volume of arms imports to China amounted about 624 million US dollars [7].

First of all it's aircraft and air defense systems. The main group of domestic goods exported to China, was weapons which other exporters ignored mostly for political reasons. Ukrainian military-technical production is very attractive for China because of its high adaptability and low price in comparison with competitors. However, the Chinese government pursues active policy of all around support of domestic military industry development. Thus, the competitive advantages of the Ukrainian military-technical production become unstable and gradually are leveled off. In view of this there is only one solution – collaboration in the projects and technologies development which are inaccessible for the military-industrial complex enterprises of the PRC.

The structure of import from China to Ukraine in 2015 is more balanced, the production means amounted to 26,14% (982,22 million US dollars) total imports; intermediate goods – 48.13% (1808,28 million US dollars, consumer goods – of 25.73% – 966,75 million US dollars).

The increase in the share of mineral raw materials is characterized by the simultaneous decrease of partners' interest from China in finished and intermediate industrial products of domestic producers, especially concerning the trade in iron and steel. In 2001, this product group was dominated in the nomenclature of the Ukrainian-Chinese export trade (378.26 million US dollars or 78% of total domestic imports). In 2015, this figure decreased to 20.1 thousand US dollars, which accounted less than 1% of total exports to China [8].

Table 1. Dynamics of business relations indicators between Ukraine and China, billion US dollars*

	2001	2002	2003	2004	2005	2006	2007	2008	2009	2010	2011	2012	2013	2014	2015
Exports	0,48	0,67	1,00	0,82	0,71	0,54	0,43	0,55	1,43	1,32	2,18	1,78	2,73	2,67	2,40
Imports	0,20	0,26	0,52	0,74	1,81	2,31	3,31	5,60	2,73	4,7	6,27	7,9	7,90	5,41	3,77
ET	0,68	0,93	1,52	1,56	2,52	2,85	3,74	6,15	4,17	6,02	8,45	9,68	10,63	8,08	6,17
Balance	0,29	0,41	0,48	0,08	-1,1	-1,8	-2,9	-5,1	-1,3	-3,4	-4,1	-6,1	-5,2	-2,7	-1,4
REI	2,48	2,58	1,93	1,11	0,39	0,24	0,13	0,10	0,52	0,28	0,35	0,22	0,35	0,49	0,64

ET – external turnover, REI – the ratio of export to import

*According to the data [4].

In accordance with the available data the share of production in total exports is now 1.58% (37.77 million US dollars); intermediate goods – 97.76% (2338,23 million US dollars), consumer goods – 0.66% (15.79% million US dollars) [9].

The technological and innovative component of the Ukrainian-Chinese trade remains at a very low level. This means that Ukraine is China's mainly raw materials exporter with low added value and the importer of finished products with relatively high added value, mainly commodities (Fig. 1).

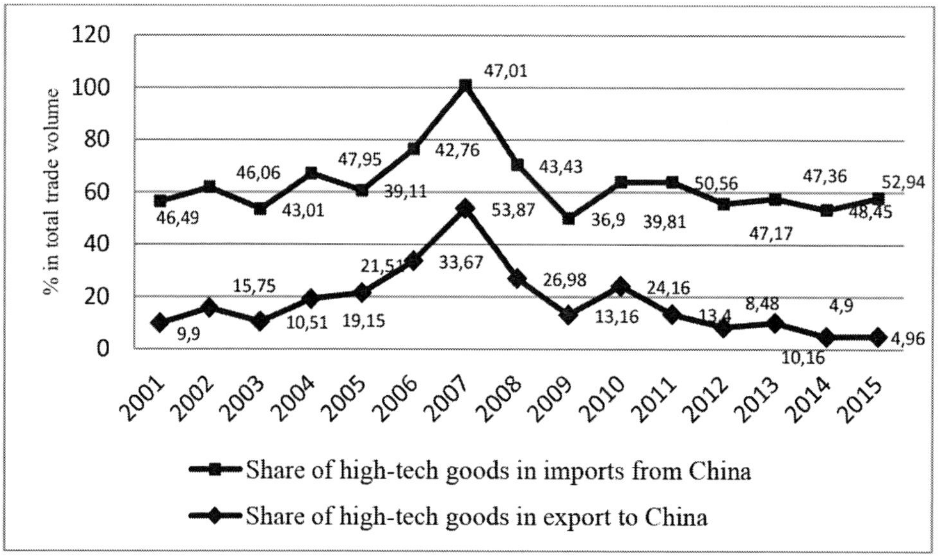

Fig. 1. Dynamics of high-tech products in business relations between Ukraine and China, 2001-2015.

It should be noted the existing trend of trade turnover growth between our countries, that is a confirmation of the thesis that China is the leading trade partner of Ukraine among the Asian countries. In 2015, the volume of Ukrainian-Chinese trade turnover made up 8.08 billion US dollars or 28.5% of Ukrainian volumes of foreign trade relations with the Asian region countries (17.26% of Ukraine's exports to Asia; 48.84% of imports from Asia to Ukraine) it's 21.24% higher than total trade turnover with the Republic of Turkey (4.86 billion US dollars), which ranks the second place among Asian countries-partners of Ukraine in 2015 [6].

However, the dynamics of trade relations development of Ukraine and China is not the only positive aspect in bilateral relations totally and in economic security in particular. This is confirmed by the figures of goods exports and imports and trade balance, the coverage of goods imports by exports, that is indicating the instability situation on the basis of imports over exports prevalence. In fact, the coverage ratio of imports by exports is less than one since 2005 (in 2015: 0.64) and balance of external trade turnover is negative (in 2015: -1, 37 billion US dollars) that does not correspond to the boundary values of the state external economic safety [6].

To search and identify the main vectors of Ukrainian-Chinese trade the predominant groups of goods, the ratio of raw materials and finished products in the exports and imports structure, the technological level of the goods turnover, the general trends of trade development should be established.

The existing Program of strategic partnership relations development between Ukraine and China for 2014-2018 should become the basis for the taking new level of interstate relations and cooperation in specific directions and branches. There is a mutual desire of the parties to increase the volume of bilateral trade, diversification of its structure, the development of long-term forms of economic cooperation, deepening of investment cooperation between Ukraine and China.

Relations between the two countries are still developing: for the 6 months of 2016 the total trade turnover has increased by 0.8% compared to 2015 and has reached 4.3 billion US dollars. However, the amount of Chinese direct investment in the Ukrainian economy is far from desirable and there are objective reasons for it. The business pools of our countries can gain plenty by collaborating and working together. Progressive advance of developing contacts between economic entities and the continued deployment of joint production bases is important for both sides.

Such branches as transportation, agriculture, infrastructure creation, aviation, space, innovation, scientific-technical and other fields are strategic for mutually beneficial cooperation development. Future-oriented strategic direction of economic cooperation with China is the direct involvement of Ukraine in the implementation of the development the Silk Road economic belt initiated by the President of China. Ukraine intends to participate actively in the concept implementation "One zone, one way", and to become a part of New "Silk Road".

Ukraine intends to use the Chinese initiative "One zone, one way" to strengthen the cooperation with China in transportation and communication so as in infrastructure branches. The advantageous geographical location of Ukraine, the country is at the crossroads of major transit routes between Europe and Asia, it has 18 ice-free ports on the Black sea (five of which are in the temporarily occupied territory), an extensive network of railways and automobile roads facilitate to this. This point is particularly relevant in connection with the EU Association Agreement and the further Ukraine integration into the transport system of corridor "Europe-Asia-Europe". Odessa sea port is the largest port in Ukraine which is simultaneously the main point of the IX Trans-European corridor.

It also should be mentioned about the Beskids tunnel reconstruction completion that is the strategic in relation of transit goods delivery towards the European countries and is the part of V pan-European transport corridor (Italy - Slovenia-Hungary-Slovakia-Ukraine).

Having all the above mentioned features and the potential Ukraine strives to become a bridge between Asia and Europe turning the existing potential into the business opportunities. Ukraine may be an important hub in logistical operations of Chinese goods to EU markets and goods of domestic and European manufacturers back to China. Ukraine together with its partners trashes out the best possible conditions for container traffic through the country ports. In the estimation of Ukrainian railway department containers transportation can take up to 15 days within the new corridor in the case of synchronization of the maritime (railway ferry) and railway sections work. Ukraine together with its partners are focused on formation new logistic chains and the integration of national transport networks into a global network of international transport.

The National Development and Reform Commission of China expect that there will be about 5,000 freight trains plying between China and Europe every year till 2020. 43 transportation hubs and 43 railway lines will be constructed. Ukraine with its seaports and well-developed railway infrastructure is a transportation gateway between East and West, North and South building its transit and transportation strategy to meet the initiatives of China on cooperation in the regions of the Adriatic, Baltic and Black seas. In this regard the work with Chinese partners first of all in creating the effective model of Ukrainian ports using is of great importance.

Now the Chinese economy faced the necessity of development model change. The current model has demonstrated (and continues to demonstrate) unprecedented pace mainly due to foreign investment and rapidly growing exports. Although this model with its role and scope (the second largest economy in the world after the United States and first by volumes of world exports), can not be quickly upgraded the Chinese government wants to minimize the risks of external shocks, the threat of which increases in the growing problems of China economic development.

Among the key challenges that will influence the content and dynamics of Ukrainian-Chinese relations, we can emphasize the following: the restructuring of China's economy; a change in the quality of economic growth, transition from extensive development to improvement the production manufacturability. At the moment a lot of Chinese enterprises, especially in the public sector, will remain uncompetitive as before and will be forced to face serious financial difficulties if the government cancels the subsidies, which include land and energy assistance now. For bilateral cooperation, the emphasis alteration in the national development of China will affect the competition growth between companies that have similar competitive advantages and are equally interested in building up the technological capabilities and to the new markets access. Keeping in mind that China has set a goal to become a leader in the production of not only goods but also services and technologies Ukraine can benefit from participation in the new products creation as a partner (or orders executive) [6].

We can predict that promotion in the Chinese market the Ukrainian engineering products and other technological multi-purpose products (and/or finished products) will occur hard. Therefore, the optimal form of consolidation of the Ukrainian manufacturers on the Chinese market may be the implementation of joint sectorial and/or regional projects to solve specific problems such as energy, environmental, scientific and technological and etc. A gradual increase in domestic demand that will make compensation for the decline in external demand increases the possibility for the Ukraine to bring the products to the Chinese market which it lacks. At a time of high savings rates consumer expenditures of households grow very slowly, so the economy remains vulnerable to fluctuations in global demand, uneven and fragile. The Chinese government will increase the purchasing power, so we can expect demand growth for Ukrainian products, which will be aimed at domestic consumption – food, agricultural raw materials and industrial products will be used for the following processing. A different component of domestic demand is the social services development– medicine, science, education. As Ukraine has considerable experience in the social services provision, cooperation with Chinese partners (on condition the availability of a systematic approach of the Ukrainian side to cooperation) can significantly stimulate the export of services – educational, recreational, health, tourism and others. It is obvious that the price that we pay for "progress" of capital intensive and export-oriented development is becoming extremely high. For example, according to expert estimates, the costs associated with the deterioration of the environmental situation in the country in 2010 amounted to at least $ 230 billion, i.e. around 3.5% of China's GDP. Ukraine may get additional opportunities to participate in environmental projects in China, in parallel implementing effective solutions for Ukrainian enterprises. It may be the solution to the debt problem at the level of regions and enterprises [7].

The beginning of reforms and local governance in complex with the financial and fiscal reforms implementation could potentially change the investment policy model of the PRC in global markets. For example, according to the National audit office of China local governments at the level of provinces, counties, urban districts and cities the total debt has already reached 11 trillion yuan (1.8 trillion US dollars). This issue potentially threatens the stability of the banking system and may lead to a wave of irrevocable bank loans, which will paralyze the banking sector and will become a serious obstacle for reforms carrying out. In light of the accumulated risks and in prospect reduce of active trade balance Chinese power can reduce the investment volumes in other countries to raise requirements for foreign projects financing, change priorities for investment. The solution of energy problems, liberalization of pricing policies. Because these problems are topical for Ukraine, the government and enterprises of the two countries have many common areas for cooperation and exchange of experience and technologies. Now the question of the pricing policies revision in the energy sector is very serious. While the Chinese economy has been growing rapidly, the price of energy was artificially underestimated, because energy is a principal component of the Chinese economic model. Anticipating a sharp rise in inflation, the government has often interfered in the pricing process, ensuring that the price of electricity, coal and other energy sources could remain at the same level. However, the energy low cost has led to the fact that the incentive to increase the efficiency of the Chinese enterprises production has almost disappeared. The transition to higher energy prices that reflect their real value, will force Chinese enterprises to improve the efficiency of their production and develop cleaner production methods. Excessive centralization of the tax system has led to the lack of finances at the regional level. This caused to a permanent crisis of local budgets, the accumulation of "bad" municipal debt and sale of land for development, the pursuit of "customs" collection from export activities. autonomy of regional authorities for the development of territories will increase as solutions to these problems exists , so now it is advisable to develop the "horizontal" relations with China at the level of regional authorities and in the context of solving regional problems.

In the midterm it becomes possible to predict the further strengthening of international economic and political competitiveness of China and improving its activity in the global and regional processes. According to Chinese experts, the country has a resource in 10-15 years to a radical modernization of the economy and China admission to technological leadership. Subject to the focused and systematic foreign policy, the Ukrainian-Chinese cooperation will contribute to the solving a similar task of the economy modernization by the Ukraine. Cooperation is advisable to focus on several directions: cooperation in the sectors of complex machine building (the defense industry, aviation and shipbuilding, space industry); to attract Chinese investment in Ukrainian agriculture (a priority is for the productive and technological projects), establishment of new enterprises producing and processing agricultural products, the supply to the Chinese market a wide range of agricultural and food products; collaboration in the social services sector, tourism development, export of educational services; the development of cooperation in the sphere of transport services (priority – opening of a direct air route Beijing-Kiev); the attraction of Chinese investment to finance infrastructure projects and so on. Proper development of the Chinese (and generally Eastern) vector of foreign economic policy will allow Ukraine not only to attract additional investment, to preserve and enhance economic, defensive and scientific-technical potential, but also to benefit from the diversification of foreign economic cooperation, since countries in the Asia-Pacific region are becoming the "locomotives" of world economic development and have a significant weight in global politics [8].

In this context, it is appropriate to strengthen the effectiveness of the Ukrainian Executive authorities concerning relations development with China which involves: creating a predictable environment for attracting investment from China into the economy of Ukraine through the legislation improvement (benefits should be given to universal regulations stimulating investment in Ukraine foreign investment and technology transfer); the communication component strengthening in the bilateral relations, the dissemination of information about Ukraine and creation a positive image in the Chinese business circles (for example, placement of direct advertising in the Chinese media, events, etc.); limiting forms of cooperation with China, providing for the transfer to the Chinese party licenses for the products manufacturing (in this case China, generally, is phasing out import of production, setting up production of the same goods on its territory); removing artificial regulatory barriers (initiating relevant measures from the Chinese side) in the business way of two countries; the language barriers weakening in communication, promoting the Chinese and English languages study as tools of economic cooperation intensification; establishing humanitarian, cultural and scientific cooperation between the public, professional and business circles of two countries [9]. The implementation achieved on the highest level of Ukrainian-Chinese agreements, the filling of strategic partnership relations between two countries practical economic content, require from Ukraine the a comprehensive and coherent policy implementation that would take into account political, economic and financial weight of China in the modern world.

4 Conclusion

The made analysis of business relations between Ukraine and China at this stage allows selecting the following options for their development: 1. The no availability of a clear government position in the points of "strong growth of Chinese goods imports " will lead to the fact that Ukraine in the nearest future will turn into a "raw materials appendage" not only for economically developed countries but for China and will also become the uncontrolled market for Chinese goods. 2. The state program development, providing for a shift in emphasis in the structure of exports to China. The basic principle of foreign policy of Ukraine should become turnover in raw materials export and cheap materials on the finished and high-tech products. But there must be a policy for such trade promotion. And the choice must be made as quickly as possible – no time to delay. Ukraine and the People's Republic of China have a real opportunity for qualitative and quantitative improvement of mutually beneficial cooperation in specific industries and economy sectors, however it is necessary to substantially revise the main provisions of the foreign policies of both countries.

References

1. L. Vlasenko, Dominant tendencies of Ukrainian-Chinese trade development. *Scientific Transactions of Kherson State University*, (2017)
2. M. Vysotskaya, Analysis of the main tendencies in the development of economic relations between Ukraine and China. *Strategy of development of Ukraine: economics, sociology*, (2013)
3. V. Levkivsky, Ukrainian-Chinese Economic Cooperation in the Context of Foreign Economic Security. *Scientific Bulletin of the Chernihiv State Institute of Economics and Management*, (2013)
4. The World Bank [online], Available at: https://data.worldbank.org/country/china (2017)
5. State Statistics Service of Ukraine [online], Available at: http://www.ukrstat.gov.ua/operativ/operativ2008/zd/o_eit/arh_o_eit_u.html (2017)
6. International trade center [online], Available at: http://intracen.org/ (2017)
7. Official site of the UN Statistics Division on International Trade. [online], Available at: https://comtrade.un.org/pb/CountryPagesNew.aspx?y=2015 (2017)
8. The National Institute for Strategic Studies [online], Available at: http://www.niss.gov.ua/articles/1359/ (2017)
9. UN Food and Agriculture Organization (FAO) [online]. Available at: http://www.fao.org/documents/card/en/c/463e6c11-c576-41f9-ae29-911bceb2924d (2017)

The Impact of the Economic Crisis on Czech Municipal Expenditures

Pavel Rousek[1*]

[1]Institute of Technology and Business, School of Expertness and Valuation, Okružní 517/10, 37001 České Budějovice, Czech Republic

Abstract. The paper focuses on the links between municipal expenditures and the economic crisis (or the economic cycle). The aim is to improve the predictions of municipal expenditures: current expenditures, capital expenditures and total expenditures. Current research shows that the predictions in times of economic growth are relatively good, but in times of economic crisis, models fail. The focus of this paper is the context of the business cycle in order to implement these macroeconomic contexts into predictive models. The tool for achieving this goal is to find strong connections, including time shifts.
Key words: prediction, economic crisis, municipal expenditures, planning

1 Introduction

Predictions of economic indicators are very important tools for public sector management. Unfortunately, some variables are not predicted with sufficient precision for now. This paper aims to show the link between the economic cycle and municipal expenditures. It should serve to predict the future development of municipal expenditures, which will subsequently serve for better planning in public sector.

2 Methods and Data

Conclusions of the paper are based on three-dimensional correlation analysis. The first dimensions are input data, which are the macroeconomic indicators of GDP, unemployment, inflation, the balance of payments, fiscal policy. The second dimension is the output data for which municipal expenditure (current, capital and total) is considered for these purposes. The third matrix dimension consists of time shifts in which input and output data can be affected (3 years forwards and 3 years backwards).

* Corresponding author: rousek@mail.vstecb.cz

DOI: 10.1051/shsconf/20173901023

2.1 Macroeconomic Indicators

Gross domestic product (GDP) is a sum of what is produced in all sectors of the national economy. The reporting unit is the Czech crown at purchase prices including taxes. In the case of goods and services which are not passed through the market and therefore their purchase price cannot be ascertained, the value is calculated by adding value (sum of compensation to employees and consumption of fixed capital) [1]. Nowcasting of Czech GDP via Dynamic Factor Model (DFM) is introduced [2]. Values of Czech GDP in current prices from 1997 to 2015 are in Table 1.

Table 1. Macroeconomic Indicators of Gross Domestic Product (GDP) [1]

Year	GDP (bn. CZK)	GDP per Capita (CZK)	GDP Growth (Percent y/y)	GDP Deflator (Percent y/y)
1997	1 953,3	189 575	-0,7	8,5
1998	2 142,6	208 120	-0,3	10,0
...	...	...	...	...
2013	4 098,1	389 900	-0,5	1,4
2014	4 313,8	409 870	2,7	2,5
2015	4 554,6	432 006	4,5	1,0

Unemployment is another monitored macroeconomic parameter. Employees are considered to be all persons† who develop any remunerated activity for at least 1 hour per week in a permanent, temporary, seasonal and occasional manner [1]. Unemployment is connected to GDP via Okun's law. Czech conditions are analysed [3]. The specific values of unemployment in the Czech Republic are shown in Table 2.

† Including persons working in their own or family business, professional members of the army and persons in the basic military and civil service, persons on maternity leave who worked before going to work.

Table 2. Macroeconomic Indicators of Labour and Unemployment [1]

Year	General Unemployment Rate (Percent)	Long-Term Unemployment Rate (Percent)	Average Gross Nominal Wage (Percent y/y)
1997	4,8	1,3	N/A
1998	6,5	1,9	N/A
...	...	...	...
2013	7,0	3,0	-0,1
2014	6,1	2,7	2,9
2015	5,0	2,4	2,7

Inflation is defined as a rise in the price level in professional literature. It characterizes the rate of depreciation of the currency within a precisely defined time period. The inflation rate is measured by the consumer price index. The development of consumer prices (cost of living) is monitored on consumer baskets based on a set of selected types of goods and services paid by the population. The total number of representatives is about 700. Industrial producer price indices are applied in the industrial, construction, market and agricultural sectors [1]. The Czech inflation values are presented in Table 3.

Table 3. Macroeconomic Indicators of Price Level and Inflation [1]

Year	Inflation Rate (Percent y/y)	Producer Price Index (Percent y/y)			
		Industrial	Construction Work	Market Services	Agricultural
1997	8,5	5,0	11,3	10,8	N/A
1998	10,7	4,8	9,3	8,8	N/A
...	...	...	...	...	...
2013	1,4	0,8	-1,1	-1,5	4,5
2014	0,4	-0,8	0,5	0,3	-3,7
2015	0,3	-3,2	1,2	0,0	-6,0

The balance of payment (BoP) records economic transactions of domestic entities (residents) with foreign entities (non-residents) over a certain period of time. The current account includes in particular flows of goods (exports and imports) and services (revenues and expenditures from transport services, tourism and other commercial and non-

DOI: 10.1051/shsconf/20173901023

commercial services). The financial account includes transactions relating to the origination, extinction, and change of ownership of the financial assets and liabilities of the government, the banking and corporate sectors and other entities in relation to foreign countries [1]. Some consequences of Czech foreign trade are analysed [4]. Czech export, import and balance of payment are in Table 4.

Table 4. Macroeconomic Indicators of the Balance of Payment (BoP) [1]

Year	Export (Percent y/y)	Import (Percent y/y)	Current Account of BoP (bn. CZK)	Current Account of BoP (Percent of GDP)	Financial Account of BoP (bn. CZK)	Financial Account of BoP (Percent of GDP)
1997	7,6	5,9	-113,0	-5,8	-90,4	-4,6
1998	9,7	5,5	-40,5	-1,9	-31,7	-1,5
...	...	...	...	...	...	...
2013	0,3	0,2	-21,8	-0,5	68,3	1,7
2014	8,6	10,0	7,9	0,2	59,4	1,5
2015	7,8	8,3	11,3	0,9	175,8	4,3

The balance of the state budget is the difference between revenues and expenditures of the state budget. The budget ends with a surplus (revenues is higher than expenditures) or a deficit (the difference between revenues and expenditures is negative). The definition of the balance of government budgets is in line with international standards and regulations of the EU institutions. According to the Maastricht criteria, the ratio of the negative balance to GDP should not exceed 3% [1]. Detailed analysis of budgets is provided [5]. The values of the Czech state budget and the Czech government budgets are presented in Table 5.

Table 5. Macroeconomic Indicators of Fiscal Policy [1]

Year	Balance of Government Budgets (bn. CZK)	Balance of Government Budgets (Percent of GDP)	Balance of the State Budget (bn. CZK)	Balance of the State Budget (Percent of GDP)
1997	-68,2	-3,5	-15,7	-0,8
1998	-99,5	-4,6	-29,3	-1,4
...	...	...	...	...
2013	-51,1	-1,2	-81,3	-2,0
2014	-83,1	-1,9	-77,8	-1,8
2015	-28,6	-0,6	-62,8	-1,4

DOI: 10.1051/shsconf/20173901023

2.2 Municipal Indicators

Current expenditures are funds spent for operation without long-term benefit, they are classified in the 5th-grade of the budget composition. Czech data of current expenditures are in Table 6. Current expenditures are for non-investment expenditure excluding budget transfers [6].

Table 6. Current Municipal Expenditures [7-9]

Year	Current Expenditures (bn. CZK)	Current Expenditures (Percent y/y)	Current Expenditures (Percent of GDP)
2000	553,3	N/A	23,3
2001	482,9	-12,7	18,8
...	...	...	...
2010	423,9	-10,1	10,7
2011	400,0	-5,6	9,9
2012	349,6	-12,6	8,6

Capital expenditures (Table 7) are classified into the 6th-grade budget line. Capital expenditures are different kinds of long-term investments such as investment purchases and related expenses, purchases of shares and pecuniary interests, and deposits in foundations and institutions, investment transfers and investment lent [6].

Table 7. Capital Municipal Expenditures [7-9]

Year	Capital Expenditures (bn. CZK)	Capital Expenditures (Percent y/y)	Capital Expenditures (Percent of GDP)
2000	55,0	N/A	2,3
2001	64,1	16,7	2,5
...	...	...	...
2010	87,7	-3,4	2,2
2011	75,8	-13,6	1,9
2012	67,4	-11,0	1,7

Total expenditures to be seen in Table 8 are the sum of current and capital expenditures.

DOI: 10.1051/shsconf/20173901023

Table 8. Total Municipal Expenditures [7-9]

Year	Total Expenditures (bn. CZK)	Total Expenditures (Percent y/y)	Total Expenditures (Percent of GDP)
2000	608,2	N/A	25,6
2001	547,0	-10,1	21,3
...	...	...	...
2010	511,6	-9,0	12,9
2011	475,8	-7,0	11,8
2012	417,0	-12,4	10,3

3 Results and Discussion

3.1 Correlation Analysis without the Time Factor

Correlation analysis without the time factor means that the same time periods are correlated.

3.1.1 Current Expenditures

The analysis of the correlation of current expenditures shows that the strongest positive dependency exists between the unemployment rate and the share of current expenditure on GDP (0.577). The strongest negative dependence is the share of the financial account‡ on GDP and the share of current expenditure on GDP (-0,733).

Substantial negative dependence can also be seen when comparing the absolute amounts of current expenditure with total GDP (-0.621) and GDP per capita (-0.640). This negative dependency means that current expenditure develops opposite the gross domestic product. There is, therefore, a visible influence of fiscal policy that stimulates the economy during the crisis and prevents its overheating at the time of expansion.

Growth rates do not appear to be a good predictor, as the dependence of these indicators only ranges from -0.332 (time development of exports versus time development of current expenditures) and 0.536 (time development of price level of market services versus time development of current expenditure).

3.1.2 Capital Expenditures

In the case of capital expenditures and investments, the strongest correlation is to total GDP (0.667) and GDP per capita (0.671). In this case (unlike current expenditures) this is a positive correlation. This means that in times of economic prosperity there are enough financial resources for long-term investments.

‡ Financial account of the balance of payment.

Similar results are in percentage analysis. There is a positive correlation with the unemployment rate (0.625) and the long-term unemployment rate (0.600) and negative correlation to the current account (-0.670) and financial account (-0.856) of the balance of payment.

Even capital expenditures are not predictable based on growth parameters (GDP growth, GDP deflator, wage growth, inflation rate, price indexes, export growth and import growth). The only two significant parameters are producer price index of market services (0.621) and agricultural producer price index (-0.686). However, there are no logical reasons for these dependencies.

3.1.3 Total Expenditures

The total sum of Czech municipal expenditures is composed of 83-91% current expenditures and 9-17% capital expenditures. Therefore, we can expect similar developments and similar correlations as dominant current (operating) expenditures.
There are some strong negative correlations:

- The share of municipal expenditures on GDP vs. Share of the financial account of the balance of payment on GDP = -0.755;
- The share of municipal expenditures on GDP vs. Share of the current account of the balance of payment on GDP = -0.695;
- There are some substantial positive correlations;
- The share of municipal expenditures on GDP vs. General unemployment rate = 0.591;
- The share of municipal expenditures on GDP vs. Long-term unemployment rate = 0.564.

The above results show that ratios are the most appropriate form of prediction variables because other types of variables show worse values. For example, absolute indicators in Czech crowns correlate negative (from -0.304 to -0.551) and percent changes in time correlate very weakly (from -0.337 to 0.582). As stated above, total expenditure can be predicted similarly to current expenditure.

3.2 Correlation Analysis with the Time Factor

Correlation analysis with the time factor means that different time periods are correlated, for example, how much the previous year GDP affects analysed years expenditures (-1 year). There is 7 analysis made from -3 years' time shift to +3 years' time shift. The overview is in Table 9:

Table 9. Overview of the Time Shifts

Time shift	Time Period of Municipal Expenditures	Time Period of Macroeconomic Indicator
-3 years	from 2000 to 2012	from 1997 to 2009
-2 years	from 2000 to 2012	from 1998 to 2010
-1 year	from 2000 to 2012	from 1999 to 2011
0 years	from 2000 to 2012	from 2000 to 2012
+1 year	from 2000 to 2012	from 2001 to 2013
+2 years	from 2000 to 2012	from 2002 to 2014
+3 years	from 2000 to 2012	from 2003 to 2015

3.2.1 Current Expenditures

The first analysis is focused on current expenditures. Really interesting results are in case of balance of payment correlation. We can see the positive correlation of the past data (0.200 to 0.518) and negative correlation of the current and future data (-0.298 to -0.637). We can use the financial account of the balance of payment from 2 years ago (-2 years) to predict future values with an accuracy of 0.518. Or the current account of the balance of payment from 2 years ago (-2 years) with an accuracy of 0.501.

Correlation of GDP and GDP per capita values is negative all the time. However, it is getting stronger and stronger in future correlations. The values are:

- -0.545 (GDP) and -0.551 (GDP per capita) in the past (-2 years);
- -0.621 (GDP) and -0.640 (GDP per capita) in the present (0 years' time shift);
- -840 (GDP and -0.850 (GDP per capita) in the future (+3 years).

As mentioned above, there is a visible influence of fiscal policy that stimulates the economy during the crisis and prevents is overheating at the time of expansion. However, this stimulation comes too late, 3 years after the crisis peak. The reason is that information about crisis comes too late. The statistical office can measure the GDP ex-post, 1 or 2 years later. Local authorities and local governments can change future budgets according to this information. So the largest changes of municipal budgets are made 3 years after the economic cycle.

Percentage values are not significant in the past (from -0.422 to 0.552). However, we can follow some connections in the present and in the future. The positive correlation of unemployment rates is getting stronger:

- General unemployment rate in the present (0 years' time shift) is 0.577 and 0.642 in the future (+3 years);
- Long-term unemployment rate in the present (0 years' time shift) is 0.550 and 0.825 in the future (+3 years).

The negative correlation of changes in the balance of payment is getting stronger as well:

DOI: 10.1051/shsconf/20173901023

- Current account share on GDP in the present (0 years' time shift) is -0.682 and -0.780 in the future (+1 year);
- Financial account share on GDP in the present (0 years' time shift) is -0.733 and -0.811 in the future (+2 years).

Percentage changes in time are not significant at all. The values are around zero, sometimes slightly positive, sometimes slightly negative.

3.2.2 Capital Expenditures

Second analysis if focused on capital expenditures. Its connection to GDP is quite strong. Especially the positive correlation between capital expenditures and GDP per capita from 2 years ago (-2 years). The correlation index is 0.707, which makes it the best value for future predictions. Total GDP is a good indicator as well. The best value is 0.688 for the same time shift (-2 years). Correlations to the balance of payment are not significant, some of them are slightly positive, some of them slightly negative.

Percentage values give weak results in the past (from -0.424 to 0.388). However, we can follow some connections in the present and in the future. There are some positive correlations of unemployment rates:

- General unemployment rate in the present (0 years' time shift) is 0.625;
- Long-term unemployment rate in the future (+2 years) is 0.695.

There are some negative correlations of changes in the balance of payment:

- Current account share on GDP in the future (+1 year) is -0.681;
- Financial account share on GDP in the future (+1 year) is -0,869.

Percentage changes in time are not significant at all. The values are around zero, sometimes slightly positive, sometimes slightly negative.

3.2.3 Total Expenditures

The third analysis is focused on **total expenditures**. Some results are similar to current expenditures because the majority (more than 80 %) of total expenditures are the current ones. Therefore, the results are similar to the analysis of current expenditures.

The strongest negative correlations:

- The share of municipal expenditures on GDP vs. Share of the financial account of the balance of payment on GDP (+2 years) = -0.816;
- The share of municipal expenditures on GDP vs. Share of the current account of the balance of payment on GDP (+1 year) = -0.789.

The strongest positive correlations:

- The share of municipal expenditures on GDP vs. General unemployment rate (+3 years) = 0.643;
- The share of municipal expenditures on GDP vs. Long-term unemployment rate (+3 years) = 0.827.

Based on these time shifted correlations we can see that municipal budgets are reason factors and macroeconomic variables are consequence factors. Based on total municipal budgets we can expect the future development of unemployment and balance of payment.

Other types of variables show worse results for predictions. For example past absolute indicators in Czech crowns correlate from -0.497 to 0.500. Past percent time changes correlate from -0.438 to 0.557. Future absolute indicators in Czech crowns correlate from -0.405 to -0.782. And future percent time changes correlate from -0.529 to 0.441.

4 Discussion

Municipal managers must consider a number of different factors in their decision-making. One of the most important factors is the risk [10]. Another factor specific for public sector is so called budget cycle [11-12]. Some authors are discussing specifics of the public sector, such as its centralization and decentralization [13]. This paper focuses on connections between macroeconomic indicators and municipal budgets. There are macroeconomic causes and macroeconomic consequences in the case of municipal budgets. Main reasons or factors influencing the municipal budgets are in Table 10.

Table 10. Factors Influencing the Municipal Expenditures

Factor	Current Expenditures	Capital Expenditures	Total Expenditures
Gross Domestic Product	Indirect proportion	Direct proportion	Indirect proportion
Balance of Payment	Direct proportion	Indirect proportion	Direct proportion

A useful tool to nowcast Czech GDP via Dynamic Factor Model (DFM) introduced Rusnák. He "demonstrated good performance for nowcasting of the expenditure components of Czech GDP" [2]. His framework "might be useful in nowcasting other variables as well" [2].

Main consequences or factors influenced by the municipal budgets are in Table 11.

Table 11. Factors Influenced by the Municipal Expenditures

Factor	Current Expenditures	Capital Expenditures	Total Expenditures
Gross Domestic Product	Indirect proportion	Direct proportion	Indirect proportion
Unemployment Rate	Direct proportion	Direct proportion	Direct proportion
Balance of Payment	Indirect proportion	Indirect proportion	Indirect proportion

5 Conclusion

This paper revealed a negative dependence between current expenditures and gross domestic product. The cause of this negative dependence is fiscal policy, which is done through current expenditures. Thus, the economy is stimulated during the crisis and it is dampening during the expansion.

The strongest correlation is between current GDP and current spending over 3 years. This three-year delay is due to mechanisms within the public sector and time lags, which are seen as one of the causes of government failures. The time delay is the result of ex-post GDP measurement, administrative processes, decision-making processes, budgeting for future periods, etc.

Furthermore, a positive correlation between capital expenditures and GDP was found. The reason for this is the fact that investments are made only with sufficient funds after ensuring the normal running of the municipality. This situation occurs at a time of economic growth.

Total expenditures consist mainly of current expenditures, so the prediction mechanism of their development is almost identical to current expenditures. The aforementioned negative correlation with GDP and the three-year delay can be applied.

Similar analysis is done in the paper Determinants of Employment and GDP Resilience in the Context of an Economic Crisis [14].

References

1. Czech Statistical Office, *Main Macroeconomic Indicators* [online], Available at: www.czso.cz (2017)
2. M. Rusnák, Nowcasting Czech GDP in real time, *Economic Modelling*. **54**, 26-39 (2016)
3. E. Mielcova, Economic Growth and Unemployment Rate of the Transition Country – The Case of the Czech Republic 1996-2009. *Ekonomie a Management*, **14**, 29-37 (2011)
4. P. Rousek, Mutual Foreign Trade between China and the Visegrad Group. *Littera Scripta*, **9**, 42-51 (2016)
5. P. Dvořák, *Veřejné finance, fiskální nerovnováha a finanční krize*, 1. Vyd. Praha: C. H. Beck, 343 s. (2008)
6. Ministry of Finance of the Czech Republic, *Decree of the Ministry of Finance No, 323/2002 Coll. of 2 July 2002 on the budget structure* [online], Available at: http://www.mfcr.cz/cs/legislativa/legislativni-dokumenty/2002/vyhlaska-c-323-2002-sb-3461 (2002)
7. Ministry of Finance of the Czech Republic, *ARIS web* [online], Available at: http://www.mfcr.cz/cs/o-ministerstvu/informacni-systemy/arisweb (2017)
8. Ministry of Finance of the Czech Republic,*ÚFIS* [online], Available at: http://www.mfcr.cz/cs/o-ministerstvu/informacni-systemy/ufis (2017)
9. Ministry of Finance of the Czech Republic, *MONITOR – Information Portal of Ministry of Finance* [online], Available at: http://www.mfcr.cz/cs/o-ministerstvu/informacni-systemy/iissp--monitor (2017)
10. H. Elsawah, I. Bakry, O. Moselhi, Decision Support Model for Integrated Risk Assessment and Prioritization of Intervention Plans of Municipal Infrastructure. *Journal of Pipeline Systems Engineering and* Practice, **7**(4) (2016)
11. T. Aronsson, D. Granlund, Federal subsidization of state expenditure to reduce political budget cycles. *International Tax and Public Finance.*,**24**, 536-545 (2017)
12. L. Matějová, M. Plaček, F. Ochrana, M. Půček, M. Křápek., Political Business Cycle in Local Government. Case Study of Czech Municipalities. *Current Trends in Public Sector Research*, **19**, 142-148 (2015)
13. N. K. Hlepas, Is it the twilight of decentralization? Testing the limits of functional reforms in the era of austerity, *International Review of Administrative Sciences*, **82**, 273-290 (2016)
14. O. Svoboda, P Applová, Determinants of Employment and GDP Resilience in the Context of an Economic Crisis: Evidence from EU Countries and Regions. *Littera Scripta*. **9**(2), 140-155 (2016)

Changes in the Concept of Public Greenery on the Basis of an Analysis of Czech Municipal Financing

Pavel Rousek[1] and *Simona* Hašková[1,*]

[1]Institute of Technology and Business, School of Expertness and Valuation, Okružní 517/10, 37001 České Budějovice, Czech Republic

Abstract. The public sector is an inseparable part of every economy. It performs a number of functions such as providing public services, environmental protection, defense, security, education, health, justice and many others. Each institution's duty is to care for its own property. Public spaces, parks and other public greenery are mostly owned by municipalities. Taking care of public greenery is a part of public spending because the source of money is municipal budgets.
Key words: green area, municipal financing, analysis, costs

1 Introduction

Green spaces play a great role in developing urban ecological and social environment. The amount of greenery, the distribution of green space in populated areas and the easiness by which city inhabitants can access such spaces are regarded as key factors of social and ecological quality of urban environments [1]. Urban green areas play a significant physical and psychological role in human health. One of the medical contributions rests in mitigating the human health impacts of rising temperatures resulting from global climate change. According to Bowler et al. [2] review evaluating the evidence on whether green space affects the air temperature of urban areas within parks, it is broadly assumed that these sites are cooler than non-green sites; the meta-analysis results in Bowler et al. [2] confirmed a park space was 0.94 °C cooler in the day on average.

Green area coverage differs considerably among cities. Fuller and Gaston [3] used standardized regression to examine the relation between urban green space coverage, city area and population density in 386 European cities; their results showed that green space within a city is primarily related to city area rather than the population density, which proofs that largely a space-filling effect is being promoted.

Besides the benefits named above, it is necessary to consider diverse real and potential costs that have to be taken into account when optimizing the net benefits from urban vegetation [4]. According to Nowak and Dwyer [4], the environmental costs can be initiated by the inappropriate landscape designs, tree selection, and tree maintenance, which

* Corresponding author: haskovas@post.cz

can result in pollen production and chemical emissions from trees and grass. Improper maintenance procedures can contribute to air pollution, excessive energy waste, poor excess waste management and water consumption. The most obvious direct costs are the explicit economic costs associated with urban green development and treatment covering, for instance, the costs of planting, maintenance, management, and removal, the costs of damage from falling tree limbs and cracked sidewalks due to tree roots [5].

The hot problem of many cities leans in a surplus of grass cuttings resulting from the urban green maintenance. Its management is usually financially demanding despite the energy saving up to date technologies; this is caused by the high acquisition costs of technologies and considerably high variable costs resulting from the human resource consumption and costs of the waste disposal. The cost reduction of waste management is discussed in a number of papers. For instance, Nakasaki, Aoki and Kubota [6] state that the progress in accelerated composting resulted in decreased costs; however, as Maroušek et al. [7] points out, the accelerated composts have the low ion-exchange capacity, hence it hardly provides any advantage in terms of its fertilization value. The grass cuttings can be utilized in biogas production [8]; in this case, though, this is associated with bad odor for which reason the approach to the waste cost reduction is difficult to implement on a large scale.

The European regulation of waste management affects the increasing necessity of municipal waste separation and recycling. In the Czech Republic, major changes have taken place based on national waste treatment legislation adopted in 2001 [9]. Nowadays a fundamental principle rests in waste recovery, which is prior to waste disposal. The practical realization of waste management differentiates economical and efficiency view. Based on scientific research [10] it was shown that the variable charges can be considered as an effective choice for the waste management regulation; the effectiveness of variable charges is greater than the effectiveness of fixed fees. Unfortunately, as the authors concluded, the key question for local representatives in the Czech Republic is not the effectiveness but the economy of waste management fees.

2 Data and Methods

Data sources for these analyses are Czech public budgets databases [11-13] led by the Ministry of Finance of the Czech Republic. Municipal expenditures are divided into current expenditures, capital expenditures, and total expenditures. Authors calculate basic data at current prices, which then convert to constant prices in 2015 or measure with gross domestic product.

This part is dedicated to public expenditures on public greenery in Czech public budgets. Budget breakdown by branch [14] states care for public greenery under subheading 3745 as part of following headings:

- 3 – Services for the population;
- 37 – Environmental protection;
- 374 – Nature and landscape protection;
- 3745 – Care for the appearance of villages and public greenery.

2.1 Case study Prague

There is a case study as a part of the paper. It attempts to evaluate efficiency, effectiveness, and efficacy (so called 3E) of Prague Municipal Expenditures for the Public Greenery. In this case, efficiency means "cheap public service delivering", effectiveness means "right public service delivering", and efficacy means "delivering public service in the right way".

3 Results and Discussion

Current technology allows lowering of public green care expenditures. "It was shown that it is possible to construct and operate newly proposed technology which utilizes the urban green cuttings with profit instead of the usual loss making. This has been achieved by maceration and intensive anaerobic fermentation followed by mechanical and heat dewatering and the subsequent pyrolysis of the fermentation residue" [7].

For this reason, this part of the paper discusses whether the cost reduction is only a theoretical option or whether it actually occurs. Those facts relate in particular to current expenditures. The starting point for verifying this thesis is the time series of Czech public expenditures for care of public greenery.

Table 1. Czech Municipal Expenditures for the Public Greenery Care (Current Prices, thousands CZK) [11-13].

Year	Current Expenditures	Capital Expenditures	Total Expenditures
2000	2 705 100	331 345	3 036 444
2001	2 717 392	456 325	3 173 716
2002	2 959 707	548 575	3 508 282
2003	3 142 977	518 189	3 661 166
...			
2007	4 202 685	1 069 699	5 272 385
2008	4 594 565	978 845	5 573 410
2009	5 230 116	1 762 026	6 992 142
2010	5 626 348	1 689 377	7 315 725
2011	5 408 944	1 210 432	6 619 376
...	...	...	...

Based on current prices, it is not possible to draw the consequences of the development over time. Data is distorted by inflation. Therefore, it can not be ascertained whether nominal growth is caused by real changes or just by raising the price level.

For this reason, authors recalculated data to permanent prices in 2015 (Table 2) and the share of Gross Domestic Product (Table 3). These are the methods commonly used to filter the undesirable influence of inflation.

Table 2. Czech Municipal Expenditures for the Public Greenery Care (Permanent Prices in 2015, thousands CZK).

Year	Current Expenditures	Capital Expenditures	Total Expenditures
2000	3 731 332	457 047	4 188 378
2001	3 580 026	601 184	4 181 211
2002	3 830 319	709 941	4 540 260
2003	4 063 435	669 946	4 733 381
...	...	...	...
2007	4 922 601	1 252 938	6 175 539
2008	5 062 662	1 078 570	6 141 232
2009	5 705 904	1 922 320	7 628 224
2010	6 047 470	1 815 824	7 863 294
2011	5 705 391	1 276 772	6 982 163
...	...	...	...

The data from Table 2 show that the development of all indicators had its minimum in 2000 or 2001. A period of gradual growth until 2009 or 2010 followed. However, in the crisis and in the years immediately following, the development turned and there was a decrease in expenditures for urban green maintenance.

In the period of growth (the first decade of the 21st century) we can see that current expenditures grew at the rate of 274 million Czech crowns per year. Current prices show annual growth of over 292 million, but an increase of 18 million is due to inflation, as noted above. Capital expenditures grew at a slower rate of about 163 million a year. For total expenditure, annual growth is 409 million (427 million nominal – 18 million)

Expressed as a percentage, growth is the following. The difference between the real and the nominal value is always due to the rise in the price level or inflation.

- Current expenditure real growth rate + 6.0% (nominal growth rate +7.6%);
- Capital expenditure real growth rate +17.3% (nominal growth rate +20.4%);
- Total expenditure real growth rate +7.3% (nominal growth rate + 9.2%).

On the other hand, in the period of decline (the second decade of the 21st century), we can see the following changes in expenditure. Current expenditures in permanent prices declined by 291 million a year (nominally -137 million), capital expenditures in permanent prices declined by almost 242 million a year (nominally -196 million), and total expenditures in permanent prices declined by 600 million a year (nominally -395 million).

If we express the values in percent change, the trend of the annual decrease has the following values:

- Current expenditures -4.9% real rate of decline / -2.5% nominal rate of decline;
- Capital expenditures -14.6 real rate of decline / -12.7% nominal rate of decline;

DOI: 10.1051/shsconf/20173901024

- Total expenditures -7.9 real rate of decline / -5.6% nominal rate of decline.

Table 3. Czech Municipal Expenditures for the Public Greenery Care (Percent GDP).

Year	Current Expenditures	Capital Expenditures	Total Expenditures
2000	0.11401	0.01397	0.12798
2001	0.10604	0.01781	0.12384
2002	0.11066	0.02051	0.13117
2003	0.11220	0.01850	0.13070
...	...	...	...
2007	0.10968	0.02792	0.13759
2008	0.11443	0.02438	0.13880
2009	0.13336	0.04493	0.17829
2010	0.14231	0.04273	0.18504
2011	0.13409	0.03001	0.16410
...	...	...	...

From the share of public green maintenance expenditures to GDP, we can see the same conclusions – growth followed by a decline. In the first decade of the 21st century, the share of these expenditures increased by 0.004 percentage point per year (current expenditures), 0.003 percentage point per year (capital expenditures) and 0.007 percentage point per year (total expenditures). In the second decade of the 21st century there was a decrease of -0.005 pp annually (current exp.), -0.005 pp annually (capital exp.) and -0.012 pp annually (total exp.).

3.1 Case study Prague

Within the Czech Republic, Prague is the capital city, the city with the largest population, the city with the largest municipal budget. Due to its area, many public spaces, parks, and gardens are also within its territory. The total area of parks and gardens reaches almost 40 square kilometers, some of them of the Prague-wide significance and other local significance. The basic overview is shown in Table 4.

Table 4. Parks of Prague-wide significance [15].

Park	Area	District
Garden complex Petřín	42.7 ha	Prague 1, Prague 5
Deer-park Hvězda	85.5 ha	Prague 6
Royal Deer-park Stromovka	84.4 ha	Prague 6, Prague 7
Park Vítkov	31.3 ha	Prague 3
Park Letná	46.5 ha	Prague 7
Others (local significance)	3 674.6 ha	Prague
Total	**3 965.0 ha**	**Prague**

In this case, study authors will focus on whether and how the reduction of municipal expenditures for public greenery care in the Czech capital is being achieved.

Table 5. Prague Municipal Expenditures for the Public Greenery Care (Current Prices, thousands CZK) [11-13].

Year	Current Expenditures	Capital Expenditures	Total Expenditures
2000	407 473	85 811	493 284
2001	432 307	88 917	521 225
2002	462 388	136 420	598 808
2003	527 848	143 635	671 483
...	...	...	...
2007	782 165	505 313	1 287 478
2008	778 895	418 147	1 197 043
2009	1 021 910	558 712	1 580 622
2010	1 085 135	681 127	1 766 262
2011	903 812	338 632	1 242 445
...	...	...	...

For the same reason, as in the case of data for the whole of the Czech Republic (influence of inflation), it is not appropriate to work with current prices.

The values of Prague municipal expenditures for the public greenery care at constant prices in 2015 and the share of these expenditures on gross domestic product of the Czech Republic follows in Tables 6 and 7.

Table 6. Prague Municipal Expenditures for the Public Greenery Care (Permanent Prices in 2015, thousands CZK).

Year	Current Expenditures	Capital Expenditures	Total Expenditures
2000	562 056	118 365	680 421
2001	569 543	117 144	686 687
2002	598 402	176 548	774 950
2003	682 435	185 700	868 134
...	...	...	...
2007	916 149	591 873	1 508 022
2008	858 250	460 748	1 318 998
2009	1 114 874	609 538	1 724 412
2010	1 166 355	732 108	1 898 463
2011	953 347	357 192	1 310 539
...	...	...	...

Concerning Prague expenditures and their comparison with the national development, we have similar results. The growth period in the first decade is changing in 2010 to decline. Perhaps the only difference is that Prague has these periods more pronounced - faster growth is followed by a deeper fall.

Period of growth:

- Current expenditures annual growth +60 mCZK or +7.6%;
- Capital expenditures annual growth +68 mCZK or +22.6%;
- Total expenditures annual growth +122 mCZK or +10.8%.

It is interesting to note that at the time of maximum there is an annual maintenance of each hectare of public greenery of almost half a million crowns. One third of this amount is capital spending and two thirds of operating spending.

Period of decline:

- Current expenditures annual decline -123 mCZK or -11.2%;
- Capital expenditures annual decline -189 mCZK or -30.5%;
- Total expenditures annual decline -312 mCZK or -18.1%.

Table 7. Prague Municipal Expenditures for the Public Greenery Care (percent GDP).

Year	Current Expenditures	Capital Expenditures	Total Expenditures
2000	0.01717	0.00362	0.02079
2001	0.01687	0.00347	0.02034
2002	0.01729	0.00510	0.02239
2003	0.01884	0.00513	0.02397
...	...	...	...
2007	0.02041	0.01319	0.03360
2008	0.01940	0.01041	0.02981
2009	0.02606	0.01425	0.04030
2010	0.02745	0.01723	0.04467
2011	0.02241	0.00839	0.03080
...	...	...	...

The share of Prague public green expenditures on GDP grew by 0.001 pp per year (operating), 0.002 pp per year (investment) and 0.003 pp per year (in total) in the first decade. The decade of decline is -0,003 pp annually (operating), -0,004 pp annually (investment) and -0,007 pp annually (in total).

4 Recommendations and Suggestions for Practice

Companies and other institutions (municipal companies and municipalities included) must consider multi-source funding. The size of the enterprise is important. "The companies that mostly finance corporate training exclusively through their own financial resources are micro-companies. In contrast, large companies finance corporate training from these resources the least" [16]. Really significant in case of human resources and corporate training are innovation activities, which can change the ratio between costs and outputs rapidly [17].

A modern method of similar predictions is the use of neural networks. "The suggested neural structures are useful in practice for an enterprise financial plan composition" [18]. Like predicting enterprise financial plans, the municipal expenditures could be predicted as well.

5 Conclusion

The examination of Czech municipal expenditures and previous or potential changes made several findings. The analysis of the time series of maintenance costs for urban green has proved that the data contain two local extremes. The minimum is in the years 2000 or 2001 and the maximum in 2009 or 2010. The above applies to a number of municipalities (including the Capital City of Prague) and municipal expenditures of the Czech Republic as a whole.

Changes in municipal expenditures are triggered by changes in the gross domestic product. The negative correlation is demonstrated by the impact of the fiscal policy that stimulates the economy during the crisis and prevents its overheating at the time of expansion, as other studies have shown. [19] Unfortunately, the new concept [7], which would save public funds, has not been implemented yet.

The analysis further revealed that the trend periods in public expenditures are different. The growth period comes after economic expansion is longer and slower. The decline period begins in the post-crisis years and is shorter, more intense and steeper.

Further research may be about how to put this information into practice. Finding a concrete formula to improve the prediction of spending would help. This would considerably improve public sector planning, where the most important planning document is the one-year budget. Expenditure frameworks with a longer future horizon are often only a formal document without any practical application.

References

1. O. Barbosa, J. A. Tratalos, P. R. Armsworth, R. G. Davies, R. A. Fuller, P. Johnson, K. J. Gaston, Who benefits from access to green space? A case study from Sheffield, UK. *Landscape and Urban Planning*, **83**(2), 187-195 (2007)
2. D. E. Bowler, L. Buyung-Ali, T. M. Knight, A. Pullin, Urban greening to cool towns and cities: A systematic review of the empirical. *Landscape and urban planning*, **97**(3), 147-155 (2010)
3. R. A. Fuller, K. J. Gaston, The scaling of green space coverage European cities. *Biology letters*, **5**(3), 352-355 (2009)
4. D. J. Nowak, J. F. Dwyer, Understanding the benefits and costs of urban forest ecosystems. *Urban and community forestry in the northeast*, 25-46 (2007)
5. J. Dwyer, The significance of trees and their management in built environments. *Trees and building sites. Proc. of an international workshop on trees and buildings*, 3-11 (1995)
6. K. Nakasaki, N. Aoki, H. Kubota, Accelerated composting of grass clippings by controlling moisture level. *Waste Manage Review*, **12**,13-20 (1994)
7. J. Maroušek, R. Zeman, R. Vaníčková, S. Hašková, New concept of urban green management. *Clean Technologies and Environmental Policy*, **16**(8), 1835-1838 (2014)
8. M. Seppälä, T. Paavola, A. Lehtomäki, J. Rintala, Biogas production from boreal herbaceous grasses – specific methane yield and methane yield per hectare. *Bioresource technology*, **100**(12), 2952-2958 (2009)
9. E. Valentova, Waste management in the Czech Republic. *Journal for European Environmental & Planning Law*, **2**(6), 502-511 (2005)
10. J. Slavik, J. Pavel, Do the variable charges really increase the effectiveness and economy of waste management? A case study of the Czech Republic. *Resources, Conservation and Recycling*, **70**, 68-77 (2013)
11. Ministry of Finance of the Czech Republic, *ARISweb* [online], Available at: http://www.mfcr.cz/en/ (2017)

12. Ministry of Finance of the Czech Republic, *ÚFIS* [online], Available at: http://www.mfcr.cz/en/ (2017)
13. Ministry of Finance of the Czech Republic, *MONITOR – Information Portal of Ministry of Finance* [online], Available at: http://www.mfcr.cz/en/ (2017)
14. J. Marhounová, Š. Nováková, Harmonization of accounting in the field of state administration and self-government [Harmonizace účetnictví v oblasti státní správy a samosprávy]. Czech Financial and Accounting Magazine *[Český finanční a účetní časopis]*, **1**, 30-43 (2007)
15. Forecast of the capital city of Prague: grow with you [Lesy hl. m. Prahy: Rosteme s Vámi] [online], Available at: http://www.lesypraha.cz/ (2017)
16. Z. Caha. Exploitation of external financial resources for corporate training purposes in the Czech Republic. *Littera Scripta,* **10**(1), 10-21 (2017)
17. M. Vokoun. Globalization and innovation: innovation activities of multinationals in the Czech Republic 1998-2010. *Globalization and Its Socio-Economic Consequences,* 2409-2416 (2016)
18. M. Vochozka, Usage of multiple perceptron neural networks in determination of the financial plan. *Littera Scripta,* **9**(3), 141-153 (2016)
19. P. Rousek, *The Impact of the Economic Crisis on Czech Municipal Expenditures* (in the printing)

China's economic interests in the "One Belt, One Road" initiative

Yakov Silin[1], *Larisa* Kapustina[2,*], *Italo* Trevisan[3], and *Andrei* Drevalev[4]

[1]Doctor of Economics, Professor, Ural State University of Economics, Head of the University, Department of Regional, Urban Economy and Management, 62, 8 Marta St., Ekaterinburg, 620219, Russian Federation
[2]Doctor of Economics, Professor, Ural State University of Economics, Head of the Marketing and International Management Department, of.465, 62, 8 Marta St., Ekaterinburg, 620219, Russian Federation
[3]Professor, University of Trento, Department of Economics and Management, via Inama 5, I - 38122 Trento, Italy
[4]Ural State University of Economics, Senior lecturer of the Marketing and International Management Department, of.755, 62, 8 Marta St., Ekaterinburg, 620219, Russian Federation

Abstract. The article examines the "One Belt – One Road" initiative of China aimed at the development of transport and logistics infrastructure on the trade route from China to Europe. The authors pay special attention to the history of the Silk Road, which serves as an ideological basis for the modern initiative. The scale of the new project allows the authors to expect that its impact on the international trade will be comparable with the contribution of the historical Silk Road to the development of the global economy as we know it. The authors analyze the prospects of the development and implementation of the initiative in terms of China's economic interests. The most significant threats associated with the initiative are identified.
Key words: OBOR, Silk Road, modern initiative, economic interests

1 Introduction

China has entered a new stage of its development. Nowadays, as in the era of the Ancient Silk Road, China leads the world trade and has positive net balance of trade with Europe and the United States of America. Europe buys about 1 billion euros worth of goods from China annually, but China imports only half of this volume from the European countries. China is now in active search of the ways to change its image, to strengthen its international standing and thus becoming a leader of the world economic relations development. The modern global distribution system has been created by other countries and is not controlled by Beijing, so China is interested in ensuring that its international political and cultural influence is growing to reflect its economic strength, beginning with the region of Southeast and Central Asia [1].

* Corresponding author: lakapustina@bk.ru

The "One Belt – One Road" (OBOR) initiative was first announced by China internationally in 2013 [2]. Its fundamental goal is to create transport corridors to facilitate access to foreign markets and increase the turnover of goods transported from China to Europe and adjacent regions through Central Asia. The region of Central Asia includes Afghanistan, Kazakhstan, Kyrgyzstan, Mongolia, Tajikistan, Turkmenistan, Uzbekistan, and also vast territories of India, Iran, China, Pakistan and Russia. The geographic center of the region is located on the border of Kazakhstan, China and Kyrgyzstan.

In 2015, the National Development and Reform Commission approved the name of the initiative as the "Silk Road Economic Belt and the 21st Century Maritime Silk Road" or a short version of the "Belt and Road Initiative" abbreviated to BRI or just B&R. According to M. Glantz, by using the image of the Ancient Silk Road, the representatives of the state authorities and business of China try to revive the memories of the era of trade prosperity in ancient China [3]. Taking historical analogies into account, it is not surprising that in some of the countries involved the initiative is commonly referred to as just the New Silk Road. However, the former reduction of OBOR has already become a recognizable global brand, and is therefore widely used in scientific literature and this article, although it is not currently used in official Chinese documents and publications.

OBOR affects various areas of development of many states and regions of the world: security, socio-cultural, political, diplomatic and civilizational aspects of their existence. At the same time officials of China prefer not to focus attention on the geopolitical goals of the initiative.

In May 2017, Beijing hosted the "One Belt – One Road" International Economic Forum, dedicated to the prospects of implementing the initiative and the new strategy of China's economic expansion. In his speech, Xi Jinping identified five fields of international cooperation [4]: political coordination, the interconnection of infrastructures, free trade, free movement of capital, and strengthening of ties between peoples. The scale of the OBOR initiative is in fact unprecedented. The projects within the initiative provide for participation of 65 countries, which account for 70 percent of the world's population. China proposes cooperation with joint investments of 4 trillion dollars in the implementation of the program of the OBOR initiative. It will enable both China and its partners to realize their potential for economic development [5]. The most active supporters of OBOR apart from China itself are Kazakhstan, Iran and Turkey.

The purpose of the study is to clarify the aims of the OBOR initiative, determine the economic interests and potential benefits of China in its effort to revive the Silk Road.

2 Evolution of the Ancient Silk Road

Over the centuries, trade routes between China and Europe have experienced periods of prosperity, decline and new revivals. The term "Silk Road" has been introduced by the German geographer F. Richthofen in his book "China", published in 1877. He is considered the first author to use the metaphor referring to the network of land caravan routes between Europe and Asia [6].

The eastern part of the Silk Road has formed as a regular caravan route with the Chinese emperor Qin Shi Huangdi in 211 BC, which founded the first dynasty of the Chinese Empire. Unified rules of trade were established throughout the territory. The problem of raids on caravans with cargo of neighboring nomad states was solved diplomatically, but in this period, there was no direct route from China to Europe. After the fall of the Qin Dynasty, trade routes collapsed.

The revival and final formation of the Silk Road is associated with the name of the Emperor Wudi (Han Dynasty), who ruled China in 141-87. BC, recaptured Hexi (Gansu corridor), extended the Great Wall of China. China has gradually gained control over all the cities of the Tarim Basin. All in all, these efforts allowed establishing direct contacts between the Chinese and Western peoples in the Fergana Valley. Chinese silk enjoyed high demand among the noble Romans and was an exotic luxury good [7]. In exchange for silk and spices of the East, Rome exported metals, dishes, glassware, but more often the imports of Oriental goods were paid for by gold and silver. China and India had a positive trade balance and received an average of 100 million sesterces per year from trade with the Roman Empire; for comparison, the total costs of having an army for the Roman Empire were about 150 million sesterces [8].

For several centuries, the trade had been flourishing between West and East, and the main route of the Silk Road had formed: from the Tarim Basin through cities of Khujand, Samarkand and Bukhara along the Amu Darya river to the Aral Sea, around the Caspian Sea to the northern slopes of the Caucasus Mountains (northern path). The southern route went from the Fergana Valley to Egypt. Approximately 120 ships were sent annually from Egypt to India [9]. Although the sea route was shorter (average delivery time was 9 months) and cheaper, the land route remained competitive due to high profitability of caravans [10]. The volume of shipments by land significantly exceeded the volume of shipping.

The factors that made the Silk Road a successful trade route in the analyzed period include the following ones:

- broad and rich markets of goods in the East and West;
- strong, well-organized imperial states;
- political and economic stability of the states along the Silk Road, which ensured the safety of traders and the safety of caravans.

The slowdown in land trade began as early as at the end of the 2nd century AD, when the Han Empire lost control of the Tarim Basin. Only by the end of the 5th century the centralized Chinese state was again created, however, the trade was far less intensive compared to the golden age. By then, the predominant role had shifted to the sea freight and at the end of the 5th century, the Silk Road land route virtually disappeared.

However, it revived centuries later. All in all, there are three most successful stages in the evolution of the Silk Road [11]:

1. 206 BC - 220 BC. In the era of the Han dynasty, diplomatic relations were established with nomadic tribes in the eastern part of the Silk Road to exclude their raids on trade caravans. The modern Chinese city of Xi'an served as the starting point of the road.
2. 618 AD - 907 AD. The Chinese Tang Dynasty controlled the trading corridor of Hexi and ensured the growth of trade in the space of Eurasia.
3. 1271 AD - 1368 AD. During the reign of the Mongol Yuan dynasty, Genghis Khan united China and built one big empire, which ensured the flourishing of the Silk Road and trans-Eurasian land trade.

Historically, the main geopolitical confrontation of the states in the space of Eurasia was set for control over caravan routes. Only a century and a half during the existence of the Silk Road, it was controlled by China (from China to Bukhara) in the 13th-14th centuries. During this period, China accounted for 58% of world GDP [12]. In the years of 1389 - 1395 Timur erased all trading cities on the coasts of the Caspian and Black Seas from the face of the earth. The Silk Road land route was abandoned and there were almost no feasible incentives to restore it. The sea trip to China from the Persian Gulf took 150 days, the overland route from the area of modern Azov up to 300 days [13]. Besides the transportation of a cargo of one ship required 1000 animals. As a result of not being competitive, in the 16th century the Silk Road finally ceased to exist.

The Silk Road played a significant role in the formation of institutions and legal support for international trade. Mediation got widely spread. Merchants usually delivered the goods to the nearest caravan-shed, which served as a hotel and a warehouse. Banks, joint-stock companies, currency exchange and checks appeared and became usual on the Silk Road. Marco Polo noted the high state guarantees to merchants: a plaque (equivalent to a modern passport or license) was given to each merchant that allowed him to move around Eurasia freely and protect him from looting [14].

The international division of labor was formed. Elephants, lions, silk, porcelain, paper, spices, nutmeg, cloves, cinnamon, tea, etc. were sent from the East. The West traded horses, silverware, agricultural crops, grapes, precious stones and metals, spirits, slaves and others. Cotton and wool products and fabrics were also imported from Central Asia to China. The development of trade along the Silk Road caused the emergence of new professions: changers, translators, camel drivers, caravan guards, customs tax collectors, and many more.

The Silk Road also provided for the transfer of technologies and innovations, as noted by Yu. V. Latov [13]. Papyrus was replaced by paper from China, which increased the circulation of manuscripts in Europe. Europe learned to produce paper only in the 13th century. Silk production technology was gradually exported from China to East Turkestan, in the 5th century - to Iran, in the 6th century - to Greece. The knowledge of the gunpowder production came from China to Europe in the 9th century. Europeans learned to use spices for canning agricultural products. And on the other hand, China has mastered winemaking through technologies that have come from the Roman Empire. It can be concluded that the cooperation of countries within the international division of labor was mutually beneficial and enriched the states of the East and the West with new goods, technologies for their cultivation and production. At the same time, China maintained its technological superiority until the 18th century.

3 China's Goals and Potential Benefits of the One Belt – One Road Initiative

China's economic interests in the 21st century are once again directed through Central Asia to Europe. China is the largest country in terms of population and the leading economy in terms of economic growth: China has become the world leader as suggested by such indicators as total exports and GDP, calculated at purchasing power parity. OBOR has emerged as a large-scale project, which ideologically goes back to the Great Silk Road uniting the countries into the Silk Road Economic Belt (SREB).

When formulating the tasks for the SREB project, the Chinese state authorities, firstly, considered the already existing economic priorities of the People's Republic of China in the region: ensuring the promotion of Chinese goods and the implementation of the "soft power" strategy through investment and trade cooperation to neutralize the fears associated with the Chinese expansion and creating a favorable public perception background for commercial projects.

China treats SREB as a comprehensive plan to promote its interests in the world as a whole, and especially in the neighboring regions. It is considered officially that these interests coincide with the interests of partners.

The initiative combines various projects aimed at development of transport and logistics infrastructure in several key geographical areas (Figure 1). Although the initiative includes up to six main routes, usually, the following three land routes of SREB are identified with the starting point in China – as the most extensive ones:

- the Northern Corridor, or the New Continental Bridge to St. Petersburg and Helsinki through Kazakhstan and Russia with the possibility of extension to Rotterdam;
- the Central corridor through Central Asia, which then branches into several directions: to Western Asia and the Persian Gulf, to Mediterranean countries (Turkey, Italy), and also to Western Europe (the port of Rotterdam);

- the Southern Corridor through the regions of South-East Asia and South Asia to the largest ports of Pakistan and Indonesia allowing easy access both to Indian and Pacific oceans.

Less commonly mentioned are the corridor China-Mongolia-Russia to the North, as well as parts of the southern corridor, with Bangladesh and Indochina becoming the end points [15].

In addition to the land routes of SREB, the OBOR initiative includes the "Maritime Silk Road of the XXI century" from China to Europe through the ports of Southeast Asia and the North of Africa.

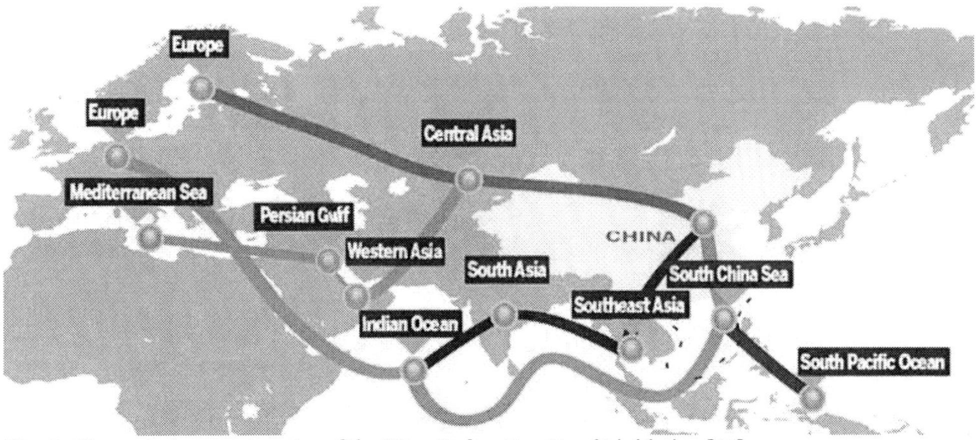

Fig. 1. The main transport routes of the "One Belt – One Road" initiative [16]

Currently, more than 60 countries are involved in negotiations. Their economic and foreign policy interests are not uniform and do not always coincide. An analysis of experts' opinions and the documents describing the China's initiative made it possible to identify the interests of China and the actual goals of OBOR from the initiators' perspective.

1. *Expanding exports, which is an important factor in the development of the Chinese economy.* The project is designed to promote Chinese goods around the world. It requires reducing the transport gap between East and West, to cut down the cost of delivering goods. It will become a valuable economic driver for the territories that transport routes will pass. The leading role among the exports belongs to high-tech machinery, light and electronic industries with high added shipped to the countries of Central Asia [17]. The region sees railways being built here, compatible with the standards of China's gauge. One of the innovative aspects of the OBOR initiative is the construction of industrial and logistics centers that will later form industrial clusters in the countries of the SREB. China has built 56 economic cooperation zones in 20 countries [4]. Chinese companies are now involved in constructing facilities of an international cross-border cooperation center in Khorgos on the border with Kazakhstan which is becoming the largest inland port [18]. Construction also takes place in China. The project of creating the third integrated duty-free zone - Urumqi, in addition to Alatanhou and Kashgar - is nearing completion. All types of communications will connect Urumqi with Almaty, Bishkek, Tashkent, Samarkand, Tehran and other cities of the Road.

2. *Expanding markets for products and services of the new economy.* Within the framework of OBOR, the digital Silk Road is being formed. It implies rapid development of information and communication, e-commerce, Internet banking, smart cities, industrial computer networks. The Chinese e-commerce giant Alibaba Cloud plans to finance 14 data centers around the world, to promote Chinese companies doing business in countries along

the Road [19]. Data centers and other facilities are supposed to be equipped with 5G communication networks that are a thousand times faster than 4G networks. This effort will likely simplify processing of goods and lead to unifying of the customs formalities in 10 countries along the new Silk Road. Further development will provide for automating some of the operations in logistics, receiving information on the delivery of cargo in real time, and the introducing unmanned automobile and railway transport.

3. *Diversifying China's land export routes and decreasing dependence on the USA as China's most important trading partner*. According to the National Bureau of Statistics of China, the main export markets in 2015 were as follows (Figure 2): the USA (18.0% of the value of exports of China), Hong Kong (14.6%), Japan (5.6%), the Republic of Korea (4.5%), Germany (3.0%), Vietnam (2.9%), Great Britain (2.6%), Netherlands (2.6%), India (2.6%), Singapore (2.3%) [20].

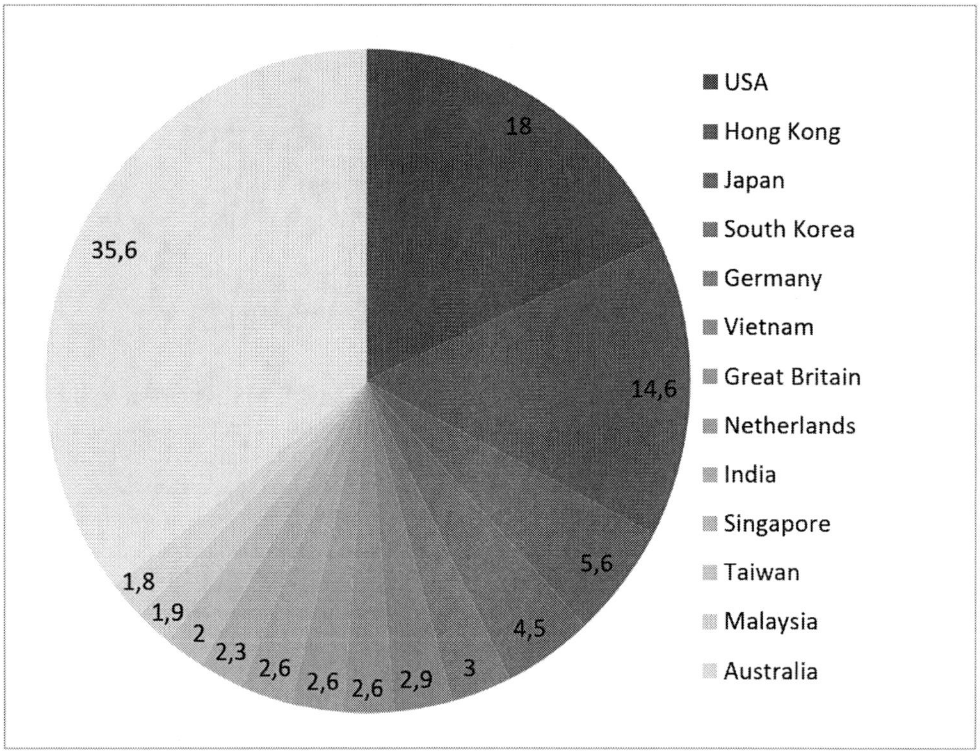

Fig. 2. Geographical Structure of China's Exports, 2015, % [20].

With most countries, China has positive balance of commodity exchange. But with a number of countries the negative trade balance remains, among them are: Japan, the Republic of Korea, Germany, Taiwan, Malaysia, Australia, and Brazil.

At the same time, the regional structure of China's total trade turnover in 2015 was characterized on the one hand by a certain decrease in the share of countries of Asia, Europe, Latin America, Africa and Oceania, and on the other hand an increase in the share of North America. China's foreign trade is 90% shipped by the sea, while the US military fleet controls the sea coast and at any time it can stop trade flows from China [21]. It can be assumed that the OBOR project has already begun to bear fruit. Thus, in the first half of 2017, compared to the same period of the previous year, the figures show impressive growth of mutual trade with the EU (by 17.4%), Kazakhstan (by 46.8%), Russia (+ 33.1%), Poland (+24, 6%), and Pakistan (+ 14.5%) [22].

4. *Ensuring access of the PRC to the natural resources, and especially fuel-and-energy sector, of the countries of Central Asia and Africa.* China proves to be successful in diversifying its import purchases of energy commodities, developing the pipeline infrastructure of Eurasia. In the framework of the new Silk Road, oil and gas pipelines from Kazakhstan and Turkmenistan are being built on a mutually beneficial basis, allowing intra-continental countries of Eurasia to expand exports beyond their traditional region [23]. Chinese companies already control 25% of Kazakhstan's oil production. The initiative includes gas and oil pipelines in the Far East of Russia and in West Africa. Pipelines are designed and modernized with the participation of Chinese companies, mainly with the goal of obtaining access to seaports. The Silk Road Fund acquired shares of Russian oil and gas companies: Yamal LNG (the share of 9.9%) and Sibur (10%) in 2016. Besides China needs uranium to ensure the increase in the capacity of nuclear power plants. Currently, 75% of China's uranium imports fall on Kazakhstan [17]. Chinese companies consume more than half of Turkmenistan's gas exports, have concluded gas and uranium deals with Uzbekistan worth $ 15 billion [3].

5. *Increasing foreign direct investment made by Chinese companies.* China moves from exporting goods to increasing outgoing investment [24]. In particular, China invested in Europe 4 times more in 2016 than European companies in China [12]. When implementing investment projects in countries along the new Silk Road, China strictly defends a controlling stake, a significant share in the ownership. It also strives to ensure the full, or at least exceeding 50% of local production, export of raw materials to be processed in China [17]. The OBOR initiative implements the PRC's global investment strategy, diversification of foreign assets, and the international expansion of large Chinese companies in the field of transport infrastructure [25]. Financing of the projects is provided from the Silk Road Fund of $ 40 billion, as well as the China Import-Export Bank and the Asian Infrastructure Investment Bank are also involved in lending.

6. *Leveling the pace of economic development of the country's regions.* Projects of OBOR are aimed at reducing the concentration of the economy in coastal areas of China and its excessive dependence on maritime transit routes. The PRC is interested in realizing the potential of the western and north-eastern provinces, which requires land transport corridors development.

7. *Reducing pressure on the Chinese labor market suffering from the slowdown of the national economy.* China is pursuing the goal of creating new jobs for its citizens, therefore, contracts for establishing joint ventures abroad involve the involvement of the Chinese labor force. Thus, in 2015, 376 thousand people got jobs outside China contributing to foreign projects. The total number of Chinese workers of various occupations outside the country exceeded one million people [20].

8. *Forming the Chinese yuan currency zone in Central and South-East Asia.* Trade flows along the new silk route will be serviced by a financial system based on yuan. The PRC seeks to create a unique financial system that will unite regional development banks and provide for an increase in settlements in national currencies, primarily in yuan [26].

China's goals related to the OBOR initiative are obvious. It is an attempt to increase exports of goods and services overcoming discriminatory restrictions on Chinese investments imposed by the US. China is trying to integrate a huge economic space with its initiative. By simplifying the logistics of export shipments, China will get a new impetus to the country's economic development. The investments made by China will potentially change the direction of the economic development of the countries of Asia and Africa. At the same time, an increase in the volume of payments in yuans will put pressure on the US dollar rate.

The active promotion of the OBOR initiative is perceived by many countries as China's attempt to claim the role of an economic leader. To support that point, People's Daily (or Renmin Ribao) newspaper has noted that Western politicians consider the OBOR initiative a geopolitical tool used by China to ensure domination in the region and create a new sphere of influence, and the United States and Europe are afraid of losing their leading positions in the world financial system due to the strengthening of the yuan [27]. The authors can agree with the opinion that in the event of threats to its foreign investments, China will have to shift from a moderate style of conducting trade negotiations that seemingly has mostly prevailed before, to a more rigid one. The "slow but steady" tactics can thus change into more proactive ways to defend national interests [28].

The security problems are not being specifically addressed officially. At the same time a significant number of Chinese experts recognize the geopolitical importance of SREB for the country. It aims to promote regional security and stability which meets China's interests in developing trade and economic ties with its neighbors [29]. The Chinese interests mostly coincide with the interests of the Central Asian countries. Just like China, they are interested in their own economic development and political stability. In addition, China does not impose political conditions in the implementation of trade and economic cooperation.

China persuades the partners of the initiative that it provides for fair joint development, offers a path of peace, prosperity, mutually beneficial freedom of trade, as well as innovative and humanitarian cooperation. The prospect of implementing the OBOR initiative is realistic, as the projects included in it are tied to existing infrastructure and logistics facilities or at least the projects being implemented now.

4 Conclusion

It can be concluded that the trade vector along the ancient Silk Road was directed from East to West. China in the past was more developed economically and technologically and had an active trade balance with Europe for 1,500 years. Luxury goods for the European nobility formed the basis of the trade. Periods of booms of Eurasian trade occurred when the Silk Road was controlled by one major regional power or less often – several powers. Thanks to the Silk Road, the international market has developed getting advanced logistics and transport infrastructure, a system of international settlements, an institute of protection of foreign merchants' property rights, trading cities and customs points.

The OBOR initiative is the largest contemporary global project, unprecedented in terms of the scope of the declared objectives as well as the number of participating countries. At the same time, the tasks are determined by the strategic vision only of one country – China. Internal economic problems can be found behind the foreign economic interests in many respects. Given the slowdown in the growth rate of the Chinese economy of the recent years, it is obvious that such a large-scale project aims to support existing and create new production facilities and jobs in China, especially in the north-western provinces.

Summarizing the goals and objectives set by the PRC, one can recognize that, they should be reviewed from the point of view of China with its centuries-old history. In this regard, the OBOR initiative is the program plan aimed to regain the role of economic leader of the macroregion and even the world economy as a whole. The look at the project from such a perspective seems to give better understanding of the representatives of the "Central State". As it seems now, they have never taken seriously the claims of other countries and regions of the world to lead the global economy and civilization.

At the same time, general reference to the historical role of the participating countries in the development of the Great Silk Road is a powerful unifying factor for the states with currently different economic and political potential, such as the countries of Central Asia,

China and Pakistan. It seems that China is sometimes exploiting this trust while investing in Central Asia. Investors often insists on using Chinese technologies and labor in foreign projects while not all Chinese companies are ready to comply with local environmental legislation, safety, immigration and other regulations.

The main driver for the development of SREB for China is its economic interests: the need to provide its own economy with raw materials, extend the markets for its manufactured products, and create jobs for labor force.

One of the main threats to the implementation of the new Silk Road may be the discontent of China's trading partners with the current legislative barriers and restrictions imposed by China on the access to the domestic market for foreign companies. Those barriers don't correspond to the priorities of free trade and capital movement declared for the OBOR initiative. The policy of creating special economic zones inside China doesn't seem to provide equal benefits of cooperation to foreign partners who face China's external economic expansion. Imbalances in trade with major partners can become an obstacle in the development of trans-Eurasian trade and international movement of capital. In addition, rebranding of OBOR to BRI or B&R initiative may lead to negative consequences for the countries involved.

There is a risk that the OBOR initiative will not be implemented in the declared scope, remaining a set of separate projects. Given the historical context of the evolution of the Great Silk Road, which was almost never homogeneous and static, such a scenario seems quite real. However, this scenario will not signify a defeat for China. On the contrary, OBOR sets a sufficiently transparent, flexible and scalable framework for the development of Chinese business in foreign markets, the use of excess production capacity and labor resources.

References

1. B. Maçães, New Western frontier, conquered by China, *POLITICO SPRL* [online], Available at: https://www.politico.eu/article/khorgas-china-kazakhstan-new-silk-road-trade/ (2016)
2. T. Fallon, The New Silk Road: Xi Jinping's Grand Strategy for Eurasia. *American Foreign Policy Interests*, **37**(3), 140-147 (2015)
3. M. H. Glantz, China's «One Belt, One Road» (OBOR) Initiative: What a Difference a «Brand» can make. *Post-Soviet Issues*, **4**(1), 8-19 (2017)
4. One Belt, One Road. *Main theses of Xi Jinping's speech* [online], Xinhua, Available at: http://news.xinhuanet.com/english/2017-05/14/c_136282982.htm (2017)
5. O. Shevtsov, "One Belt – One Road" – the most ambitious geo-economic project. *Soyuznoe Veche*, **21**, (2017)
6. UNESCO, *The Silk Roads Projects CCT/ CPD/ DIA/ 2008/ P1/61* [online], Available at: http://unesdoc.unesco.org/images/0015/001591/159189E.pdf (2008)
7. W. K. Klingman, *The first century*. New York, HarperCollins (1990)
8. L. Casson, New light on maritime loans: P.VINDOB. G 40822. *Zeitschrift für Papyrologie und Epigraphik*, **84**, 195-206 (1990)
9. Strabo, *Geography (Γεωγραφικά – Gheographikà), Book II*, 118
10. CH. Scarre, *The Penguin Historical Atlas of Ancient Rome*, Harmondsworth, Penguin (1995)

11. J. Wuttke, The history of the Silk Road. J. Wuttke, Xi Jinping's Silk Road is under threat from one-way traffic. *Financial Times* [online], Available at: https://www.ft.com/content/61c08c22-3403-11e7-99bd-13beb0903fa3 (2017)
12. Yu. V. Latov, The Great Silk Road is the prologue of the world economy and globalization (timed to the 2130th anniversary of its "opening"). *Historical and Economic Research*, **11**(1), 123-140, (2010)
13. Yu. V. Latov, The Great Silk Road. China – Britain Business Council, One Belt One Road: new opportunities in China and beyond. *A role for UK companies in developing China's new initiative*, (2015)
14. J. Catanzaro, R. Qi, C. Jia, B. Han, *Silk Road initiative connects countries on path of prosperity*, The Washington Post China Watch [online], Available at: http://www.telegraph.co.uk/news/world/china-watch/business/silk-road-initiative-china/ (2015).
15. P. Minakir, Low hanging fruit. *Expert*, **20**, 30-32 (2017)
16. S. Korzhumbaem, On the both sides of Khorgos. Rossiyskaya Gazeta, 115, **7281**, *International Radio of China*, 32-34 (2017)
17. K. Volkov, Quantum prospects: The Silk Road will be digital – not only economic. Rossiyskaya Gazeta, 115, **7281**, *International Radio of China*, 38-39 (2017)
18. The economy and foreign trade of China, *Trade Mission of the Russian Federation in the People's Republic of China*, 7-8 (2015)
19. E. Chernyh, Russia and China are getting ahead of America on the New Silk Road. Komsomolskaya Pravda 13.05.2017, 2, (2017)
20. Review of China's Foreign Trade in the First Half of 2017. *General Administration of Customs People's Republic of China* [online], Available at: http://english.customs.gov.cn/statics/36233f04-6390-45ad-b7b7-8d824031263c.html (2017)
21. H. Karahan, Economic relapse of history: The Silk Road, Yeni Safak, (2015)
22. P. Kober. Moscow, look, with whom we deal with. *Expert-Ural*, **23**, 8-11 (2017)
23. A. B. Galagan, Yu. A. Savinov, Evolution of China's economic development model. *Russian Foreign Economic Bulletin*, **6**, 32-47 (2016)
24. V. V. Ivanov, N. I. Bochkareva, Financing of Russia's projects within Silk Road Economic Belt. *Russian Foreign Economic Bulletin*, **8**, 38-48 (2016)
25. China's "Belt and Road" initiative poses no threat. People's Daily, (2017)
26. S. Pale, "New Silk Road": Pros and Cons for Russia, *New Eastern Outlook* [online], Available at: https://journal-neo.org/2015/07/10/new-silk-road-pros-and-cons-for-russia/ (2015)
27. A. V. Lukin et al., Chinese global project for Eurasia: formulation of the problem, Analytical report, 6, (2016)

Innovation as a source of country's global competitiveness growth

Beata Sofrankova[1,*], *Dana* Kiselakova[1], and *Veronika* Cabinova[1]

[1]University of Presov in Presov, Faculty of Management, Department of Finance, 080 01 Presov, Slovakia

Abstract. In the context of ensuring the long-term prosperity of each country, the more and more emphasis is placed on the mutual comparison of the nation´s competitiveness in the European or global area, so every country monitors and supports the development of its global acceptability. The aim of this paper was to analyze the development of Slovakia's competitiveness through the Global Competitiveness Index (GCI) during the period 2006-2016 and to reveal the interrelationships among the GCI indicator and its individual pillars. Of the total 12 pillar scores, a statistically significant correlation with the GCI of Slovakia was confirmed in the case of 7 pillars, whereas the strongest positive relationship was proved for 1st Pillar: Institutions. The level of Slovakia's spending on R&D belongs to one of the lowest within the EU countries. Over the years 2006-2015, the average R&D intensity fluctuated around 0.68%, whereas in 2015 it reached the value of 1.18%. The top leader in R&D expenditure per capita within the V4 countries is the Czech Republic followed by Hungary, Poland and Slovakia which in 2015 significantly increased its R&D spending to the level of 171.7 € per capita and ranked 2nd in the V4 countries.

Key words: innovation, competitiveness, global, performance

1 Introduction

Competitiveness and innovation performance represent two interrelated economic categories that reflect their synergistic effect in different areas of economic and social life [1]. Through innovation, products and services become competitive not only in the domestic but also in the global economy. Nowadays, technology and innovation activities become the main source of economic growth and competitiveness in the local as well as international business environment. This fact is becoming topical and discussed issue, especially in the developed EU countries.

International competitiveness represents one of the alternative performance economic indicators which allows monitoring of all important factors that affect not only economic performance but also many social aspects and social maturity of the country. However,

* Corresponding author: beata.sofrankova@unipo.sk

despite all the definitions, in literature there is no clearly defined attitude to the issue of competitiveness as well as its measurement and the way of expression [2].

1.1 Background in empirical studies and problems

A market economy is a competitive economy, thus competition is a crucial part of economic activity. A number of economic analyses and research experience show that the national economy development is undoubtedly dependent on the success of domestic producers and their skills to compete effectively in international markets [3].

An inevitable part of assessing the country's economy or economic groupings is to monitor their competitiveness and growth. In general, the most common opinion is that the competitiveness is mainly based on non-tax factors such as stability of the economic environment, enforcement of rights, development of infrastructure, educational level of the population, etc. [4].

The success and growth of each country in today's competitive environment depends not only on national policy, but also on broad public policy, the social qualities of the regions, the creativity and the talents of the country's citizens, and above all by supporting these regional factors [5].

Over the years assessment methodologies relating to global competitiveness have been adjusted to correspond with the current globalization trends in the world economy. Relevant information about several aspects of competitiveness from the perspective of attractiveness of the country for foreign investment and doing business provide a number of reputable international institutions – World Bank (WB), World Economic Forum (WEF), International Institute for Management Development (IMD), The Heritage Foundation (HF), Eurostat and many others. These institutions assess and present internationally comparable results called "competitiveness rankings". They represent a way of multi-criteria evaluation through various international aggregated indexes [6].

Many empirical studies are concerned with analysing the global competitiveness of world economies based on the Global Competitiveness Index (GCI). Most authors focus on identifying factors that affect the overall score of the GCI indicator, whether in positive or negative way, and formulate specific solutions to improve the current state of analysed countries. Author as Chudarkova [7] examined the development of national competitiveness of the Visegrad group countries, while the Czech Republic was identified (during the analysed years) as the most competitive economy. The analysis of changes and current position in competitive rankings of this informal group of Central European countries was also performed by Ruzekova, Kastakova and Zatko [8]. However, the results were compared with other multi-criteria indexes World Competitiveness Index (WCI) and Doing Business Index (DBI). To demonstrate differences between evaluations of national competitiveness by applying various approaches, author as Necadova [9] focused on comparing the evaluation of competitiveness based on results in the Global Competitiveness Report with traditional macroeconomic indicators. Confirmation of the positive relationship between the size of EU economies and their achieved level of competitiveness measured by the GCI indicator was the subject of study Necadova [9]. Author as Balcarova [10] studied whether the differences in competitiveness among individual EU economies by using the GCI and WCI indicator decrease or not. Based on results, the convergence was confirmed only in the case of using the GCI indicator.

In recent years, the role of innovation as a tool for increasing the global competitiveness has been intensively discussed in literature. Authors Streimikiene and Fil'a and Kucera [11, 12] in their studies stated that research and development is one of the main drivers of economic growth and creation of modern knowledge based economy that guarantees the efficient use of natural resources and the reduction of negative impact on environment. The analysis of relationship and its intensity between innovation and national competitiveness of EU countries tested through econometric analyses was provided by Iosif [13]. She focused on comparison of Innovation Union Scoreboard (IUS) dimensions and their corresponding indicators in relation to the GCI indicator and assessed the effect of innovation´s determinants on the international competitiveness of countries. Identification of the usage of information and communication technology (ICT) on international competitiveness of European countries was the subject of study Zoroja [14]. The findings revealed that ICT has the highest impact on the GCI indicator and especially on Efficiency enhancers´ sub-index. As reported by Tiruneh and Hekelova [15] the economies with an advanced level of higher education, training and higher level of innovation activities tend to reach a higher level of the GCI indicator compared to countries with lower levels of education and innovation performance.

One of the most significant factors influencing the realization and development of innovation activities is undoubtedly the amount of public investment released to develop country's innovative potential. Author as Necadova [16] examined changes, common and different features in R&D expenditure in the business sector within the Czech and Slovak Republic and stressed the increasing importance of these financial resources. Authors as Chudarkova [7] discuss about the initial assumptions for starting process of supporting innovation potential and performance in the Czech and Slovak Republic. This process depends on the quality and quantity of human resources and on the intensity of R&D spending, however, these necessary conditions are not a guarantee of the growth and development of the national and regional innovation capacity of these countries.

Authors as Chudarkova [7] studied the impact of government grants for innovation activities in countries that represent a similar level of economic development – Turkey and Poland. The results confirmed the positive effect of government grants on development in innovation performance of companies in both countries. The analysis of R&D expenditure impact and development in the Central and Eastern European countries was also provided by Szarowska [17]. She found out that higher government R&D expenditure by 1% will increase country´s economic growth by 2%. Authors as Czarnitzki [18] also emphasize the importance and positive impact of government support by analyzing the effects of a specific government-sponsored commercial R&D program from various angles. Authors as Aerts and Schmidt [19] analyzed if the public R&D grants crowd out private R&D investments in Flanders and Germany. The results of empirical study disproved the hypothesis what was also confirmed by the findings of [20, 21, 22, 23, 24] identified as the most significant factors influencing the innovation performance of countries (except for R&D expenditure) also the number of researchers, number of patents issued, connection of labor market and vocational training system, internationalization of R&D activities or the degree of appropriate cluster policy implementation.

In the paper we focused on revealing the mutual influence of the individual pillars of the GCI index in Slovakia and finding the strongest and weakest pillars influencing the growth of competitiveness of Slovakia. In the next part of this article, we wanted to find out the relationship between the level of R&D expenditure in the V4 countries (Slovak Republic, Czech Republic, Hungary and Poland) and the level of competitiveness this countries.

2 Data and methods

The aim of this paper was to analyze the development of Slovakia's competitiveness through the Global Competitiveness Index (GCI) during the period 2006-2016 and to reveal the interrelationships among the GCI indicator and its individual pillars. The paper deals with the analysis and comparison of the development of Slovakia's competitiveness based on the GCI indicator during the years 2006-2016 and revealing the significant relationship among the overall score of GCI indicator and its individual pillars. In further analyzes we focused on the development of R&D expenditure in Slovakia and EU countries (28) over the period 2006-2015 as well as their impact on individual pillars of GCI indicator.

In order to achieve stated aims of our study we formulated the following research questions:

RQ1: Is there a statistically significant relationship among the individual pillars of the Slovakia's GCI indicator?

RQ2: Is there a statistically significant relationship among the amount of V4 countries's R&D expenditure and the score of individual pillars within the overall GCI indicator?

The analyses were prepared on the basis of secondary data drawn from the annual Global Competitiveness Reports and GCI database completed by the World Economic Forum. In the analytical part of presented paper we compared the development of Slovakia´s position within the Visegrad group and European Union (28) based on the average scores of the GCI indicator. By correlation analysis we examined the interconnections among the scores of individual pillars and the overall GCI indicator in Slovakia but at first we focused on the development of average GCI pillar scores.

In order to process the secondary data we used the statistical programs STATISTICA and MS EXCEL. From the statistical tools there were the correlation matrix and Kendall´s test employed to determine the strength of interdependencies among the selected indicators, which

Based on secondary data gathered from the Global Competitiveness Reports for the years 2006-2016, we focused on the development of Slovakia´s GCI indicator compared with its average scores reached in the Visegrad group countries and European Union (28).

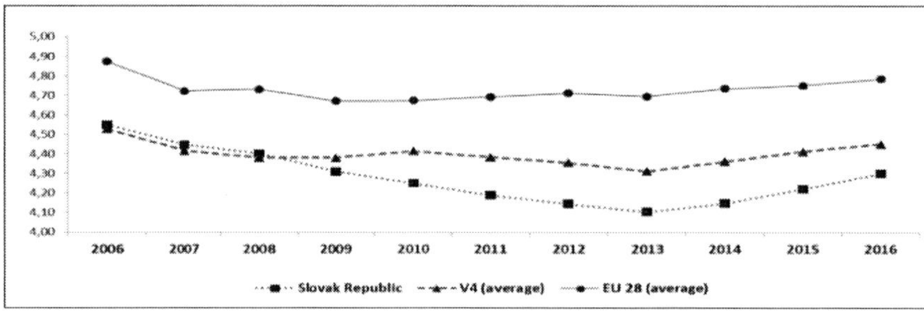

Fig. 1. GCI indicator development analysis of the Slovak Republic over the years 2006-2016

When analyzing the development of the Slovakia´s GCI indicator we can notice a negative trend from 2006 to 2013. During the years 2006-2008 Slovakia reached better results of GCI indicator than its average value for all V4 countries. The highest score of Slovakia´s GCI indicator (4.55) was achieved in 2006, vice versa, the lowest score was recorded in year 2013 (4.10) and it has increased till 2016 to the level of 4.22. A similar trend was also observed in case of V4 countries. In 2006, the highest GCI average score (V4) was recorded at the level of 4.53 and the lowest value (4.31) was achieved by Visegrad group countries in 2013. In the case of EU countries (28) was the development a bit different, because the worst average results were reached in years 2009 and 2010 (4.67). Despite the fact that since 2010 the GCI indicator (EU) has recorded an increasing trend, the maximum score was achieved in 2006 (4.87) and in 2016 it was only 4.79.

The next part of presented paper was aimed at the development analysis of average GCI pillar scores in Slovakia during the period 2006-2016. The graphic development of values is presented in the following figure.

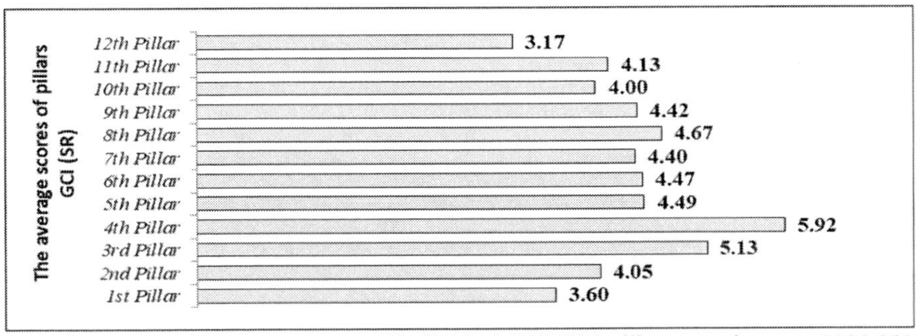

Fig. 2. Development of the average scores of individual GCI (SR) pillars over the years 2006-2016
Source: own processing by data of Global Competitiveness Reports 2006-2016

Based on results of the average scores of individual GCI (SR) pillars it can be concluded that the highest value on average (5.92) was reached within 4th Pillar: Health and primary education, vice versa, the lowest average score (3.17) was recorded in case of 12th Pillar: R&D Innovation. Insufficient results were achieved by Slovakia also for 1st Pillar: Institutions with its average score of 3.60.

3 Results and discussion

The following analysis is focused on revealing the interrelationships among the overall GCI score and its individual GCI pillars within the Slovak Republic. Significant results of the correlation analysis using the Kendall´s test are shown in the next table.

Table 1. Result of the correlation analysis among the GCI (SR) indicator and indexes of 12 pillars

Correlations among scores of Pillar of GCI (SK) Marked correlations are significant at p* < 0.05 (p < 0.01)**							
Kendal Tau	**1st**	**2nd**	**3rd**	**6th**	**8th**	**11th**	**12th**
GCI (SK)	0.7818**	-0.5138*	0.4630*	0.5872*	0.5636*	0.6853**	0.5138*

Source: own processing in STATISTICA

Out of the 12 scores of individual pillars, a statistically significant relationship with the GCI indicator of Slovakia was confirmed for 7 pillars. At the significance level $p < 0.05$ we recorded a significant relationship for 5 pillars, while at the significance level $p < 0.01$ was significant relationship detected in the case of 2 pillars. Statistically significant relationship was not proved for the following pillar: 4th Pillar: Health and primary education, 5th Pillar: Higher education and training, 7th Pillar: Labor market efficiency, 9th Pillar: Technological readiness and 10th Pillar: Market size.

The highest positive relationship was confirmed in the case of 1st Pillar: Institutions (τ = 0.7818) followed by 11th Pillar: Business sophistication (τ = 0.6853) and 6th Pillar: Goods market efficiency (τ = 0.5872). The negative relationship was detected only for 2nd Pillar: Infrastructure at the level of τ = - 0.5138. For 12th Pillar: R&D Innovation, the same strong dependency (τ = 0.5138) was revealed as for the 2nd Pillar: Infrastructure, but this dependency was confirmed as positive.

Based on previous analyzes it may be asserted that 1st Pillar: Institutions reached the lowest average score within the evaluation of 12 pillars forming the GCI indicator whereas for this pillar was also confirmed the highest relationship with the GCI indicator. The highest average score within 12 analyzed GCI pillars was reached for 4th Pillar: Health and primary education, but the statistically significant relationship between this pillar and the overall GCI score was not proved.

Considering the worst average values attained in the case of 12th Pillar: R&D Innovation we focused on a more detailed analysis of mentioned pillar and its graphic development presented in the following figure.

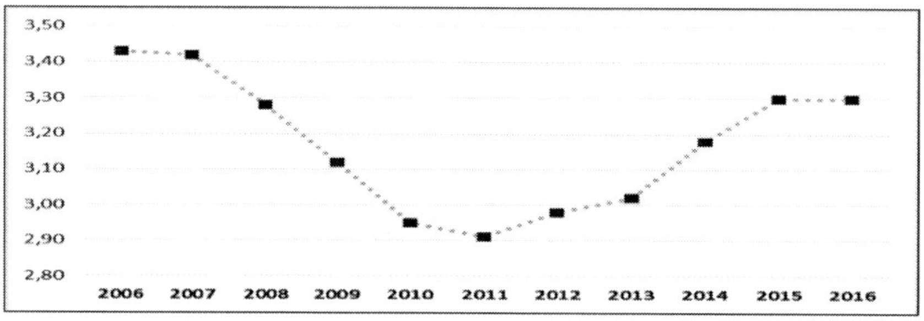

Fig. 3. Development of 12th Pillar: R&D Innovation over the years 2006-2016
Source: own processing by data of Global Competitiveness Reports 2006-2016

In terms of the development of 12th Pillar: R&D Innovation we can notice a decreasing trend in years 2006-2011, whereas the highest value of this pillar was noticed in 2006 at the level of 3.43 and lowest value (2.91) was recorded in 2011. Since 2011, the score of 12th Pillar started to increase gradually till 2015 when it reached the level of 3.30 and remained stable also in the following year 2016. The most significant decline (- 5.76%) was attained in 2010 and the highest increase (+ 5.03%) in 2014.

Following part of presented paper addresses the development of R&D expenditure within the Slovak Republic and European Union (28).

3.1 Development analysis of R&D expenditure in Slovakia

R&D expenditure as a percentage of GDP is used as an indicator of an economy's relative degree of investment in generating new knowledge. R&D intensity of Slovakia belongs to the lowest within the EU countries. The average R&D expenditure (% of GDP) moves on the level of 0.68% during the years 2006-2015. The development of mentioned indicator is shown in the following figure.

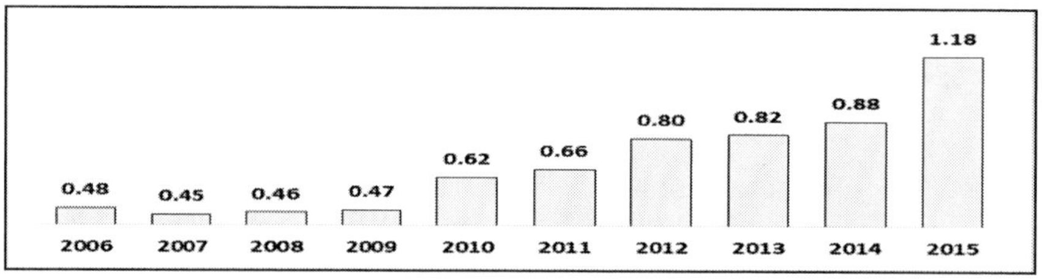

Fig. 4. Development of R&D expenditures (% of GDP) in Slovakia over the years 2006-2015
Source: own processing by data of Slovak Office of the Slovak Republic

When assessing the development of the share of R&D expenditure in Slovakia's GDP, we can notice a gradual increase, whereas the analyzed share fluctuated around 0.5% from 2006 to 2009, in 2010 and 2011 it increased to the level around 0.6% and in the following period it was almost around 0.8%. In 2015, the share of R&D expenditure increased by more than 146% compared to the 2006.

DOI: 10.1051/shsconf/20173901026

Although there has been a slight increase of R&D expenditure (% of GDP) over the last period, the Slovak Republic is not able to achieve such results as most European countries. However, the level of innovative activity of countries cannot be judged solely in terms of R&D expenditure. It is important to set up their appropriate amount, structure and interconnection with qualified human resources, innovation strategy and business environment.

Research and development activities can be funded from public resources, private resources, resources from abroad or other resources. It is really interesting to analyze the R&D expenditure structure in terms of public financial resources and the private ones (see Figure 5).

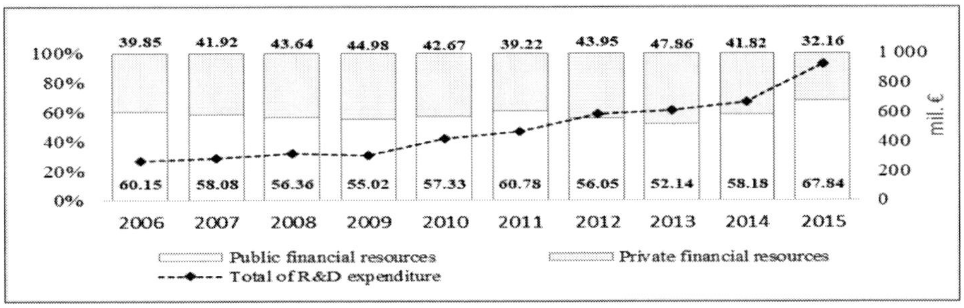

Fig. 5. Structure of R&D expenditure in the Slovak Republic over the years 2006-2015
Source: own processing by data of Slovak Office of the Slovak Republic

Since 2006, the amount of Slovakia's R&D expenditure has increased year by year. This positive trend started in 2006 with the total R&D expenditure of 267.65 million € and till 2015 these financial resources reached its highest year-on-year increase and also the highest value of 927.27 million €. The highest year-on-year decrease (- 13.47 million €) was recorded only in the year 2009. Looking at the structure of R&D expenditure and its development over the whole analyzed period, it is obvious that public financial resources represented higher share of total resources than the private ones. The average share of public financial resources reached the value of 58.2% and the remaining share of 41.8% belonged to private resources. The highest share of public financial resources (67.84%) was recorded in 2015, vice versa, the lowest share (52.14%) in 2013 when private resources achieved their maximum values.

In the next part of analysis we dealt with the development comparison of R&D expenditure per capita during the years 2006-2015 within the Slovak Republic, V4 countries (Slovak Republic, Czech Republic, Hungary, Poland) and European Union countries (28).

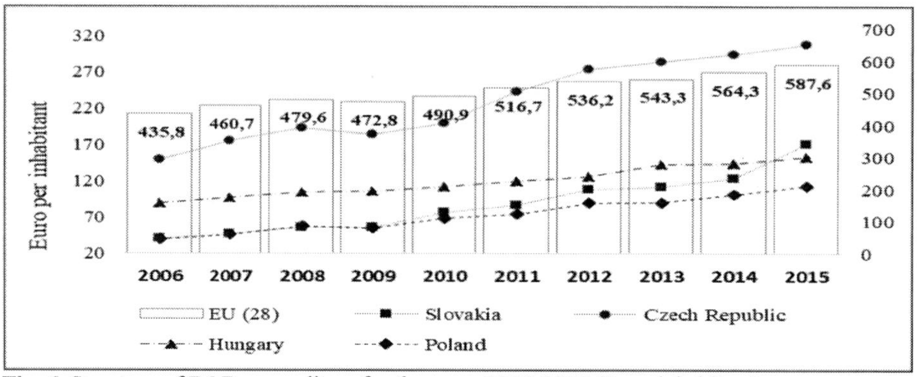

Fig. 6. Structure of R&D expenditure for the years 2006-2015 (€ per inhabitant)

EU countries (28) invested 435.8 € per capita in R&D activities in 2006. However, gradually ascending trend has led to reaching the level of 587.6 € per capita in 2015 that represented an increase of 151.8 € per capita in absolute terms (+ 34.8%). Over the analyzed years 2006-2015, the Czech Republic has a dominant position among the Visegrad group countries. It allocated on average 230.78 € per capita in R&D activities, Hungary ranked 2nd (119.87 € per capita), the Slovak Republic 3rd (88.01 € per capita) and in the rankings of average R&D financing Poland recorded the lowest values (73.72 € per capita). Looking at the growing trend in all V4 countries we can notice significantly different and more positive development of R&D expenditure per capita in the Czech Republic. A slightly different positive trend of mentioned financial resources is detected in the case of Slovakia which in 2006-2009 achieved values almost the same as Poland. In 2015, R&D spending in Slovakia reached 171.7 € per capita representing an increase of 38.4% compared to 2014. As a result, Slovakia exchanged its position with Hungary and moved up 1 place to reach 2nd position within the V4 countries.

In the following part of the paper, we focused on examining the existence of a statistically significant relationship among the amount of R&D expenditure and individual pillars of the GCI indicator within the V4 countries.

Table 2. Result of the correlation analysis among the R&D expenditure and indexes of 12 pillars

Kendall Tau correlations among scores Pillar of GCI (SK, CZ, PL and HU) and R&D Expenditure (SK, CZ, PL and HU). Number of observations = 10 for each countries Marked correlations are significant at p* < 0.05												
Pillars	**1st**	**2nd**	**3rd**	**4th**	**5th**	**6th**	**7th**	**8th**	**9th**	**10th**	**11th**	**12th**
R&D E (SK)	**-0,78***	**0,63***	-0,20	0,09	-0,39	**-0,58***	**-0,82***	**-0,82***	0,42	0,16	-0,45	-0,20
R&D E(CZ)	-0,24	0,24	0,09	0,11	0,11	**-0,52***	-0,45	-0,22	**0,78***	0,11	-0,45	**-0,64***
R&D E (HU)	**-0,90***	**0,72***	**0,56***	-0,16	-0,02	0,11	**-0,56***	**-0,94***	**0,54***	-0,23	**-0,82***	-0,39
R&D E (PL)	0,33	**0,73***	0,00	0,18	**0,60***	**0,67***	-0,43	0,09	**0,85***	**0,80***	0,07	-0,11

Source: own processing in STATISTICA

Based on the results of correlation analysis we may assert that the positive relationship among the Poland's R&D expenditure per capita and individual pillars of the GCI indicator was confirmed in the case of 5 pillars (2nd, 5th, 6th, 9th and 10th Pillar). For the Czech Republic, which is the leader of V4 countries, a statistically significant relationship was noticed for 3 pillars (6th, 9th and 12th Pillar), however, in case of 2 pillars there was a negative dependency revealed. The Slovak Republic demonstrated a statistically significant positive relationship only for 2nd Pillar and the negative dependency was recorded in the case of 1st, 6th, 7th and 8th Pillar. According to the revealed results of analyzing the interdependence among the R&D expenditure and individual pillars of the GCI indicator, a deeper and more detailed analysis as well as a larger range of input data is needed. This topic will be the subject of our further research studies.

4 Conclusions

Every year many international organizations, international research institutes or foundations compile multi-criteria evaluations of competitiveness which lead to the compilation of worldwide rankings where countries defend not only their overall position, but also position in selected areas. Those organisations with other multi-criteria evaluation systems of competitiveness provide countries a certain benchmark view on selected aspects and their position in comparison with other countries and analyse macroeconomic performance of the country. Each country follows its range of important economic aspects and features that compares with other states at the global level.

This paper presents the development analysis of Slovakia's competitiveness based on the GCI indicator over the years 2006-2016 as well as the detection of relationships among the individual pillars of the GCI indicator and its overall score. The results revealed that the development of Slovakia's GCI indicator had a decreasing trend till 2013 and in the following years a gradual increase was recorded. In 2016, Slovakia reached the overall GCI score at the value of 4.30. According to the results of average scores we can state that the highest value on average (5.92) was achieved within 4th Pillar: Health and primary education, vice versa, the lowest average score (3.17) was recorded for 12th Pillar: R&D Innovation. Within the 12 pillars forming the structure of the GCI indicator, the statistically significant relationship with the GCI of Slovakia was revealed in the case of 7 pillars, whereas the strongest positive relationship was confirmed for 1st Pillar: Institutions. The negative dependency (τ = - 0.5138) was detected only for 2nd Pillar: Infrastructure.

In next analyses we also focused on the development of R&D expenditure in V4 countries during the period 2006-2015 and their influence on individual GCI pillars.

While assessing the development of R&D expenditure (% of GDP of Slovakia) over the years 2006-2015, the gradual increase was recorded. The analyzed share fluctuated around 0.5% in 2006, in the following years it was almost around 0.8% and in 2015 the share of R&D expenditure increased by more than 146% compared to the first analyzed year 2006. The amount of Slovakia's R&D expenditure recorded a positive increasing trend which started in 2006 at the level of 267.65 million € and till 2015 these financial resources reached the highest value of 927.27 million €. Most R&D expenditure is secured by public financial resources (approximately 60%) and the remaining share is made up of private ones (about 35-40%). A substantial part of public resources financing R&D activities secure only basic research with no linking to the economic performance of the country. Within the V4 countries, the Czech Republic was identified as a leader in R&D expenditure per capita, followed by Hungary, Poland and Slovakia with one exception in 2015 when R&D spending in Slovakia attained 171.7 € per capita and reached 2nd position.

In conclusion we can say that we found statistically significant relationship among the 7 individual pillars of the Slovakia's GCI indicator. The statistically significant relationship among the amount of V4 countries' R&D expenditure and the score of individual pillars within the overall GCI indicator were confirmed for each V4 countries (Slovak Republic and Poland - 5 pillars, Hungary – 7 pillars, Czech Republic – 3 pillars)

References

1. E. Ivanova, M. Kordos, Competitiveness and innovation performance of regions in Slovak Republic. *Marketing and Management of Innovations*, **8**(1),145-158, (2017)
2. E. Ivanova, M. Kordos, J. Habanik, Microenterprises' entrepreneurs' attitudes to managing financial risks. *Actual Problems in Economics*, **5**(167),39-49, (2015)
3. I. Kravcakova – Vozarova, R. Kotulic, E. Sira, V4 countries' agricultural sector evaluation in terms of competitive advantage. *Economic Annals-XXI*, 5-6, 60-63 (2015)
4. A. Schultzova, *New Challenges in Public Finances 2016. Collection of Scientific Writings on Tax Policy*. Bratislava: Ekonóm EU Bratislava (2016)
5. E. Lapinova, M. Varga, B. Sarkanova, *Reprodukce lidského kapitálu* (RELIK 2016): vzájemné vazby a souvislosti. *Mezinárodní vědecká konference*. Praha: Vysoká škola ekonomická v Praze, 325-337, (2016)
6. P. Ochotnicky, B. Lajzova, D. Kiselakova, Analysis of macroeconomic factors for the establishment of industrial parks and their effects on regional development: empirical study from Slovakia. *Asian Economic and Financial Review*, **4**(9), 1220-1236 (2011)
7. S. Chudarkova. In: *10th International Scientific Conference on Economic Policy in the European Union Member Countries*, 94-104, (2013)
8. V. Ruzekova, V. Kastakova, E. Zatko, Comparison of the V4 Economies According to International Competitiveness Indices and the Basic Pillars of the Knowledge Economy. *3rd International Conference on European Integration (ICEI)*, 838-848 (2016)
9. M. Necadova, Concepts of Firm and National Competitiveness and Changes in Competitiveness of Visegrad Group Countries. *9th International Days of Statistics and Economics*, 1169-1179 (2015)
10. P. Balcarova, Competitiveness of the European Union Countries: Converging or Diverging Tendency? *18th International Conference on Enterprise and the Competitive* Environment, 34-45 (2015)
11. D. Streimikiene, Business Infrastructure for Sustainability in Developing Economies. *Amfiteatru Economic*, **16**(37), 965-979 (2014)
12. M. Fiľa, J. Kucera, Innovation management and corporate sustainability 2015. *3rd International Conference on Innovation Management and Corporate Sustainability* (IMACS), 39-51 (2015)
13. A. E. Iosif, Management challenges for Sustainable Development. *8th International Management Conference: Management challenges for Sustainable Development*, 670-679 (2014)
14. J. Zoroja, Fostering Competitiveness in European Countries with ICT: GCI Agenda. *International Journal of Engineering Business Management*, **7**(18), (2015)
15. M. W. Tiruneh, E. Hekelova, Central Bohemia University International Conference 2016. *CBU International Conference on Innovations in Science and Education (CBUIC)*, 132-141 (2016)
16. M. Necadova. In: *12th International Conference on Hradec Economic Days (HED): Economic Development and Management of Regions*, 324-330, (2014)
17. I. Szarowska, European Financial Systems 2017. *14th International Scientific Conference on Economic Policy in the European Union Member Countries*, 687-698 (2016)

18. D. Czarnitzki, C. Lopez-Bento, Value for money? New microeconometric evidence on public R&D grants in Flanders. *Research Policy*, **42**(1),76-89 (2013)
19. K. Aerts, T. Schmidt, T, Two for the price of one?: Additionally effects of R&D subsidies: A comparison between Flanders and Germany. *Research Policy*, **37**(5), 806-822 (2008)
20. X. Gonzalez, C. Pazo, Do public subsidies stimulate private R&D spending? *Research Policy*, **37**(3), 371-389 (2008)
21. B. Svagzdiene, J. Kuklyte, The Analysis of Factors which have Impact for Summary Innovation Index in Germany, Estonia and Lithuania. *Transformations in Business & Economics*, **15**(2), 784-799 (2016)
22. R. Wiedenhofer, C. Friedl, L. Billy, D. Olejarova, Application of IC-models in a combined public-private sector setting for regional innovation in Slovakia. *Journal of Intelectual Capital*, **18**(3), 588-606 (2017)
23. J. Rahko, Internationalization of corporate R&D activities and innovation performance. *Industrial and Corporate Change*, **25**(6), 1019-1038 (2016)
24. M. Balog, Development Factors of Cluster Organizations in the Slovak Republic. *Journal of Economics*, **64**(2),148-168 (2016)

Business process projection in relation to the internationalization of the external environment of industrial companies

Jan Váchal[1,*], *Petra* Pártlová[1] and *Jarmila* Straková[1]

[1]The Institute of Technology and Business, Okružní 517/10, 37001 České Budějovice, Czech Republic

Abstract. Highly intensive internalization of the world economy requires the business sector to have different view of strategic management and decision making, together with the quality and content of business processes. Changes in the environment of enterprises significantly influence their function and behaviour, causing a need to significantly change enterprise architecture, including the design of development strategies. The paper focuses on analysing the processes of industrial enterprises, dealing with their importance and profit, classified by size. The result revealed that process activities, such as input logistics and marketing, are dominant in micro-enterprises, and such activities are important in profitable enterprises. In middle-sized enterprises, there is a more significant difference between marketing/sales and services without any relations to profit. However, there is a requirement to intensify these activities in non-profitable enterprises. Large enterprises revealed a significant influence of the scientific and technological development and also, to a lesser extent, of marketing, sales and purchase. Non-profitable enterprises should also focus on improving the quality of such processes.
Key words: internationalization, enterprises, business process, business environment

1 Introduction

Internationalizing the world economy in the age of its regionalism and realignment exerts intense pressures and actions which have a profound impact on business environments and their conduct. The business environment, in the manufacturing and industry sector, in particular, is subject to constant changes. These systematic changes require continual and recurrent theoretical examination of the corporate growth, carried out by many authors. Dagnino, King and Tienari [1] defined the human capital, strategy, management of human resources, innovation and skills as five main factors of the significant growth. This process is supported by the fact that companies work in conditions of discontinual and turbulent environment [2, 3], i.e. unpredictable, volatile and disconnected environment [4]. Thus, the strategic management constitutes a central organ of the organization management within the dynamic and competition environment. The strategic management consists of three

* Corresponding author: vachal@mail.vstecb.cz

processes, such as strategic planning, strategy implementation and strategic control [5]. The most evident trends influencing business environment, processing and implementation of business strategy are suggested in literature [3, 6-12]. As a matter of fact, a good and flexible decision-making, speeding up information flows and effective use of modern information technologies are the key requirements for a successful modern company management [13]. In addition, activities and strategies of international companies (multinational enterprises, MNE) are more global than a perspective of the regional strategy shows [14, 15]. Hence, a further research should suggest a detailed comparison of MNE approach towards the strategy on different levels (National, regional and global), based on the value chain. Works which claim that a business strategy does not depend on the ability to predict changes [16, 17] present a development milestone in projecting and implementing business strategies in conditions of the new economy. Moreover, a wide vision of potential customers and their needs within following three to five years is essential. The literature also notes that a thorough analysis is of a crucial importance. Therefore, there is not a sound reason to draw up a business strategy in advance. It may last a couple of years to find a really good strategy. Porter's model of five powers constitutes one of the basic pillars of the functional strategy, i.e. an essential characteristic of five competitive powers, their mutual configuration and a value chain analysis, including branches and value chain adapted to the requirements of a particular enterprise [18]. One of the premises of successful managerial strategic decision-making is a strategic situational analysis [19], which means the evaluation of factors of business environment, i.e. changes in customers' requirements, competitors, suppliers and development of macroeconomic factors including characteristics of internal enterprise resources. Similar definitions of situational analysis with only small differences in the descriptions of micro environment, mid-environment and macro-environment are suggested by a number of authors [20-29]. A strategic situational analysis means a complex approach to capture important factors influencing business activities in their mutual relations. Sedláčková [30] emphasizes the important role of the SA in strategic planning and decision-making at the top level of management. Furthermore, she points out individual strategic methods of the business environment and its internal resources, which play an important role in the business potential. Dedouchová [31, 32] in her works focuses on strategic analyses, shows relations between a strategy, organizational structure and business management system and deals with the conflicts of interests within their production and implementation. Companies often present themselves as socially responsible. Actions proving their social engagement are not confined only to the society in which companies exist, but they also achieve a huge impact on a wider environment [33, 34]. The social responsibility of companies (corporate social responsibility, CSR) is often associated with a managers' effort to foster and maintain good relationships with the interested parties (employees, suppliers, bankers etc.). From this point of view, the social responsibility constitutes a very important component of the business strategy. The SA is also defined as a means for projecting a business strategy based on realizing its inner potential, possibilities and limits of a further development upon considering the impact of mid-environment and external environment [13, 14]. Studying the available literature, we may accept the assumption that using analytical methods is highly effective in the frame of strategic business management, or more precisely, formulating a viable strategy; however, their structure and content should be considered since some of them do not have a universal character, and therefore they cannot be applied in a unified way. Follow-up analyses should be used as a support means for creating a company-wide strategy.

2 Methods

The research was based on a wide-range questionnaire survey of the business sector in the Czech Republic. The survey is still carried out. The results as presented in the paper include a sample of 176 enterprises, 70% of which are profitable. Of the total, there were 74 micro-enterprises, 70 small enterprises and 32 medium-sized enterprises. The size structure of the sample corresponds to the SME structure in the Czech Republic. In order to analyse the research questions, such enterprises that did not answer or the answers were unreliable due to a link to follow-up question were left out of the sample. The statistics used Fisher´s exact test and a two-sample test. The two-sample test revealed the p-value and two-sided 95% interval estimate of ratio differences. For Fisher´s test, p-value is reported as it is the test statistics itself. Both tests were carried out at the significance level of 0.05. The research analysed questions related to testing the significance of business processes in terms of their significance for profitability classified by the size of the enterprises (micro-enterprises and small enterprises, medium-sized and large enterprises.

3 Results and discussion

3.1 Micro enterprises and small enterprises

In regard to a small number of enterprises in the micro-enterprise group, micro-enterprises and small enterprises were grouped together.

Table 1. Processes in micro-enterprises and small enterprises and their impact on profitability (in %)

Process	**non-profit**	**profit**	**Difference profit – non-profit**
Input logistics	13.6	37.7	24.1
Production	59.1	81.1	22.0
Output logistics	13.6	32.1	18.4
Marketing and sale	22.7	45.3	22.6
Basic and accompanying services	18.2	32.1	13.9
Purchase	31.8	32.1	0.3
Scientific-technological development	0.0	5.7	5.7
Human resources management	18.2	20.8	2.6
Enterprise infrastructure	4.5	15.1	10.5

Tested differences bigger than 10% are marked in bold. With the view of a large number of aggregates of profit and non-profit enterprises, results of the two-sample test on ratio differences are shown. Significant differences are marked in bold.

DOI: 10.1051/shsconf/20173901027

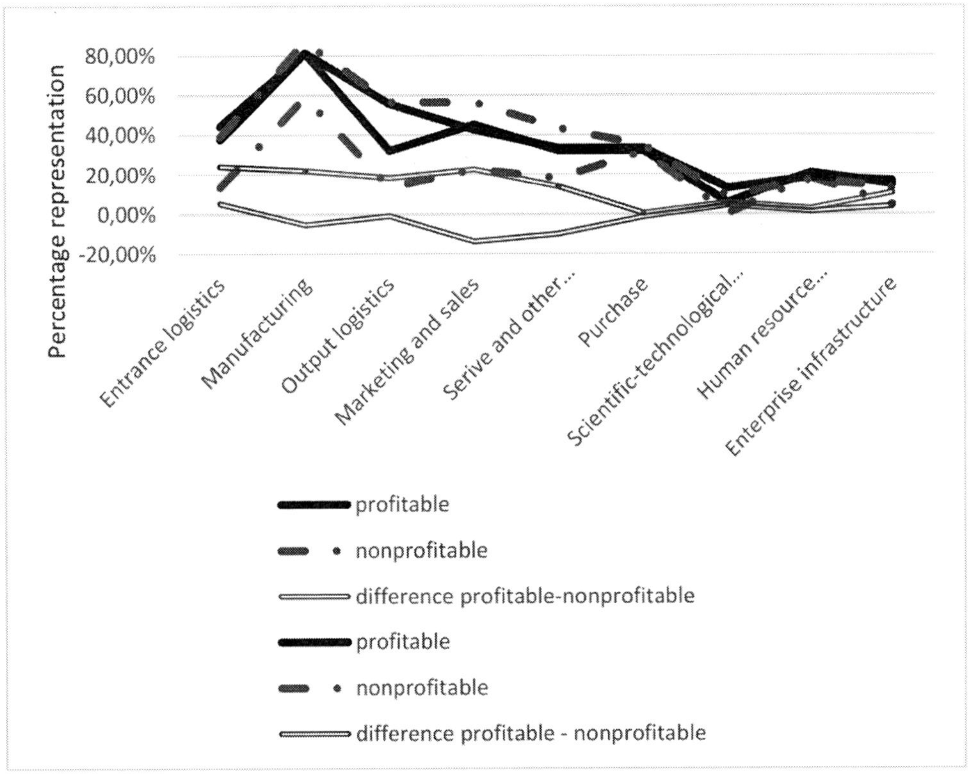

Figure 1. Processes in micro-enterprises and small enterprises and their impact on profitability

Table 2. Testing individual processes in regard to their significance and impact on profitability

Process	Testing statistics	P-value	Interval estimate of the ratio difference
Input logistics	2.436	0.0148	(0.047; 0.435)
Manufacturing process	1.871	0.061	(-0.010; 0.451)
Output logistics	1.895	0.058	(-0.006; 0.375)
Marketing and sale	2.005	0.045	(0.005; 0.446)
Basic and accompanying services	1.332	0.183	(-0.065; 0.343)
Enterprise infrastructure	1.592	0.111	(-0.024; 0.235)

The results shown in the table indicate significant differences on the confidence level of 95% only in processes in the area of input logistics and marketing and sale. In this case, the ratio of profit enterprises, which focus on the process, is significantly higher than the non-profit ones. The test did not reveal any statistically significant ratio differences in other areas. P-value is only little higher than 0.05 level and interval estimates, of which the low limit is negative but very close to zero, indicating a very close result, which means that choosing only a little lower confidence level would show a difference. Therefore, profit enterprises show higher ration than the non-profitable.

3.2 Middle-sized enterprises

Table 3. Processes in middle-sized enterprises and their impact on profitability (in %)

Process	non-profit	profit	Difference profit - non-profit
Input logistics	39.1	44.4	5.3
Production	87.0	81.5	-5.5
Output logistics	56.5	55.6	-1.0
Marketing and sale	56.5	42.6	**-13.9**
Basic and accompanying services	43.5	33.3	**-10.1**
Purchase	34.8	33.3	-1.4
Scientific-technological development	8.7	13.0	4.3
Human resources management	17.4	18.5	1.1
Enterprise infrastructure	13.0	16.7	3.6

Tested differences bigger than 10 % are marked in bold. With the view of a large aggregate number of profit and non-profit enterprises, results of the two-sample test on ratio differences are shown.

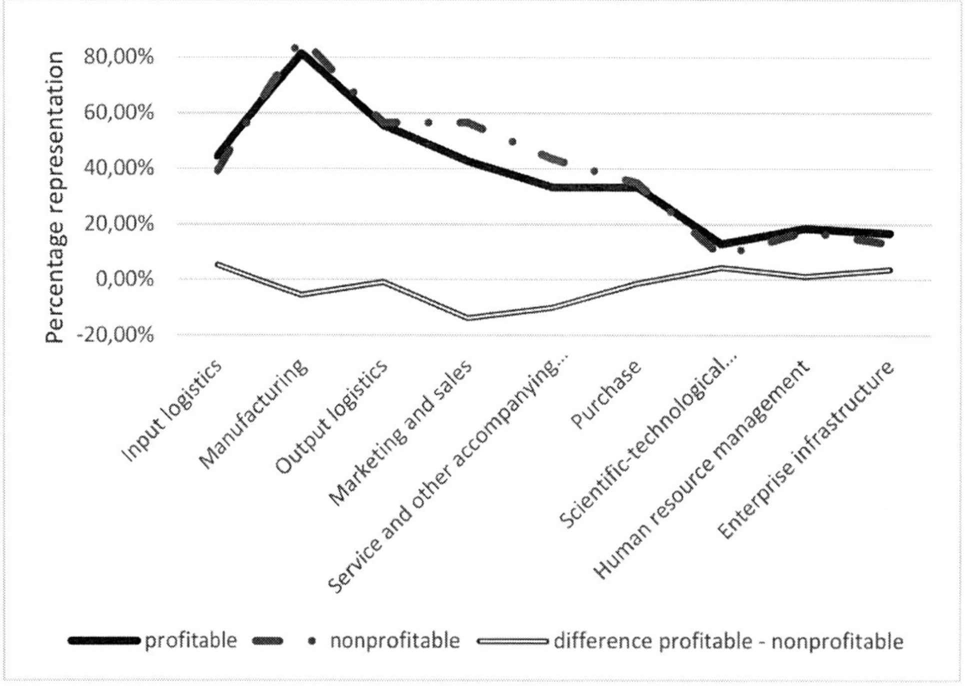

Figure 2. Processes in middle-sized enterprises and their impact on profitability

DOI: 10.1051/shsconf/20173901027

Table 4. Testing of individual processes in regard to their significance and impact on profitability

Process	Testing statistics	P-value	Interval estimate of the ratio difference
Marketing and sale	-1.129	0.259	(-0.381; 0.102)
Basic and accompanying services	-0.834	0.404	(-0.340; 0.137)

A significant difference in the result of the group of medium-sized enterprises was indicated only in two cases, i.e. "Marketing and sale" and "Basic and accompanying services". Although the statistic test did not show any significant difference, even very convincingly (p-values are multiplied higher than 0.05 level and interval estimates show zero, while extreme limits are far from zero), the opposite difference from the small enterprises is interesting, i.e. non-profit enterprises have a higher ratio of a central focus on these processes than the profitable ones (therefore, there are the negative ratio differences). It can, therefore, be assumed that non-profit enterprises feel the need to improve and enhance the activity of the aforementioned activity.

3.3 Large enterprises

Table 5. Processes in large enterprises and their impact on profitability (in %)

Process	profit	non-profit	Difference profit - non-profit
Input logistics	52.6	36.4	16.3
Production	86.8	81.8	5.0
Output logistics	55.3	45.5	9.8
Marketing and sale	57.9	27.3	30.6
Basic and accompanying services	31.6	9.1	22.5
Purchase	39.5	54.5	-15.1
Scientific-technological development	34.2	0.0	34.2
Human resources management	39.5	27.3	12.2
Enterprise infrastructure	18.4	27.3	-8.9

Tested differences bigger than 10% are highlighted in the table. In regard to the extent of the aggregate of non-profit enterprises, p-value of Fisher's exact test is suggested as the main result. The results of the two-sample test on the ratio difference and the interval estimate for this difference are given for guidance only. Statistically, significant differences are highlighted in the following tables.

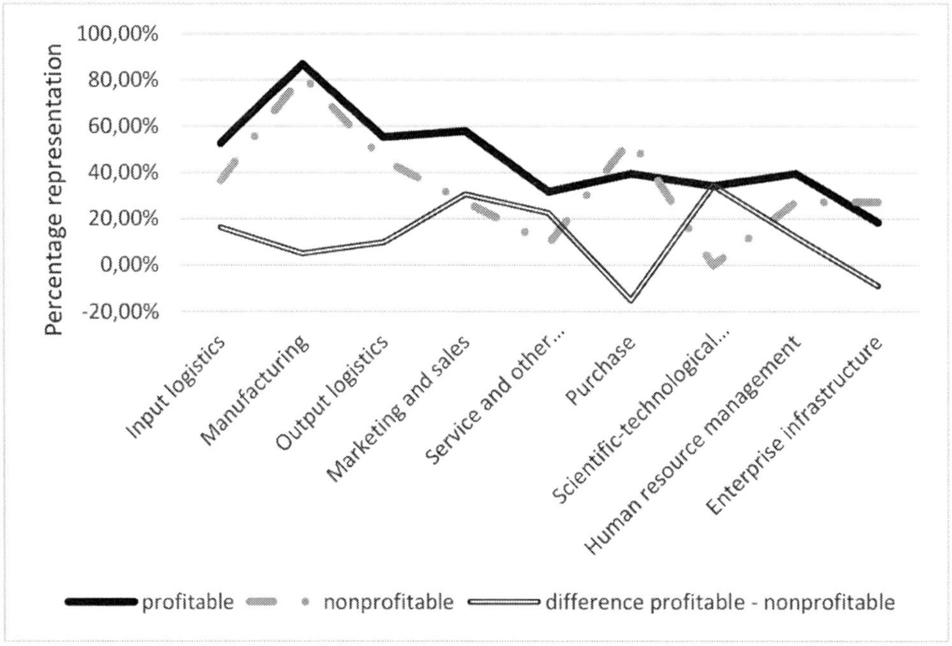

Figure 3. Processes in large enterprises and their impact on profitability

Table 6. Testing of individual processes in regard to their significance and impact on profitability

Process	P-value of Fisher's exact test	Two-sample test on ratio differences		
		Testing statistics	P-value	Interval estimate of the ratio difference
Input logistics	0.496	0.979	0.327	(-0.163; 0.488)
Marketing and sale	0.095	1.959	0.0502	(-0.00023; 0.613)
Basic and accompanying services	0.246	1.957	0.0503	(-0.00029; 0.450)
Purchase	0.493	-0.887	0.375	(-0.483; 0.182)
Scientific-technological development	**0.024**	Double-choice test on ratio differences was not carried out as a result of zero frequency of non-profit enterprises in the group.		
Human resources management	0.724	0.782	0.433	(-0.184; 0.428)

In regard to the small number of non-profit enterprises, we cannot rely on the results of the two-sample test for ratio differences, it is necessary to take Fisher's Exact Test as a conclusion. The test in all areas showed a statistically significant difference between profit and non-profit enterprises, with a difference of more than 10%, except in the field of scientific and technological development. There was a decisive influence in this area of the zero frequency in the group of non-profit enterprises. The marketing and sales situation was closest to the statistical significance.

Although the questionnaire finds a difference in ratio over 30%, the test of significance was rejected. The tightness of the conclusion, however, indicates that this area is worth the attention as well as the area of purchase. Although the test did not prove the statistical significance of the difference, it shows that the ratio in non-profit enterprises is greater than in profitable enterprises.

4 Conclusion

The representation of processes in the business system on the general level depicts system framework for procedural business management; on the other hand, it introduces the corporate philosophy of the senior management. It is clear that senior management must take into account both the current competitive ring in the relevant sector and in the relevant size category and the prediction of the future development, regarding priorities and intensity.

The paper focused on these particular questions and outputs from the research can be summarized as follows.

Process significance in the area of "Input logistics" and "Marketing and sales" was analysed in micro-enterprises and small enterprises. The enterprises which prioritize these processes proved to be highly profitable.

The aggregate of middle-sized enterprises indicated a process significance only in two processes, i.e. "Marketing and sales" and "Basic and accompanying services. The fact that non-profit enterprises achieved a higher ratio of focusing on these processes than profitable is also interesting. It might be assumed that non-profit enterprises feel the need to improve and enhance the suggested activities, and profit enterprises consider these processes as sufficient and stable.

Large enterprises showed a dominant relation in "Scientific-technological development", which is in compliance with the expectation and commonly declared principles in business and management. The managers should also pay attention to the area of "Marketing and sales" and "Purchase". However even in this case, the same rules as in middle-sized enterprises are applied, i.e. business processes need improving.

Researchers consider these results only as the first entry to the system analysis of business processes. Apart from practical outcomes, valuable methodological experience has been gained; mainly in a view of the approach towards such burning and important issue as a projection of the process enterprise system.

References

1. G. B Dagnino, D. R. King, J. Tienari, The Strategic Management of Dynamic Growth. *Long Range Planning*, **50**, 427-430 (2017)
2. P. F. Drucker, *Driving in turbulent times [Řízení v turbulentní době]*. Prague: Management Press (1994)
3. P. F. Drucker, *Management challenges for the 21st century*. New York: HarperBusiness (1999)
4. J. Truneček, *Knowledge Management [Management znalostí]*. Prague: C. H. Beck (2004)
5. Y. Salamzadeh, M. Y. Nia, M. R. Markovič, A. Salamzadeh, Strategic management development: the role of the learning) school in the promotion of the competence of managers. *Economy and Society*, **21**, 1-23 (2016)
6. P. F. Drucker, *Management: the future begins today [Management: budoucnost začíná dnes]*. Prague: Management Press (1992)
7. P. F. Drucker, *The Daily Drucker: 366 Days of Insight and Motivation for Getting The Right Things Done [Drucker na každý den: 366 zamyšlení a podnětů, jak dělat správné věci]*. Prague: Management Press (2006)

8. P. Kotler, *Kotler on Marketing: How to Create and Control New Markets [Marketing podle Kotlera: jak vytvářet a ovládnout nové trhy]*. Prague: Management Press (2000)
9. Z. Pitra, *Improving business performance [Zvyšování podnikatelské výkonnosti firmy]*. Prague: Ekopress (2001)
10. J. Jirásek, Strategy: *The Art of Business Victory [Strategie: umění podnikatelských vítězství]*, 2.nd ed. Prague: Professional Publishing (2003)
11. F. Šmída, *Strategies in Business Practice [Strategie v podnikové praxi]*. Prague: Professional Publishing (2003)
12. Z. Souček, *The 21st century company: we will overtake the best [Firma 21. století: předstihneme nejlepší]*. Prague: Professional publishing (2005)
13. J. Straková, J. Váchal, P. Pártlová, *Strategic Management: Study Script [Strategický management: studijní skripta]*. České Budějovice: The Institute of Technology and Business in České Budějovice [České Budějovice: Vysoká škola technická a ekonomická v Českých Budějovicích] (2017)
14. T. Blumentritt, W. Danis, Business strategy types and innovative practices. *Journal of Management,* **18**, 274-291 (2006)
15. R. Mudambi, J. Puck, A Global Value Chain Analysis of the 'Regional Strategy' Perspective. *Journal of Management Studies,* **53**, 1076-1093 (2016)
16. M. E. Porter, Recent Developments in Competitiveness and Strategy. *Porter Prize Conference, Hitotsubashi University. Graduate School of International Corporate Strategy*, Tokyo, Japan (2012)
17. M. E. Porter, *Shared Value and Strategy*. Shared Value Leadership Summit, New York (2015)
18. J. Magretta, *Michael Porter clearly and comprehensibly [Michael Porter jasně a srozumitelně]*. Prague: Management Press (2012)
19. M. Keřkovský, O. Vykypěl, *Strategic Management: Theory for Practice [Strategické řízení: teorie pro praxi]*, 2.nd ed. Prague: C. H. Beck (2006)
20. J. Jakubíková, *Strategic management [Strategický management]*. Prague: Grada (2008)
21. M. Dedouchová, *Business strategy [Strategie podniku]*. Prague: C. H. Beck (2001)
22. T. Mallya, *Fundamentals of Strategic Management and Decision Making [Základy strategického řízení a rozhodování]*. Prague: Grada (2007)
23. M. Šulák, E. Vacík, *Strategic management in enterprises and projects [Strategické řízení v podnicích a projektech]*. Prague: The university of finance and administration [Praha: Vysoká škola finanční a správní] (2005)
24. J.G. Frynas, K. Mellahi, *Global strategic management*, 2nd ed. New York: Oxford University Press (2011)
25. E. Kislingerová a kol., *Business behavior in a globalizing environment [Chování podniku v globalizujícím se prostředí]*. Prague: C. H. Beck (2005)
26. J. Veber, *Environmental management [Environmentální management]*. Prague: Oeconomica (2004)
27. E. Gomes, *Mergers, acquisitions, and strategic alliances: understanding the process*. New York: Palgrave Macmillan (2011)
28. E. Nigel, D. Campbell, *Strategic management for travel and tourism*. Repr. Oxford: Butterworth-Heinemann (2002)

29. M. E. Porter, M. R. Kramer, Strategy and Society: The Link between Competitive Advantage and Corporate Social Responsibility. *Harvard Business Review,* **84** (2006)
30. H. Sedláčková, *Strategic analysis [Strategická analýza]*. Prague: C. H. Beck (2000)
31. M. Dedouchová, *Business strategy: teaching texts for engineering studies [Strategie podniku: učební texty pro inženýrské studium*]. Prague: College of Economics [Praha: Vysoká škola ekonomická] (1995)
32. M. Dedouchová, *PE412 - Enterprise Strategy [PE 412 - Strategie podniku]*. Prague: C. H. Beck (2002)
33. J. Cox, Corporate social responsibility and social capital. *Journal of Banking and Finance,* **60**, 252-270 (2015)
34. A. B. Carroll, K. M. Shabana, The Business Case for Corporate Social Responsibility: a review of concepts, research and practice. *International Journal of Management Reviews,* **12**, 85-105 (2010)

Effect of the economic outturn on the cost of debt of an industrial enterprise

Marek Vochozka[1,*]

[1]Institute of Technology and Business, School of Expertness and Valuation, Okružní 517/10, 37001 České Budějovice, Czech Republic

Abstract. The cost of debt is referred to as the key factor determining profitability. It is a decisive factor in decision making of the management, especially in strategy development. The purpose of this paper is to establish the relationship between the volume of debt and the economic outturn of industrial enterprises. Using artificial neural networks, the relationship between interest costs and three profit categories is examined. Data of 5622 Czech processing enterprises in the years 2015-2017 are used. Multilayer perceptron neural networks and neural networks of basic radial functions are used for processing. A total of 10,000 neural structures are generated for each cost-interest relationship and the corresponding profit, of which 5 are retained, showing the best results. The results indicate that in all cases of profit there is no dependence between the interest and the amount of profit generated. Profiting companies do not get debt cheaper than other businesses.
Key words: economic outturn, debt, artificial neural networks, profit

1 Introduction

Capital refers to financial assets or financial values of assets such as cash and cash held on deposit accounts, as well as tangible machines and production facilities used at different workplaces, for example in factories. In addition, capital includes equipment such as buildings used to produce and store manufactured goods [1]. The company's financial management is a decision-making process with the sole purpose of achieving the company's main goal. Estimates of capital costs are therefore of great importance in this area [2]. The use of information on capital costs in the company's decision-making process is closely related to the assessment of the company's financial management using market value added [3].

Capital costs have a major impact on the key decisions of the Board of Directors on the scope of investment plans, the setting of objectives, the growth of capital, the creation of an optimal capital structure and other areas of financial management in the company such as capital budgets, takeover processes, etc. It is also a parameter in the calculation of return on investment and a useful tool for a number of other analyses [4].

* Corresponding author: vochozka@mail.vstecb.cz

The average capital cost of an enterprise can be expressed using the WACC model, the weighted average cost of capital. The weighted average cost of capital consists of part of the cost of the debt and partly of the cost of equity [5].

The total cost of capital in % can be expressed as follows [5]:

$$WACC = kd * (1 - t) * D/C + ke + E/C \quad (1)$$

Where:

kd = costs of debt,

ke = equity costs,

t = rate of profit tax (per decimal number),

D = market value of debt,

C = value of total capital,

E = market value of equity.

Debt costs are a key component for managerial decision-making and are an important factor determining profitability. Debt costs determine the availability of capital for future investments and the profitability of current operations. Debt costs are therefore a decisive factor in strategy development [6]. The cost of debt (debt costs) in % can be expressed as follows [7].

$$Kd = r * (1 - t) \quad (2)$$

Where:

r = contractual interest rate,

t = the marginal tax rate for businesses.

Each capital is valued at its cost, so it has its own costs. For example, we can consider the return of equity (E) as the desired return of the company's owners. The interest on the loan is stated as the cost of debt. Equity is generally referred to as more expensive capital than debt [8].

We can express the cost of equity in % as follows [9]:

$$re = rf + \beta * (rm - rf) \quad (3)$$

Where:

rf = risk-free interest rate,

β = relative risk coefficient,

rm = average return on the capital market.

The relationship between the volume of debt and the economic outturn will be examined in this paper by Artificial Neural Networks (ANNs). The concept of a neural network has been developed in biology and psychology, but its use goes to other areas, such as business and economics. At present, ANNs are widely used to address possible future problems. ANNs try to copy processes in the human brain and nervous system using computer devices [10].

The big advantage is that ANN's relationships may not be determined in advance because this method itself creates relationships through the learning process. They are especially valuable where inputs are highly correlated, missing, or there are non-linear systems [11]. Artificial neural networks are currently extensively used in various technical applications and become a common method for providing various solutions in optimization, regression and problem estimation [12]. Despite their simplified design, they have the ability to mimic human qualities in problem solving through education and generalization [13].

The neural network consists of an input layer, an output layer, and one or more intervening layers, also referred to as hidden layers [14]. Hidden layers can capture a non-linear relationship between variables. Each layer consists of multiple neurons that are attached to the neurons in the adjacent layers. These networks contain many mutually interacting nonlinear neurons in several layers. They can capture relatively complex phenomena [15].

The purpose of this paper is to establish the relationship between the volume of debt and the economic outturn of industrial enterprises.

We also set the hypothesis to support the achievement of the goal: The cost of debt depends on the economic outturn.

2 Data and methods

The paper examines the data of 5622 enterprises operating in the manufacturing industry in the Czech Republic between 2015 and 2017. The data comes from the Albertina database. Complete statements of the financial statements of all these companies are available. For the analysis, however, only some data are important, namely the amount paid using debt (interest expense), operating profit, interest and tax profit and pre-tax profit. We will examine the relationship between cost and three profit categories.

Operating profit (economic outturn) relates to the enterprise's core business. It is possible to influence it managerially and its creation directly affects the future existence of the company.

Profit before Interest and Tax (EBIT) is the so-called Managerial Profit Category. Its aim is to reduce the impact of businesses that are not highly influenced by corporate interests - cost and corporation tax.

Profit before tax (EBT) is also a profit category that attempts to shield the impact of corporate income tax on its interpretation.

The tool by which we examine the relationship will be neural networks. Specifically, they will be Multilayer Perceptron Networks (MLP) and Neural Networks of Basic Radial Functions (RBFs). We will be using the program Statistica by DELL in version 12. The data will be imported from the MS Excel file, which is the output of the Albertina database. The software will use Data Mining - neural networks. In particular, we will need a part called Regression. First, we select data - it will always be a cost-dependent variable. As an independent variable, we understand the selected profit category. Subsequently, we divide the data into three sets: training, testing and validation. The ratio will be 70:15:15. On the set of training we will look for suitable neural networks that describe the relationship between the two variables. On the other two datasets, we then verify that the neuronal structures found are adequate and faithfully describe the situation.

We will generate 10,000 neural structures, of which we will keep the 5 best in terms of performance and error. For MLP we use a minimum of 2 and a maximum of 8 neurons in a hidden layer. We use the following functions to activate the inner neuron layer as well as to activate the outer layer of neurons:

- Identity,
- Logistics function,
- Hyperbolic tangents,
- Exponential function,
- Sine function.

DOI: 10.1051/shsconf/20173901028

In the case of RBF, at least 21 and up to 30 neurons will be hidden in the hidden layer. In order to confirm or refute the established hypothesis, we determine that the variables are in relation if their correlation coefficient is at least 0.7.

3 Results

As has been determined by the methodology, we have dealt with the relationship between cost and interest categories. These were used by three. This will correspond to the breakdown of the results obtained.

3.1 Operating profit (economic outturn)

Table 1 provides basic statistical characteristics of the input variables in the breakdown between the training, texting, and validation data sets. At the same time, the characteristics for the whole set are given.

Table 1. Data statistics - Operating profit and interest expense

Characteristics	Operating outturn - ths. CZK	Price of bank loans
Minimum (Training)	-507427	0.000147
Maximum (Training)	7356017	0.984321
Average (Training)	16102	0.084567
Standard Deviation (Training)	145721	0.117928
Minimum (Testing)	-211884	0.000702
Maximum (Testing)	765533	0.947368
Average (Testing)	12590	0.080872
Standard Deviation (Testing)	49279	0.119176
Minimum (Validation)	-190930	0.000164
Maximum (Validation)	965185	0.878788
Average (Validation)	11370	0.073653
Standard deviation (Validation)	80721	0.098862
Minimum (Overall)	-507427	0.000147
Maximum (Overall)	7356017	0.984321
Average (Overall)	14866	0.082377
Standard Deviation (Overall)	124833	0.114635

The operating result is expressed in thousands of CZK. The price of bank loans, or rather of paid debt is expressed by index. For example, it is clear from the table that the highest interest paid is up to 98.4%. Average interest is then almost 8.24%.

10,000 neural structures were generated. Of these, five of the best characteristics were retained. These are listed in Table 2.

DOI: 10.1051/shsconf/20173901028

Table 2. Overview of preserved neural networks (operating profit)

	Network name	Train. Perf.	Test. Perf.	Valid. Perf.	Train. error	Test. error	Validat. error	Train. algor.	Error function	Hidden layer act.	Output act. funct.
1	RBF 1-24-1	0.06914	0.04175	0.15998	1759133	2079149	1349410	RBFT	Sum.qt.	Gauss	Identity
2	RBF 1-25-1	0.06666	0.03323	0.15737	2389524	2180104	2533730	RBFT	Sum.qt.	Gauss	Identity
3	RBF 1-28-1	0.12625	0.04290	0.16755	2147330	2203807	2274527	RBFT	Sum.qt.	Gauss	Identity
4	RBF 1-21-1	0.07213	0.04567	0.14818	5610786	5981590	5496185	RBFT	Sum.qt.	Gauss	Identity
5	RBF 1-22-1	0.01193	0.01018	0.21063	74510380	55042374	88967422	RBFT	Sum.qt.	Gauss	Identity

The table shows that all retained networks use the RBF architecture. All of them also use the Gauss curve to activate the hidden neuronal layer and then to identify the external layer of the neurons.

However, network performance is scarce. This is illustrated by table number 3.

Table 3. Correlation coefficients of preserved neural networks (operating profit)

Neuron network	Price of bank loans		
	Training	Testing	Validation
1.RBF 1-24-1	0.069142	0.04175	0.159975
2.RBF 1-25-1	0.066663	0.033227	0.157366
3.RBF 1-28-1	0.126247	0.042897	0.167553
4.RBF 1-21-1	0.072133	0.04567	0.148184
5.RBF 1-22-1	0.011929	0.010175	0.210628

The performance of neural networks is measured by the correlation coefficient. The table provides correlation coefficients for all sets of preserved networks. We are ideally searching for a network that has a correlation coefficient of 0.7 and higher and ideally similar performance in the training, test and validation data sets. However, this does not show any of the preserved neural networks. Given the performance of individual networks, we can say that there is no dependence between operating profit and interest expense (the cost of debt).

A very important characteristic is the residue. Ideally, there is no prediction of any residues. In this case, it means the minimum and maximum long distance predictions from reality. Thus, we can say that in the category of operating profit, the hypothesis was reversed.

3.2 Profit before interest and taxes (EBIT)

Table 4 provides basic input statistics.

DOI: 10.1051/shsconf/20173901028

Table 4 Data statistics - EBIT and cost interest

	EBIT	Price of bank loans
	Input variable	**Output (goal)**
Minimum (Training)	-508543	0.000147
Maximum (Training)	7266527	0.984321
Average (Training)	16152	0.084567
Standard Deviation (Training)	146454	0.117928
Minimum (Testing)	-88568	0.000702
Maximum (Testing)	769040	0.947368
Average (Testing)	13316	0.080872
Standard deviation (Testing)	55622	0.119176
Minimum (Validation)	-202041	0.000164
Maximum (Validation)	916084	0.878788
Average (Validation)	10906	0.073653
Standard deviation (Validation)	71125	0.098862
Minimum (Overall)	-508543	0.000147
Maximum (Overall)	7266527	0.984321
Average (Overall)	14940	0.082377
Standard deviation (Overall)	125764	0.114635

The interest (or the cost of the debt) is the same as table 1. The change is then in the column marked EBIT, which offers the second variable to be examined. For comparison, we select one variable, the total is lower for EBIT than for operating profit. It can be deduced that the particular firm in question (certainly the same business) generated a loss of financial, (a) or extraordinary activity.

The stored networks are in Table 5. Again, of the 10000 generated networks, the five best characteristics were left.

Table 5. Overview of preserved neural networks (EBIT)

	Network name	Train. perf.	Test. perf.	Valid. perf.	Train. error	Test. error	Validat. error	Train. algor.	Error function	Hidden layer act.	Output act. funct.
1	RBF 1-23-1	0.11208	0.056685	0.13056	827743	935330	890193	RBFT	Sum.qt.	Gauss	Identity
2	RBF 1-21-1	0.090216	0.061057	0.127725	7780315	7752146	8343863	RBFT	Sum.qt.	Gauss	Identity
3	RBF 1-24-1	0.144422	0.066246	0.126835	1291421	1261820	1340023	RBFT	Sum.qt.	Gauss	Identity
4	RBF 1-25-1	0.090728	0.034356	0.142812	1016761	1115185	1045455	RBFT	Sum.qt.	Gauss	Identity
5	RBF 1-22-1	0.091924	0.089913	0.154548	830338	855661	875923	RBFT	Sum.qt.	Gauss	Identity

The result is very similar to operating profit. Again, it is only RBF. To activate the inner layer of neurons, they use Gaussian function, to activate the output layer of neurons identity.

The performance of retained networks is shown in table 6.

Table 6. Correlation coefficients of preserved neural networks (EBIT)

	Price of bank loans		
	Training	**Testing**	**Validation**
1.RBF 1-23-1	0.11208	0.056685	0.13056
2.RBF 1-21-1	0.090216	0.061057	0.127725
3.RBF 1-24-1	0.144422	0.066246	0.126835
4.RBF 1-25-1	0.090728	0.034356	0.142812
5.RBF 1-22-1	0.091924	0.089913	0.154548

Even for EBIT, net power is very small, albeit slightly better than operating profit. Still, we can not talk about any dependence. Also in this case, the hypothesis was refuted.

3.3 Profit before tax (EBT)

Table 7 lists the statistical characteristics of the input variables.

Table 7. Data statistics - EBT and cost interest

Characteristics	**Profit before tax (+/-) - ths. CZK**	**Cost of debt resources**
Minimum (Training)	-587766	0.000051
Maximum (Training)	7137185	0.195682
Average (Training)	14984	0.018553
Standard deviation (Training)	143964	0.018175
Minimum (Testing)	-91844	0.000144
Maximum (Testing)	760073	0.124312
Average (Testing)	12310	0.01737
Standard deviation (Validation)	54257	0.015643
Minimum (Validation)	-230073	0.00009
Maximum (Validation)	914086	0.131193
Average (Validation)	9652	0.017667
Standard deviation (Validation)	67963	0.017038
Minimum (Overall)	-587766	0.000051
Maximum (Overall)	7137185	0.195682
Average (Overall)	13783	0.018243
Standard deviation (Overall)	123588	0.017536

We have recently reviewed the EBT. We select one of the EBT characteristics, namely the maximum. Again, it is probably the same business that generated the maximum value of both EBIT and operating profit. In this case, the maximum is slightly lower (by value of cost interest). However, it is from this value that we can deduce the low indebtedness of the business or the low cost of the debt resources used by it.

Again, 10,000 neural networks were generated. These were retained again. These are listed in Table 8.

DOI: 10.1051/shsconf/20173901028

Table 8. Overview of preserved neural networks (EBT)

	Network name	Train. perf.	Test. perf.	Valid. perf.	Train. error	Test. error	Valid. error	Train. algor.	Error function	Hidden layer act.	Output act. funct.
1	MLP 1-2-1	0.25249	0.21798	0.25279	0.00016	0.00012	0.00012	BFGS 9999	Sum.qt.	Logistic	Logistic
2	MLP 1-2-1	0.23701	0.20376	0.23193	0.00016	0.00012	0.00013	BFGS 10000	Sum.qt.	Logistic	Exponential
3	MLP 1-2-1	0.25352	0.21917	0.25520	0.00016	0.00012	0.00012	BFGS 4640	Sum.qt.	Logistic	Logistic
4	MLP 1-2-1	0.23639	0.20013	0.22591	0.00016	0.00012	0.00013	BFGS 462	Sum.qt.	Tang	Logistic
5	MLP 1-2-1	0.24321	0.21050	0.23675	0.00016	0.00012	0.00012	BFGS 5034	Sum.qt.	Logistic	Logistic

The result differs from the previous two categories of profits. Only MLPs are now preserved. To activate the hidden layer of neurons they use a logistic function, in one case a hyperbolic tangent. The logic function is used to activate the outer layer of the neurons, only exponential functions. All networks were created using the Quasi-Newton algorithm (but it differs in its variant).

Table 9 shows the performance of individual preserved networks.

Table 9. Correlation coefficients of preserved neural networks (EBT)

	Cost of debt resources		
	Training	**Testing**	**Validation**
1.MLP 1-2-1	0.252488	0.217982	0.252787
2.MLP 1-2-1	0.237011	0.203762	0.231934
3.MLP 1-2-1	0.253517	0.219174	0.255196
4.MLP 1-2-1	0.236392	0.20013	0.225914
5.MLP 1-2-1	0.243214	0.210502	0.23675

Correlation coefficients are more than double that of EBIT. They are even similar to each set of data. Nevertheless, they are far below the required levels of 0.7. Even in this case we can say that the studied quantities are indifferent to each other. Even in this case, the hypothesis was refuted.

4 Conclusion

The aim of this paper was to determine the relationship between the volume of debt and the economic outturn of industrial enterprises.

A hypothesis was also set to support the rate of completion of the goal.

The relationship of the cost of debt and profit was analyzed using three profit categories - operating profit, EBIT and EBT. In theory, it was possible to believe that the relationship between profit and the amount of interest exists. Theoretically, it could be assumed that the higher the economic outturn, the lower the interest on the lent debt. We could have guessed that a successful profitable firm would have easier access to debt. At the same time, the cost of debt will be lower for such a company, as the company has lower credit risk. However, the correlation has not been confirmed. In all cases of profit we can state that there is no dependence between interest and the amount of profit generated.

Profiting companies do not get debt cheaper than other companies (those that do not generate profit or even generate a loss).

References

1. V. Wiegel, The prospectus directive and the prospectus ordinance is a dogmatic, economic and comparative law [Die Prospektrichtlinie und Prospektverordnung eine dogmatische, ökonomische und rechtsvergleichende]. In: *Writings on European and International Private, Banking and Commercial Law [Schriften zum Europäischen und Internatinalen Privat, Bank und Wirtschaftsrecht]*, Berlin, De Gruyter Recht, (2008)
2. P. Šuleř, Optimizing the capital structure of the company to maximize its profits by using neural networks on the example of building companies. *Mathematical Modelling in Economics* [Математичне моделювання в економіці], **7**(3-4), (2016)
3. J. Bertomeu, E. Cheynel, Disclosure and the Cost of Capital: A Survey of the Theoretical Literature. *Abacus-A Journal of Accounting Finance and Business Studies*, **52**(2), 221-258 (2016)
4. A. Michalak, The Cost of Capital in the Effectiveness Assessment of Financial Management in a Company. *Oeconomia Copernicana*, **7**(2), 317-329 (2016)
5. M. Z. Frank, T. Shen, Investment and the weighted average cost of capital. *Journal of Financial Economics*, **119**(2), 300-315 (2016)
6. M. Mirea, A. V. S. Comanescu, Financing decision based on the weighted average cost of capital. *Proceedings of the International Economic Conference on Integrative Relations between the European-Union-Institutions-and-the-Member-States*, Sibiu, Romania, 265-269 (2008)
7. T. Lindner, J. Muellner, J. Puck, Cost of Capital in an International Context: Institutional Distance, Quality, and Dynamics. *Journal of International Management*, **22**(3), 234-248 (2016)
8. A. Kumar, R. Sharma, *Financial management: Theory and Practice*. Vishla Enclave, New Delhi: Atlantic Publishers & Distributors Pvt Ltd, (2001)
9. M. Abudy, S. Benninga, E. Shust, The cost of equity for private firms. *Journal of Corporate Finance*, **37**(2), 431-443 (2016)
10. J. Dzuričková, R. Fabinyová, B. Mihalčová, The Opportunity Cost of Equity Capital. *Procedia Economics and Finance*, **23**(2), 1492-1498 (2015)
11. D. Enke, S. Thawornwong, B.C. Fidan, The use of data mining and neural networks for forecasting stock market returns. *Expert Systems with Applications*, **29**(4), 927-940 (2005)
12. H. T. Pao, A comparison of neural network and multiple regression analysis in modeling capital structure. *Expert Systems with Applications*, **35**(3), 720-727 (2008)
13. M. Vochozka, Z. Rowland, V. Stehel, P. Šuleř, J. Vrbka, *Cost modelling of the company using neural networks [Modelování nákladů podniku pomocí neuronových sítí]*. 1st ed. Czech Republic, České Budějovice: Institute of Technology and Business (2016)
14. H. Altun, A. Bilgil, B.C. Fidan, Treatment of multi-dimensional data to enhance neural network estimators in regression problems. *Expert Systems with Applications*, **32**(2), 599-605 (2007)
15. Z. Rowland, J. Vrbka, Optimization of a company´s property structure aiming at maximization of its profit using neural networks with the example of a set of construction companies. *Mathematical Modelling in Economics* [Математичне моделювання в економіці], **7**(3-4), (2016)

16. F. M. Tseng, H.-Ch. Yu, G. H. Tzeng, Combining neural network model with seasonal time series ARIMA model. *Technological Forecasting and Social Change*, **69**(1), 71-87 (2002)

Enterprise Value Generators in the Building Industry

Marek Vochozka[1] *and Veronika* Machová[2,*]

[1]Institute of Technology and Business, School of Expertness and Valuation, Okružní 517/10, 37001 České Budějovice, Czech Republic
[2]University of Economics, Faculty of Business Administration, Nám. W. Churchilla 4, 13067 Praha 3 – Žižkov, Czech Republic

Abstract. Value generators mark factors that influence the given enterprise´s success most. Thus, they refer to activities and abilities that increase profitability, decrease risk, and support the company´s growth. The aim of this contribution is to identify value generators in a building enterprise. The main presumption, however, is that the enterprise value will be measured by the EVA Equity indicator (Economic Value Added for the shareholders – owners). Data to be analysed come from the Albertina database. They include complete financial statements of building enterprises that operated on the market between 2006 and 2015. The data is organized into a table the EVA Equity in each enterprise is calculated per each year of its operation on the market. The table is subsequently imported into the Statistica software which searches for the extent to which the EVA Equity indicator is dependent on the individual items of financial statements. The result is a created adequate methodology and identification of value generators in building industries from 2006 to 2015 in the Czech Republic. The following variables are marked as the most significant: economic growth per current accounting period, equity, bank loans and bailouts, trade receivables, and current assets.
Key words: value generators, EVA Equity, enterprise value, artificial neural networks

1 Introduction

Value generators are factors that have the greatest impact on the company´s success, and are specific for different industry branches and enterprises [1]. A value generator may be described as an activity or an ability that adds value to a product, service, brand or the enterprise itself. Specifically, value generators refer to activities or abilities that increase profitability, decrease risk and support growth in accordance with strategic aims. Such aims may include a growing shareholder value, competition advantage and customer determination [2].

The main aim of every business activity is to create value for the owner. With regard to the complexity of nowadays organizations and the circumstances of growing customer

* Corresponding author: machova@mail.vstecb.cz

demands in which they compete, it is necessary for the company aims focused on value to be efficiently managed at all organization levels. A well considered strategy may be only successful if the company aims are transparent, comprehensible for workers, and controllable [3].

Nowadays, most global markets are being globalized, and competition is becoming more exponential. Companies always try to be as competitive as possible. They use different tools to do so. One of them is modern manager style called 'Value-Based Management'. Its aim is to focus on creation of values for the company. Establishing this type of management is not easy because the company has to identify the managers correctly, as they are the ones responsible for value creation in the company [4].

Value-Based Management (VBM) may be defined as an integrated managing approach based on several elements including managing systems [5]. Specifically, VBM should offer a general frame via which managers may define, implement and promote strategies when creating values within a company, and provide the owners with resources for monitoring and harmonizing manager actions [6]. The key VBM element is metrics based on values that combine costs on invested capital and company profitability [7]. These metrics and a related set of linked value controls are meant to replace traditional accounting measures, such as key financial indicators in a company, and serve as the main company performance indicators [8]. Based on this, metrics based on values should contribute to creating strategies, allocating sources and determining financial aims, and thus harmonizing the inner aims with creating shareholder value [9]. Meanwhile, integration of metrics based on values within a compensation system of enterprises should still support harmonization of well motivated managers [10].

Nowadays, knowledge assets are generally understood as organization sources that play a key role in creating the company's value. Mechanisms through which these assets participate in creating a value, are still not well known. This contribution draws on this need and suggests a model for identifying knowledge assets of an enterprise that significantly contribute creation of values. The model is basically designed as a cognitive map, and it draws on the combination of knowledge gathered from literature about strategic management [11].

Lately, more and more experts believe that the main aim of a stable enterprise is mainly growth of value for owners, shareholders. There are several options of measuring such growth. The EVA Economic Value Added is a very useful and significant tool for judging enterprise performance. It combines such factors as economy, accounting, and market information [12]. According to Morad and Balu [13], this indicator tries to express the company's real economic profit. Its significant advantage, as opposed to usual approaches, is the ability to evaluate the enterprise's economic performance comprehensively, and take into account the amount of investment risk [14]. According to Kollar and Klieštik [15], the EVA indicator represents the most advanced tool for measuring enterprise performance based on the principle of value management. According to Bluszcz and Kijewska [16], the EVA indicator depends on the amount of capital invested, and the economic dimension. Frank and Shen [17] claim that economic dimension is expressed via the weighted average costs on capital, and the return on invested capital.

The aim of this contribution is to identify the generators of a building enterprise value.

However, the assumption is that the enterprise value will be measured via the EVA Equity indicator.

2 Data and Methods

Data meant for analysis will come from the Albertina database. Building enterprises that operated on the market during 2006 to 2015 will be concerned. In total, the dataset will contain records of 65,656 enterprises. Data of their complete financial statements will be available (except attachments). That means that we will use balance sheet, profit and loss sheet, and statement of cash flow data. The data will be organized into a table which will be lined according to companies and the years of their activity. The columns will then contain individual information from Financial Statements.

In the next step, the value of Economic Value Added for shareholders (owners) will be calculated for every enterprise in each year of its operation on the market, i.e. within the parameters of EVA Equity.

This is first presumed by the Weighted Average Costs of capital [18]:

$$WACC = r_f + r_{LA} + r_{entrepreneurship} + r_{FinStab} \tag{1}$$

Where:

$WACC$	represents the Weighted Average Costs of Capital,
r_f	Risk-free yield,
r_{LA}	function of indicators characterizing the size of the enterprise,
$r_{entrepreneurship}$	function of indicators characterizing the creation of production power,
$r_{FinStab}$	function of indicators characterizing relationships between enterprise property and sources of their financial backing.

Subsequently, the value of alternative rate of equity will be stated [18]:

$$r_e = \frac{WACC * \frac{UZ}{A} - (1-d) * \frac{U}{BL+B} * \left(\frac{UZ}{A} * \frac{E}{A}\right)}{\frac{E}{A}} \tag{2}$$

Where:

WACC	Weighted Average Costs of capital,
UZ	Money Sources (Corrupt Sources) (Equity and Interest-bearing Loan Capital),
A	Assets,
E	Equity,
BL	Bank Loan,
B	Bonds,
$\frac{U}{BL+B}$	Interest Rate. The 'I' mark may also be used,
d	Income Tax Rate (in the previous models the 't' mark was used).

The Economic Value Added for shareholders will be derived from the following relationship [18]:

$$\textbf{EVA Equity} = (ROE - r_e) * VK \tag{3}$$

Where:

ROE represents Return on Equity.

Subsequently, the table will be imported into the DELL Statistica Software, version 12. We are going to look for the extent to which the EVA Equity indicator is dependent on the individual items of financial statements.

DOI: 10.1051/shsconf/20173901029

We will use the data mining tool, automated neural networks (ANN). We will deal with regression. EVA Equity will be determined the dependent variable. Financial statement data will represent the independent variables. The data file will be divided into three sets – training, testing and validational. The first group will contain 70% of input data. Based on the training data set neural structures will be generated. The left over two data sets 15% will be always left. Both groups will serve to verify the reliability of the found neural structure, respectively the found model. We will generate 1,000 neural networks. Out of them, 5 proving the best characteristics, will be kept[1].The hidden layer will contain a minimum of two neurons, a maximum of 150. In case of a radial basic function, the hidden layer will contain a minimum of 21 neurons, 80 the most. For a multilayer perceptron network, the following distributional functions will be considered in the hidden and in the output layer:

- Linear,
- Logistical,
- Atanh,
- Exponential,
- Sinus.

Neural networks predicting the EVA Equity based on input data will be the result. The model will take into account only such variables that will influence the resulting value of EVA Equity in reality. We will choose such a neural structure that will describe the relationship as correctly as possible (i.e. proving the best performance in the training, testing and validational data set, minimal error in each data set, and a clear economic interpretation). We will be subsequently interested in the value of neurons entering a network chosen in that way. Moreover, we will carry out sensitivity analysis to determine which variables enter the calculation and which of them influence the result significantly, at the same time. Value generators will be the result of sensitivity analysis.

3 Results

Table No.1 offers five of the best generated and preserved neural networks.

Table 1. Neural Networks with the best characteristics

	Network	Training Perf.	Testing Perf.	Valid. Perf.	Training Error	Testing Error	Valid. Error	Training Algorithm	Error Function	Hidden Layer Act.
1	MLP 28-19-1	0.987825	0.979665	0.98839	5810288	8348213	6624405	BFGS 59	Square Sum.	Exponential
2	MLP 28-10-1	0.992142	0.976071	0.98913	3762468	9837310	6252567	BFGS 96	Square Sum.	Tanh
3	MLP 28-21-1	0.990369	0.979875	0.987803	4605288	8283818	6963926	BFGS 131	Square Sum.	Logistical
4	MLP 28-14-1	0.989948	0.977133	0.987428	4803759	9439238	8521111	BFGS 74	Square Sum.	Exponential
5	MLP 28-17-1	0.987636	0.978024	0.987556	5904161	9012839	7575515	BFGS 65	Square Sum.	Exponential

The table clearly proves that multilayer perceptron networks have the preserved, thus the best characteristics. In all cases Quasi-Newton was used as a training algorithm, nevertheless, always in its different variant. Smallest square method was considered an error function. The hidden neuron layer is in case of preserved neural networks activated

---

[1]We will get oriented via the smallest squares method. Network generation will end if there is no improvement, i.e. a decrease in value when adding the squares. Thus, we will keep those neural structures where the total of square residues compared to the real gold development will be the lowest (in an ideal case, equal to zero).

via exponential function, respectively via hyperbolic tangens and logistical function. Logistical function is the output activation one. Number of neurons in the first layer is very promising information. In all cases, there are 28 neurons. If all of them are represented by the same variables, we may boldly identify, with a relative high exactness, the enterprise value generators in building industry.

Relevance of generated networks is given by Table No. 2.

Table 2: Performance in Preserved Neural Networks

Neural Network	Training	Testing	Validational
1.MLP 28-19-1	0.987825	0.979665	0.98839
2.MLP 28-10-1	0.992142	0.976071	0.98913
3.MLP 28-21-1	0.990369	0.979875	0.987803
4.MLP 28-14-1	0.989948	0.977133	0.987428
5.MLP 28-17-1	0.987636	0.978024	0.987556

In the given table we may observe the performance of individual networks in all data sets (training, testing, and validational). In the optimal situation, we are looking for the highest performance value (correlational coefficient), and at the same time an equal value for all data sets. Probably the best characteristics are offered by the first network, MLP 28-19-1. It is however necessary to add that differences between individual networks are minimal.

For a better guess of a correct result, prediction parameters carried out with the use of individual networks are given in Table No. 3.

Table 3. Prediction Parameters

Prediction Parameter	1. MLP 28-19-1	2. MLP 28-10-1	3. MLP 28-21-1	4. MLP 28-14-1	5. MLP 28-17-1
Minimal Prediction (Training)	-625821	-593672	-565909	-618678	-627985
Maximal Prediction (Training)	1285241	1296644	1280179	1310115	1285661
Minimal Prediction (Testing)	-439848	-420398	-485825	-472289	-419033
Maximal Prediction (Testing)	716318	773790	712872	731896	741961
Minimal Prediction (Validational)	-534500	-489643	-487837	-488074	-540623
Maximal Prediction (Validational)	931984	966569	919711	962717	932213
Minimal Residues (Training)	-151009	-79944	-115456	-118709	-123061
Maximal Residues (Trénovací)	158204	114905	122328	139520	165594
Minimal Residues (Testing)	-206335	-221582	-192218	-215316	-199480
Maximal Residues (Testing)	60896	60488	106207	69333	64274
Minimal Residues (Validational)	-171069	-133646	-204284	-166636	-182821
Maximal Residues (Validational)	68236	81589	68456	100494	72700
Minimal Standard Residues (Training)	-63	-41	-54	-54	-51
Maximal Standard Residues (Training)	66	59	57	64	68
Minimal Standard Residues (Testing)	-71	-71	-67	-70	-66
Maximal Standard Residues (Testing)	21	19	37	23	21
Minimal Standard Residues (Validational)	-66	-53	-77	-57	-66
Maximal Standard Residues (Validational)	27	33	26	34	26

Even Table No. 3 shows clearly that differences in prediction are minimal. This is proven not only by the extreme prediction values, but also the extreme residue values.

Further, sensitivity analysis was carried out. Results of the analysis are given in Table No. 4.

Table 4. Sensitivity Analysis

Indicator	1. MLP 28-19-1	2. MLP 28-10-1	3. MLP 28-21-1	4. MLP 28-14-1	5. MLP 28-17-1	Average
Economic Result per Accounting Period (+/-)	45,52534	57,62013	54,98645	47,86791	43,15851	49,83167
Equity	3,096897	3,525898	5,279423	3,107585	3,478305	3,697622
Bank Loans and Bailout	2,592221	3,381039	3,513716	2,641688	2,582952	2,942323
Trade Receivables	1,745256	1,761609	2,929951	1,689505	1,755744	1,976413
Fixed Assets	1,307376	3,434348	1,739905	1,746764	1,189961	1,883671
Current Assets	1,285155	3,014548	1,373655	1,255774	1,292558	1,644338
Short-term Receivables	1,582573	1,572459	2,128731	1,253316	1,516588	1,610733
Operating Profit	1,539484	1,580329	1,939547	1,252003	1,512833	1,564839
Total Obligations	1,373689	1,26703	1,993261	1,538607	1,301354	1,494788
Long-term Receivables	1,443037	1,684167	1,475288	1,326547	1,472512	1,48031
Tangible Fixed Assets	1,611922	1,462031	1,185565	1,288986	1,607434	1,431187
Foreign Sources	1,78316	1,238692	1,181723	1,678969	1,19326	1,415161
Personal Costs	1,070623	1,112373	1,528621	1,06989	1,05861	1,168023
Value Added	1,069547	1,289792	1,273848	1,083258	1,083973	1,160084
Costs on Sales + Power Consumption	1,13081	1,2096	1,130011	1,083609	1,024377	1,115682
Power Consumption	1,048284	1,158429	1,203669	1,027318	1,048093	1,097159
Consumption of Material and Energy	1,051657	1,172924	1,108193	1,074561	1,052695	1,092006
Total Assets	1,058739	1,198038	1,073814	1,027691	1,089817	1,08962
Current Assets	1,048995	1,086747	1,11206	1,05345	1,103385	1,080927
Long-term Intangible Assets	1,145869	0,985621	1,064498	1,137256	1,058307	1,07831
Supplies	1,014047	1,102265	1,181804	1,052269	1,038923	1,077862
Basic Capital	1,095442	1,010118	1,077835	0,987304	1,133611	1,060862
Trade Obligations	1,121967	1,024943	1,072587	1,070067	1,013501	1,060613
Business Margin	1,034675	1,008112	1,170859	0,997513	1,036753	1,049582
Sale Revenue of Own Products and Services	1,066979	1,085189	1,011392	1,048065	0,996401	1,041605
Sale Revenue of Sold Goods and Performance	1,040813	1,034926	1,029347	1,03945	1,00828	1,030563
Performance	0,999527	1,070207	1,062063	0,995275	1,007639	1,026942
Operating Production Revenues	1,001808	1,078419	1,006702	1,013017	0,993135	1,018616

The table shows that the same variables entered the calculation in all cases. Although the order of importance differs in individual networks, the difference is not too significant. First, there is, logically (and as expected), economic result. Further, Equity appears (again as expected), bank loans and bailout, trade receivables, long-term assets, and current assets. Other positions of this notional rank have been taken by variables that do influence the value creation in an enterprise, but their meaning is not too big.

4 Conclusion

The aim of this contribution has been to identify value generators within a building enterprise. An adequate methodology was created, and value generators within building enterprises from 2006 to 2015 were generated for the Czech Republic. In total, 28 quantities were identified, entering the process of enterprise value creation, which is measured via the EVA Equity indicator.

The following variables were identified as the most significant: the profit or loss for the current accounting period, equity, bank loans and bailout, trade receivables, long-term assets and current assets.

Result potential is important. Another research may follow. At this point, it is suitable to identify the amount of individual variables´ influence on the EVA Equity, and at the same time to determine the relationship of these variables towards EVA Equity. In the following step it is suitable to decompose the partial indicators, and incorporate them within the tactical and operative aims of the given enterprise. The strategic aim is given. It is the growth of value for shareholders.

References

1. B. Andrea, B Coulson, C. Professor, A. Dr, R. Hogan, J. D. Evans, Does the strategic alignment of value drivers impact earnings persistence? Sustainability Accounting. *Management and Policy Journal*, **6**(3), 374-396 (2015)
2. V. Kazlauskiene, C. Christauskas, Business Valuation Model Based on the Analysis of Business Value Drivers. *Engineering Economics*, **57**(2), 23-31 (2008)
3. I. Zareba, Strategic and financial drivers of business value creation. *Business and non-profit organizations facing increased competition and growing customers´ demands*, Muszyna, Poland, 403-416 (2014)
4. J. Šalaga, Value-based management and value drivers in world market. *Proceedings of the 15th International Scientific Conference Globalization and its socio-economic consequences*, Rajecke Teplice, Slovakia, 743-747 (2015)
5. M. Burkert, R. Lueg, Differences in the sophistication of Value-based Management – The role of top executives. *Management Accounting Research*, **24**(1), 3-22 (2013)
6. M. C. Daraban, An empirical view on value theory and value-based management. *CBU International Conference Proceedings*, **4**, 26-34 (2016)
7. D. Ch. Ittner, D. F. Larcker, T. Randall, Performance implications of strategic performance measurement in financial services firms. *Accounting, Organizations and Society*, **28**(7-8), 715-741 (2003)
8. V. Stehel, M. Vochozka, Company Management by Using INFA Method. *Proceedings of the 12th International Academic Conference*, (2014)
9. J. Šalaga, V. Berzáková, Summary and comparison of values indicators of value based management. *Economic and management spectrum [Ekonomicko-manažerské spektrum]*, **8**, 48-54 (2014)
10. S. Firk, S. Schrapp, M. Wolff, Drivers of value creation – The role of value-based management and underlying institutions. *Management Accounting Research*, **33**, 42-60 (2016)
11. F. D. Sole, F., D. Carlucci, G. Schiuma, Mapping Intellectual Capital Dynamics to Identify Company's Value Drivers. *Proceedings of the European Conference on Intellectual Capital*, 423-434 (2009)
12. I. Issham, A. Samad, M. Fazilah, Y. S. HWA, A. A. Kamil, Economic value added (EVA) as a performance measurement for GLCs vs Non-GLCs: Evidence in Bursa Malaysia. *Prague Economic Papers*, (2), 168-179, (2008)
13. B. Morard, F. O. Balu, Developing a practical model for calculating the economic value added. *Economic Computation and Economic Cybernetics Studies and Research*, **3**(3), 1-16 (2009)
14. V.Stehel, M. Vochozka, The analysis of the economical value added in transport. *Nase More*, Dubrovnik: University of Dubrovnik, **63**(3), 185-188 (2016)

15. B. Kollar, T. Kliestik, Approach in credit risk models. *Proceedings of the 4th International Conference on Applied Social Science* (ICASS 2014), **51**, 150-155 (2014)
16. A.Bluszcz, A. Kijewska, Factors Creating Economic Value Added of Mining Company. *Archives of Mining Sciences*, **61**(1), (2016)
17. M. Z. Frank, T. Shen, Investment and the weighted average cost of capital. *Journal of Financial Economics*, **119**(2), 300-315 (2016)
18. I. Neumaierová, I Neumaier, Financial analysis of industry and construction in 2007 [Finanční analýza průmyslu a stavebnictví za rok 2007]. *Analysis of the Ministry of Industry and Trade [Analýzy MPO]*, **1**, 1-187 (2008)

Innovation activities of firms: the case of 8 cis waves in the Czech industry between 2001 and 2014

Marek Vokoun[1,*]

[1]Institute of Technology and Business in České Budějovice, Department of management, Okružní 517/10, 370 01 České Budějovice, Czech Republic

Abstract. Eight Czech CIS waves between 2001-2014 were analyzed. An adjusted (incomplete) CDM model was used to estimate effects of characteristics of firms (firm size, government funding, market orientation, or foreign ownership) on the innovation process (decision to innovate, R&D intensity, sales from innovated goods). Public support had a very limited or even negative effect on innovation inputs and outputs. This analysis suggests that there is no relationship between R&D expenditures and sales of innovated goods. The coefficients are quite similar for new-to-the-market innovators and new-to-the-enterprise innovators. The orientation of the world market is a highly important factor of new-to-the-market innovators. The ability to capture sales of innovated goods is higher for smaller firms, but only for the new-to-the-firm innovators. The analysis also showed differences between innovators in service sector and manufacturing sector. Cooperation is a positive factor of innovation intensity in the manufacturing sector; however on average, there is zero effect of cooperation in the service sector. World market orientation is a key factor of service companies.
Key words: innovation activities, firms, CDM model, Czech market, CIS

1 Introduction

The paper analyses innovation activities of firms on the Czech market using 8 Community Innovation Surveys (CIS), which were carried out in 2001, 2003, 2005, 2006, 2008, 2010, 2012, and 2014. The surveys on innovation activities of firms were carried out in the transition years of the Czech economy into a modern developed country. Czech economy is one of the most open economies in the world. In the selected period, firms were joining the European Union markets in 2004, had experienced economic boom in 2005-2007, and economic crisis in 2008-2012.

The objective is to analyze the long-term relationships between innovation processes and firms' characteristics. The study focuses on the repressors such as size of the firm, government funding of research and development (R&D), and innovation input efficiency of innovation projects.

This paper explores new data from the years 2012 and 2014 and compares the manufacturing industry to the service sector. It uses two distinct definitions of innovators to

* Corresponding author: marek.vokoun@mail.vstecb.cz

observe the differences between them. One is describing innovators as those who introduced an innovation new to the market and in second case innovators as those who introduced an innovation new to the firm. The method uses three stages of the innovation process suggested by innovation modeling literature. It uses the selection equation to distinguish between innovators and non-innovators and then models innovation input by R&D expenditures and innovation output using the sales of innovated goods and services.

2 Current literature

Newest research focuses on the relationship between innovation and growth, and on the role of public support. Young firms are entering markets and usually try to utilize new technologies, cooperation, and in-house R&D activities to surpass the competition. However, the relationships between innovation and growth, R&D and young firms, and efficiency of public support of R&D are not as straightforward as one would expect.

On average, public support of R&D didn't have a significant impact on additional private investments in Italy and a complete crowding-out effect was observed there. The positive effect was observed only in the group of smaller firms; a strong negative effect was in larger firms [1].

The European Union Framework programmes (FPs) were created to promote and support cooperative R&D activities. The impact of FPs on firms' labour productivity is higher in comparison to randomly selected unsupported companies in the analysed years 1998-2002 [2].

There is a crowding-out effect of private R&D on innovation activities of firms, which is caused by the public support of R&D expenditures and innovation projects. The problem is that overall innovation activity is not affected by public support; supported firms only substitute their own resources by the government funding. This is the case especially for open innovation activities. Providing information resources and facilitating collaboration are examples of non-monetary public support. This non-monetary support seems to exert a stronger influence on firm's openness than monetary support which leads to a crowding out effect, i.e. does not foster open innovation in firms that are already innovative [3].

The relationship between innovation and firm growth is not straightforward. Firm growth is only explained with number of employees, capital, and sometimes with location or industry variables. It should be also analysed using specific unobservable factors related to the human and social capitals of firms. Growth of firms is observed in advanced technological environments where entrepreneurial firms are skilled at exploiting knowledge spillovers and investment is oriented at organisational flexibility [4].

Company growth is independent on R&D intensity and R&D investment. The Gibrat's Law (proportional rate of growth of a firm is independent on its absolute size), is rejected in their paper based on a sample of Portuguese manufacturing firms during the period from 1990 to 2001. The causal link and dynamics are more complex and primarily focus on the investment in fixed assets, which has a positive effect on the growth of firms; for small and high-tech firms such an effect is even more noticeable [5].

Company growth depends not only on the number of employees but also on the frequency of R&D (continuous, non-periodic) and the type of innovation (process, product, organizational, and marketing innovation). Continuous product and process innovators grow in the markets, however occasional process innovators grow even more rapidly [6].

The drivers of the innovative performance of young firms in the creative sectors are based on higher R&D intensity; greater emphasis is placed on internal R&D activities that focus more on product innovation than on organizational or process innovations. It is also based on networking activities and more frequent use of formal or informal methods of intellectual rights protection in comparison to firms in other industries [7]. Not all young firms invest in R&D in Spain between 2004 and 2010. It is because they are exporters and they have competitive goods and no need for intensive R&D expenditures and innovation projects as non-exporters [8].

There are other factors as well. Cooperation is a factor of innovation intensity and leads to more innovation activities of firms. Cooperation strategies that incorporate knowledge intensive business service firms (outsourcing) have a considerably greater impact on introducing innovation on the market [9]. Lack of qualified human capital is a barrier for starting innovation activities and is also a serious impediment to continuous innovator firms [10]. New management techniques are most common type of innovation in service companies in Spain [11]. The results also suggest that small-sized firms are more likely to be innovators than medium and large service companies.

The entrepreneurial opportunities of innovation ideas can be measured simply by the mean expected value of a new idea and its variance that depends on the risk appetite of the economic agent and market environment. Authors developed this opportunity based framework into the knowledge spill-over theory of entrepreneurship [12]. I agree with the thesis that too strong intellectual property protection reduces intra-temporal knowledge spill-overs, innovation, and growth since all the rents are appropriated by incumbent firms and entrepreneurship and innovation activities are too uncertain.

3 Materials and Methods

Eight Community Innovation Survey (CIS) waves (2001, 2003, 2005, 2006, 2008, 2010, 2012 and 2014) were used for the analysis. Data from the Czech Statistical Office (CZSO) questionnaires were used and connected. They are labelled "TI" and "P5-01." There was a data loss (empty cells, zero sales, innovators with zero R&D expenditures) but it was only for a few observations. The firm level innovation data and financial statements were not joined because CZSO no longer provides the opportunity to join data from two questionnaires. There was also a problem with discontinuity of some questions and only a few control variables were possible to create.

Table 1. Summary statistics for CIS waves in the Czech manufacturing industry.

Variable	Obs.	Mean	Std. Dev.	Min	Max
Innovator (new to the market)	42663	0.16	0.36	0	1
Innovator (new to the firm)	42663	0.28	0.45	0	1
Total R&D expenditures	42387	11616.37	146349.80	0	1.51E+07
Sales form innovated goods per employee	17996	2635.98	41875.90	0	4015001
Number of employees	34913	259.99	1205.67	1	81662
Manufacturing	42663	0.46	0.50	0	1
Services	42663	0.40	0.49	0	1
Foreign ownership	7130	0.40	0.49	0	1

The data sample features 16% of innovator firms (Tab. 1), which introduced innovations new to the market. The smallest firm has one employee; there are also large firms with more 10,000 employees. As described before, the Czech manufacturing industry is characterized by the foreign presence. In the data sample, there are 40% of multinational firms.

The estimation strategy is similar to the one used [13] [14]. The first step (decision) is a Heckman procedure, which is also known as the generalized Tobit II procedure. The panel version using random effect Probit estimation is, however, used in the first equation and panel regression with Mill's ratio in the second equation. The second step is a fixed effects estimator of the innovation output.

Table 2. Modified CDM model.

Heckman procedure	$r_{it}^{*}\begin{cases} 1\ if\ r_{it} = (X_{1it}\beta_1 + \rho_i + \varepsilon_{it_1}) > 0 \\ 0\ otherwise\ (r_{it} \leq 0) \end{cases}$
	$k_{it}^{*} = \ln(k_{it})\vert(r_{it} > 0) = X_{2it}\beta_2 + \rho_i + \varepsilon_{it_2}\ with\ Df(k_{it}) = (0,\infty)$
Innovation output	$t_{it}^{*} = \ln(t_{it})\vert(k_{it} > 0) = X_{3it}\beta_3 + \rho_i + \alpha k_{it}^{*} + \varepsilon_{it_3}\ with\ Df(t_{it}) = (0,\infty)$

$Xn_{it}\beta_n$'s (with n = 1, 2, 3, and 4) are vectors of explanatory variables and ε_{itn}'s (with n = 1, 2, 3, and 4) are random-error terms. The error terms are assumed to be independent of the exogenous variables. In the first (I.) and the second part (II.), the error term can be estimated as a system. The system of four econometric equations (Tab. 3) allows for arbitrary correlation among the four disturbances. The vector of parameters to be estimated is denoted βn (with n = 1, 2, 3, and 4) and the single parameters to be estimated are α and γ.

The first equation (r_{it}*) accounts for selection into R&D activities. Here the interest is focused on the probability of a firm i to engage in continuous R&D in a year t. This is specified as a probit model, i.e. $P(r_{it}{*}>0) = \Phi(X_{1it}\beta_1)$, where r_{it}* equals 1 if firm $_i$ is an innovator and reports R&D expenditures. The second linear equation ($_{ki}$*) describes innovation input, which relates the log of internal and external R&D expenditures (without machinery expenditures) to the number of employees in a firm i, conditional of being a process innovator and doing R&D. The ρ_i represents fixed or individual effects. In both equations, there is a number of potential determinants ($Xn_{it}\beta_n$'s) such as a firm's size, foreign ownership, being part of a group of companies, etc. The third linear equation (t_{it}*) models the innovation log of sales of goods and services to the number of employees. Here the aim is at the input-output elasticity α and other explanatory variables ($Xn_{it}\beta_n$s) describing the behavior and market determinants of innovators.

4 Results

Eight Czech CIS waves were analyzed. The goal was to assess the differences between industries and two definitions of innovators. First six models in Table 2 and Table 3 compare new-to-the-firm and new-to-the-market innovators. The probability to innovate is positively influenced by the size of the company and being a part of a group of companies. For new-to-the-market innovators the coefficients are quite similar and slightly lower in effect. In case of employees, the coefficient size is about 0.03 lower, in case of being part of a group the coefficient is about 0.12 lower, and industry variables are also lower. Substantial differences are observable in market orientation. New-to-the-market innovators have about 0.2 higher probability to innovate in comparison to the new to the firm innovators. The probability to innovate is higher in the manufacturing industry in comparison to services and other industries such as trade. The differences between these two sectors are described in literature [15] and it is suggested to analyze these sectors separately.

New-to-the-market innovators differ in innovation intensity (R&D expenditures per employee) when compared to new-to-the-firm innovators. Central Government support, EU funding, and cooperation positively affected the amount of R&D expenditures per employee for new-to-the-firm innovators. For new-to-the-market innovators, there is a higher R&D intensity in the manufacturing and service industries in comparison to other

DOI: 10.1051/shsconf/20173901030

industries. New-to-the-market multinational innovators are on average spending more per employee than domestic firms. Higher R&D intensity is typical for smaller innovators and it is stronger for new-to-the-market innovators.

There is a negative effect of the firm's size on the new-to-the-market innovators. This effect is not present in new-to-the-firm innovators. In both cases, Framework Programmes did not contribute to higher R&D intensity and decreased the innovation output of participated firms. Other public funding sources did not contribute to higher innovation output. The contribution is not significant because the input output elasticity ratio, i.e. the effect of R&D expenditures on sales of innovated goods and services, was not statistically significant.

There are differences between new-to-the-market innovators in service and manufacturing sector (see Table 4 and Table 5). The probability to innovate is higher for larger companies in the manufacturing industry. The situation in the service sector is similar but the effect of firm size is lower. Significant difference is also in market orientation. World market orientation is a key factor of service companies. The probability is higher than in manufacturing industry.

The innovation intensity is highly and negatively dependent on the firm size in the service industry; the effect in manufacturing is weaker. Cooperation plays a dominant role in manufacturing and it is not statistically significant in the service sector. Public funding has an impact on R&D intensity in the manufacturing sector. The only insignificant source was funding from the EU Framework Programmes. The funding from the central government in the service sector was the only significant source. Local government funding contributed to higher sales of innovated goods in the service sector.

5 Conclusion

This paper analyzed 8 CIS waves in Czech Republic in the period between 2001-2014. Adjusted (incomplete) CDM model was used to estimate variables in the panel of 24,000 observations about firms. Two definitions of innovators and two industries were selected to analyze effects of regressors (firm size, government funding, market orientation, and foreign ownership) on the innovation process (decision to innovate, R&D intensity, sales from innovated goods).

The coefficients are quite similar and the differences between new-to-the-market innovators and new-to-the-firm innovators are not substantial. We can however observe slightly lower coefficients in new-to-the-market innovators. World markets orientation is a highly important factor of being a new-to-the-market innovator. The ability to capture sales of innovated goods is higher for smaller firms, but only for the new-to-the-firm innovators.

The first analysis showed differences between innovators in the service sector and the manufacturing sector. Cooperation is a positive factor of innovation intensity. It plays a dominant role in the manufacturing sector; however on average, there is zero effect of cooperation in service sector. World market orientation is a key factor of service companies.

This analysis explored the dataset of 8 CIS waves and suggests that the effect of R&D expenditures on sales of innovated goods and services is not statistically significant. Public support had a very limited and even negative effect on innovation input and output as well. This analysis points to the fact that the investment in fixed assets is the most significant factor of company growth and value.

References

1. R. Bronzini, E. Iachini, Are Incentives for R&D Effective? Evidence from a Regression Discontinuity Approach. *American Economic Journal: Economic Policy,* **6**, 100-134 (2014)
2. L. Aguiar, P. Gagnepain, European cooperative R&D and firm performance: Evidence based on funding differences in key actions. *International Journal of Industrial Organization,* **53**, 1-31 (2017)
3. M. Cano-Kollmann, R.D. Hamilton III, R. Mudambi, Public support for innovation and the openness of firms' innovation activities. *Industrial and Corporate Change,* **26**, 421-442, (2017)
4. D. B. Audretsch, A. Coad, A. Segarra, Firm growth and innovation. *Small Business Economics,* **43**, 743-749 (2014a)
5. B. Oliveira, A. Fortunato, Firm growth and R&D: Evidence from the Portuguese manufacturing industry. *Journal of Evolutionary Economics*, **27**, 613-627 (2017)
6. M. Deschryvere, R&D, firm growth and the role of innovation persistence: an analysis of Finnish SMEs and large firms. *Small Business Economics*, **43**, 767-785 (2014)
7. A. Protogerou, A. Kontolaimou, Y. Caloghirou, Innovation in the European creative industries: a firm-level empirical approach. *Industry and Innovation,* **24**, 587-612 (2017)
8. D. B. Audretsch, A. Segarra, M. Teruel, Why don't all young firms invest in R&D? *Small Business Economics,* **43**, 751-766 (2014b)
9. D. Doloreux, R. Shearmur, Innovation Strategies: Are Knowledge-Intensive Business Services Just Another Source of Information? *Industry and Innovation*, **20**, 719-738 (2013)
10. P. D'Este, F. Rentocchini, J. Vega-Jurado, The Role of Human Capital in Lowering the Barriers to Engaging in Innovation: Evidence from the Spanish Innovation Survey. *Industry and Innovation*, **21**, 1-19 (2014)
11. A. Trigo, The Nature of Innovation in R&D- and Non-R&D-Intensive Service Firms: Evidence from Firm-Level Latent Class Analysis. *Industry and Innovation,* **20**, 48-68 (2013)
12. Z. J. Acs, D. B. Audretsch, E. E. Lehmann, The knowledge spillover theory of entrepreneurship. *Small Business Economics,* **41**, 757-774 (2013)
13. F. Castellacci, *How does competition affect the relationship between innovation and productivity? Estimation of a CDM model for Norway* (MPRA Paper 27591. B.m.: University Library of Munich, Germany) (2009)
14. B. Crepon, E. Duguet, J. Mairesse, Research, Innovation and Productivity: An Econometric Analysis at the Firm Level. *Economics of Innovation and New Technology,* **7**, 115-158 (1998)
15. J. E. Ettlie, S. R. Rosenthal, Service versus Manufacturing Innovation, *Journal of Product Innovation Management*, **28**, 285-299 (2011)

DOI: 10.1051/shsconf/20173901030

Appendix

Table 3. All industries, new market innovators, Czech Republic 2001-2014.

All industries New market innovators 8 CIS Waves 2001-2014	**(1) Innovator (0/1) df/dx**	**(2) Log of R&D expenditures per employee**	**(3) Log of sales of innovated goods per employee**
	PROBIT RE	OLS FE	OLS FE
Log of employees	0.259***	-0.594***	-0.245*
	(0.02)	(0.20)	(0.15)
Being part of a group	0.353***		
	(0.04)		
Foreign company	-0.200***	0.331*	0.166
	(0.04)	(0.18)	(0.17)
Manufacturing	0.780***	1.321***	0.769
	(0.06)	(0.36)	(1.08)
Services	0.507***	1.060**	0.016
	(0.07)	(0.52)	(0.88)
National markets	0.755***		
	(0.06)		
European markets	0.715***		
	(0.06)		
World markets	0.771***		
	(0.07)		
Linear trend	0.065***	0.004	Year dummies
	(0.01)	(0.02)	
Lack of personnel	0.327***		
	(0.04)		
Lack of information	0.096**		
	(0.04)		
Prior innovation	-0.253***		
	(0.04)		
Cooperation		0.264**	0.032
		(0.11)	(0.09)
Funding – local GOV		0.264	0.045
		(0.26)	(0.22)
Funding – central GOV		0.474***	-0.025
		(0.11)	(0.10)
Funding – EU		0.135	0.203
		(0.13)	(0.13)
Funding – UE FP		0.042	-0.308*
		(0.15)	(0.18)
Log of R&D expenditures			0.025
per employee			(0.03)
Constant	-133.757***	-2.294	10.258***
	(13.05)	(39.27)	(1.25)
Observations	23802	3378	3109
Adjusted R2	n/a	4.28 %	71.26 %

Robust Standard errors (bootstrapping) in parentheses, * $p < 0.10$, ** $p < 0.05$, *** $p < 0.01$

DOI: 10.1051/shsconf/20173901030

Table 4. All industries, new to the firm innovators, Czech Republic 2001-2014.

All industries New firm innovators 8 CIS Waves 2001-2014	**(4) Innovator (0/1) df/dx**	**(5) Log of R&D expenditures per employee**	**(6) Log of sales of innovated goods per employee**
	PROBIT RE	OLS FE	OLS FE
Log of employees	0.286***	-0.465***	-0.258
	(0.01)	(0.13)	(0.18)
Being part of a group	0.472***		
	(0.04)		
Foreign company	-0.273***	0.202	0.054
	(0.04)	(0.14)	(0.15)
Manufacturing	0.832***	0.299	0.567
	(0.05)	(0.39)	(0.85)
Services	0.526***	0.627	-0.134
	(0.05)	(0.55)	(1.01)
National markets	0.472***		
	(0.04)		
European markets	0.575***		
	(0.04)		
World markets	0.578***		
	(0.05)		
Linear trend	0.055***	0.003	Year dummies
	(0.01)	(0.01)	
Lack of personnel	0.488***		
	(0.03)		
Lack of information	0.066*		
	(0.04)		
Prior innovation	-0.297***		
	(0.03)		
Cooperation		0.264**	0.032
		(0.11)	(0.09)
Funding – local GOV		0.221	0.036
		(0.17)	(0.19)
Funding – central GOV		0.443***	-0.037
		(0.10)	(0.09)
Funding – EU		0.177*	0.133
		(0.10)	(0.11)
Funding – UE FP		0.108	-0.309***
		(0.10)	(0.11)
Log of R&D expenditures per employee			0.025
			(0.03)
Constant	-114.682***	-0.049	10.422***
	(11.32)	(26.30)	(1.34)
Observations	24289	5399	3351
Adjusted R2	n/a	3.52 %	69.70 %

Robust Standard errors (bootstrapping) in parentheses, * $p < 0.10$, ** $p < 0.05$, *** $p < 0.01$

DOI: 10.1051/shsconf/20173901030

Table 5. Services, new to the firm innovators, Czech Republic 2001-2014.

Services New firm innovators 8 CIS Waves 2001-2014	**(7) Innovator (0/1) df/dx**	**(8) Log of R&D expenditures per employee**	**(9) Log of sales of innovated goods per employee**
	PROBIT RE	OLS FE	OLS FE
Log of employees	0.098***	-1.175***	-0.854
	(0.03)	(0.30)	(0.68)
Being part of a group	0.564***		
	(0.07)		
Foreign company	-0.218***	0.132	0.656
	(0.07)	(0.24)	(0.46)
National markets	0.654***		
	(0.07)		
European markets	0.461***		
	(0.09)		
World markets	0.814***		
	(0.11)		
Linear trend	0.034***	0.064**	Year dummies
	(0.01)	(0.03)	
Lack of personnel	0.692***		
	(0.06)		
Lack of information	-0.022		
	(0.07)		
Prior innovation	-0.361***		
	(0.07)		
Cooperation		0.354	0.166
		(0.23)	(0.29)
Funding – local GOV		-0.240	0.554*
		(0.37)	(0.33)
Funding – central GOV		0.555**	-0.022
		(0.25)	(0.20)
Funding – EU		-0.118	0.137
		(0.22)	(0.35)
Funding – UE FP		-0.202	-0.593*
		(0.28)	(0.34)
Log of R&D expenditures			-0.068
per employee			(0.07)
Constant	-71.127***	-120.230**	14.408***
	(24.53)	(51.06)	(3.49)
Observations	6277	1156	675
Adjusted R2	n/a	8.38 %	69.03 %

Robust Standard errors (bootstrapping) in parentheses, * $p < 0.10$, ** $p < 0.05$, *** $p < 0.01$

DOI: 10.1051/shsconf/20173901030

Table 6. Manufacturing, new to the firm innovators, Czech Republic 2001-2014.

Manufacturing New firm innovators 8 CIS Waves 2001-2014	**(10) Innovator (0/1) df/dx**	**(11) Log of R&D expenditures per employee**	**(12) Log of sales of innovated goods per employee**
	PROBIT RE	OLS FE	OLS FE
Log of employees	0.452***	-0.297*	-0.129
	(0.02)	(0.18)	(0.15)
Being part of a group	0.365***		
	(0.05)		
Foreign company	-0.374***	0.213	-0.100
	(0.05)	(0.17)	(0.14)
National markets	0.552***		
	(0.07)		
European markets	0.503***		
	(0.07)		
World markets	0.548***		
	(0.08)		
Linear trend	0.056***	-0.009	Year dummies
	(0.01)	(0.02)	
Lack of personnel	0.392***		
	(0.04)		
Lack of information	0.106**		
	(0.04)		
Prior innovation	-0.241***		
	(0.04)		
Cooperation		0.223***	-0.078
		(0.08)	(0.09)
Funding – local GOV		0.324**	-0.261
		(0.13)	(0.17)
Funding – central GOV		0.427***	-0.118
		(0.10)	(0.08)
Funding – EU		0.246**	0.168
		(0.12)	(0.12)
Funding – UE FP		0.180	-0.169
		(0.15)	(0.13)
Log of R&D expenditures per employee			0.039
			(0.03)
Constant	-115.555***	22.705	10.075***
	(13.92)	(35.88)	(0.82)
Observations	13294	3850	2459
Adjusted R2	n/a	3.31 %	73.12 %

Robust Standard errors (bootstrapping) in parentheses, * $p < 0.10$, ** $p < 0.05$, *** $p < 0.01$

 DOI: 10.1051/shsconf/20173901031

Modern trends of development of the world economy and financial competitiveness of enterprises

Nadezhda Volkova[1,*] *and Valeria* Shagun[1]

[1]Saint Petersburg Polytechnic University Peter the Great, 195251, Saint-Petersburg, Russia

Abstract: The article considers modern trends in the development of the global economy, the dynamics of the activity of the domestic economy. The statistical data of the share of loss-making enterprises on the domestic market are analyzed. The importance of competitiveness and financial stability of enterprises in modern conditions is considered, the relationship between competitiveness and financial stability is indicated. The notion of financial competitiveness is formulated. Financial competitiveness is analyzed from the point of view of enterprise management and on the parameters for assessing the financial stability of enterprises. Methods for assessing the financial competitiveness of enterprises have been identified. The primary calculation of financial competitiveness indicators for PJSC "SF Almaz" was carried out based on the selected methods. The analysis of the obtained calculation results is carried out. Methods are proposed to ensure financial competitiveness of Russian companies.
Key words: modern trends, competitiveness, economy, finance

1 Introduction

In modern conditions, the domestic economy is characterized by a prolonged economic decline in activity, not only at the moment, but also in relation to the previous year, without justifying the forecasts of the Government and the Central Bank. This is due to a number of reasons of external and internal nature, which exert their influence on all spheres of social reproduction. There is a decline in GDP, partial stagnation of unemployment and inflation. According to the Ministry of Economic Development of Russia, the decline in GDP for the first half of the year was -0.9% to the corresponding period of 2015 [1]. The main impact on the partial slowdown in the economic recession, according to the Ministry of Economic Development of Russia, was provided by industrial production, transport, and agriculture. Construction and retail trade have a negative impact on the economic situation. These problems can presumably in the near future create a serious threat to macroeconomic stability and sustainable economic growth.

According to Rosstat, the share of enterprises that operate in Russia at a loss in January-May-May 2016 increased by 0.2% to 32.2% compared to the same period in 2015. The balanced financial result of organizations, that is, the amount of profit less losses, for the

* Corresponding author: nadine_13@mail.ru

five months amounted to 4.3 trillion rubles. More than 34 thousand organizations earned 5.2 trillion rubles, another 16.3 thousand suffered losses in the amount of 902.4 billion rubles [2]. Figure 1 shows a diagram of the proportion of loss-making enterprises in Russia by industry. The information base for drawing up the chart was provided by the Federal State Statistics Service [2].

The diagram shows that the positive dynamics for the reporting period is represented by enterprises in the spheres of agriculture and hunting (17.4% were unprofitable), followed by fisheries and fish farming (20.4%), retail trade, repair of auto transport and household goods (23.2%) and construction (32.1%).

The most unprofitable was the work of enterprises in the sectors of production and distribution of electricity, gas and water (45.5%), mining (38.6%), real estate and rent (35.3%). Last year, the share of unprofitable organizations in the country was 28.1%.

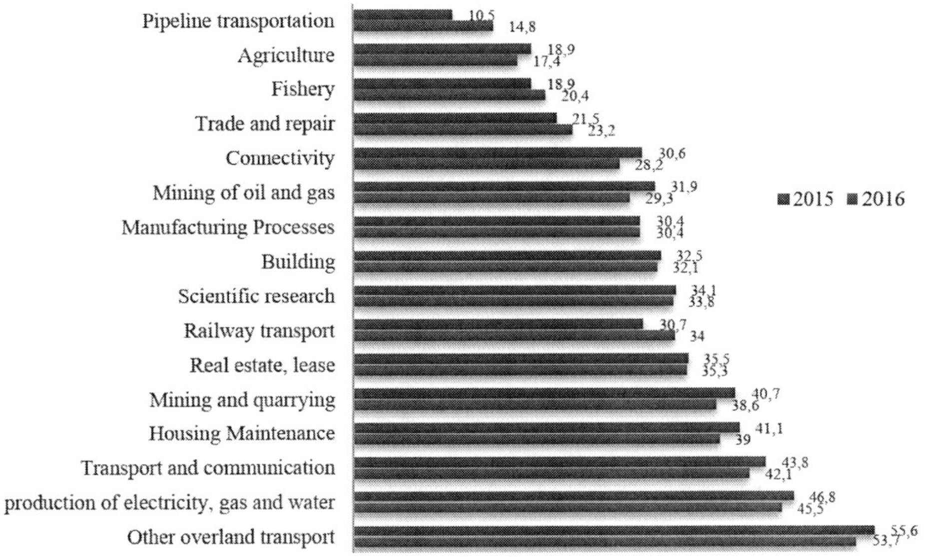

Fig. 1. Share of loss-making enterprises in Russia by industry (%)

As can be seen from Figure 1, the share of loss-making enterprises is declining quite slowly, but their share is high. This situation is a consequence of the crisis situation in the country and the fact that the enterprises do not match the conditions for the development of the world economy. The growth of the world economy has stabilized at a level of 3.4% per year, which is well below the average annual growth rate in the last ten-year period, when even taking into account the crisis phenomena of 2008-2009. Growth was higher than 4% per year [3]. In the context of globalization, this dynamic has had an impact on the Russian economy. A consequence for the domestic economy was the outflow of foreign investment from previously attractive industries, such as the resource-extracting industry, energy, wholesale and retail trade [4]. After the aggravation of the geopolitical situation, many states began to refuse to import Russian oil and gas, switching to other sources of purchasing these resources. This negatively affected the investment attractiveness of the industry, since the profits of these industries mainly depend on oil prices.

The manifestation of the negative effect on the economy on the part of the construction industry is characterized by a high percentage of commercial credit. As a result, expensive loans, the uncertain economic situation in the country and weak demand have an impact on investment in the domestic industry.

In the context of globalization, the discrepancy with new, dynamically changing processes and trends, forms the demand for a new generation of enterprises that must adequately, flexibly and quickly react to changing conditions of their functioning. Due to the impact of the global financial crisis on the economic situation in the world, the study of the role and significance of the enterprise's financial competitiveness, as well as its evaluation, is of particular interest. Given these trends, as well as the requirements for accelerating economic growth, the main task of Russia's modern development is to increase the competitiveness of the Russian economy [5].

2 Actuality

The development of enterprises in the modern domestic economy, namely, in the conditions of the continuous fall of the ruble and the strengthening of the dollar and the euro, the instability in oil prices in 2014-2015, competitive advantages in many ways determine the success or failure of enterprises. To ensure the survival of the enterprise in such conditions it is necessary to be able to reliably assess the financial competitiveness of both its enterprise and existing potential competitors.

The notion of "financial competitiveness" can be interpreted as the organization's ability to finance its sustainable development while ensuring a normal return on investment, profitability and an increase in the business price. The problem of assessing the financial competitiveness of economic entities is considered in the works of G.B. Kleiner, J. Brigham and L. Gapenski, P.V. Baranova, L.E. Bassovsky, I.T. Balabanov and a number of other authors. At the same time, defining its various aspects, all of them agree that financial competitiveness is not limited only to current profitability and solvency, but is an organic part of the overall competitiveness of the organization, reflecting its strategic advantages.

The financial competitiveness of an enterprise is characterized by the use of financial resources and capital, the fulfillment of obligations to the state and other economic entities. Analysis, evaluation and management of financial competitiveness can be considered in two aspects:

1. management aspect (Fig.2);
2. Functional aspect (Fig. 3);

It should be noted that the presented parameters are a quantitative characteristic of the properties and attributes of the financial competitiveness of the organization.

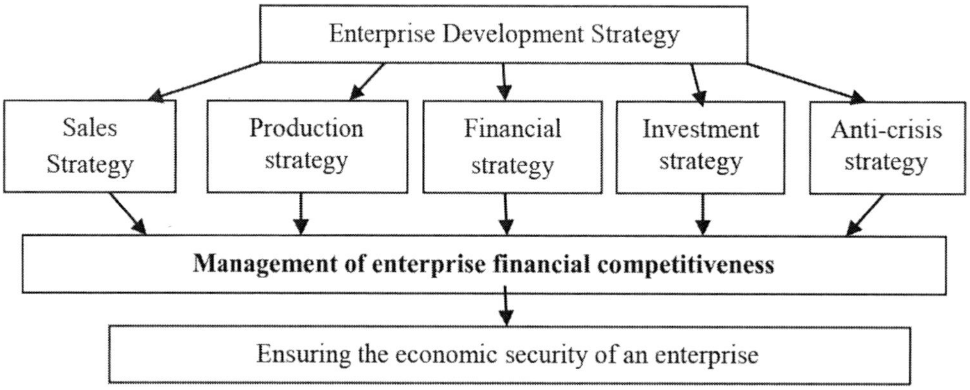

Fig. 2. Financial competitiveness from the position of enterprise management

The enterprise development strategy is the establishment of goals, objectives, identification of the main directions of its development and analysis of the opportunities to achieve these goals. The choice of strategy is largely determined by the market situation

DOI: 10.1051/shsconf/20173901031

and the organization's capabilities at the initial stage. Based on the strategy, it is necessary to develop a list of activities and actions aimed at achieving the set goals. From the correct and competently chosen composition of products, work or services, the success of the enterprise in competition can depend [6].

Sales and production strategies of the enterprise are the most important parts of the complex of the enterprise development strategy. The marketing strategy ensures the simultaneous development of the existing market and the search for new markets for the enterprise, as well as increasing the competitiveness of the enterprise. The production strategy is characterized by the development of actions to create and implement competitive products on the market.

The parameters of the financial and investment strategy, designed to increase the investment attractiveness of the enterprise and reduce the risk of loss of investment, occupy an important place in the strategic planning system. These strategies, as important components of the financial and investment policy, examine the laws of the development of financial and investment relations in the market, form the main strategic goals and directions for investment, determine the choice of sources, forms and methods of financing for the implementation of the strategy, justify the forms and ways of ensuring sustainable growth in the market.

An anti-crisis strategy is also an important part of the strategic development of the enterprise. An anti-crisis strategy should be implemented not only at the stage of strategic planning, but also be monitored in the current planning. If operational measures to overcome the economic crisis are carried out in isolation from strategic goals, this may lead to a short-term improvement in the financial situation, but it will not allow to remove the root causes of the crisis phenomena. The anti-crisis strategy is aimed at regulating business in the conditions of a recession in the industry, a steady decline in the basic financial performance indicators and the threat of bankruptcy. Untimely implementation of anti-crisis measures has a negative impact on financial competitiveness and, accordingly, on the economic security of the enterprise.

Enterprises that did not create an effective system for protecting their economic interests risk significantly worsen their economic situation and stop their activities. Therefore, the issues of ensuring economic security will always be paramount for the enterprise.

The main factor determining the state of economic security is the company's possession of sustainable competitive advantages. These benefits should be in line with the strategic objectives of the business entity.

The economic security of an enterprise is defined as:

• protection against various economic crimes (theft, fraud, falsification, industrial espionage, etc.);

• the state of the enterprise's security as an object of economic relations from internal and external threats;

• the presence of such competitive advantages that are due to the conformity of the material, financial, personnel, technical and technological potentials and organizational structure of the enterprise to its strategic tasks, which is intended to provide it with a certain level of economic security. The realization of advantages and potential gives the enterprise the opportunity to ensure economic security. Such an interpretation of the economic security of the enterprise is best combined with the category of the enterprise's financial competitiveness;

• the state of efficient use of resources, where the possibility to avoid the threat of economic security is based on economic concepts of achieving the goal or, in other words, on the resource-functional approach;

• the implementation of the protection of the economic interests of the enterprise, its human and intellectual potential, information, technology, capital and profit, which is

DOI: 10.1051/shsconf/20173901031

provided by a system of special legal, economic, organizational, information-technical and social measures [7].

It is important to consider financial competitiveness not only from the point of management of the enterprise, but also in the functional context of the enterprise's financial competitiveness. Figure 3 shows the parameters of the company's financial competitiveness, grouped according to the functional areas: sustainability, liquidity, solvency.

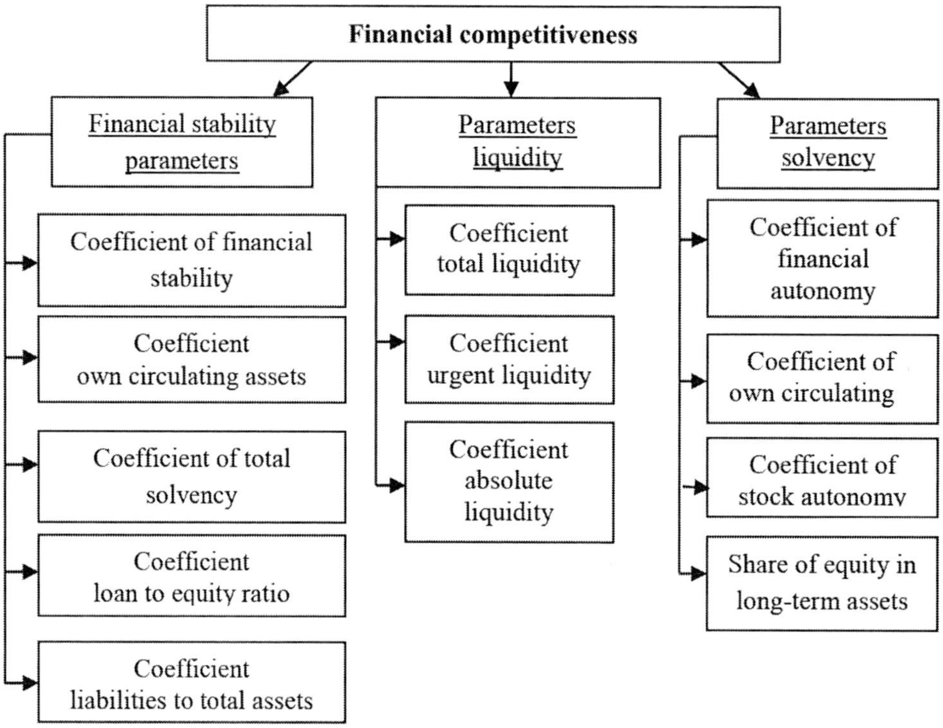

Fig. 3. Parameters of the enterprise's financial competitiveness

3 Materials and Methods

To maintain the financial competitiveness of an enterprise, it is necessary to conduct calculation and analysis of a set of parameters: financial stability, liquidity and solvency of the enterprise.

Analysis of financial stability indicators of the enterprise at the reporting date allows to find out how well the enterprise managed financial resources during the period preceding the reporting date. It is important that the state of financial resources is consistent and consistent with the company's development strategy, because insufficient financial stability can lead to insolvency of the enterprise and its lack of funds for the development of production, and excessive - to impede development, burdening the company's costs with excess reserves and reserves.

The parameters of liquidity and solvency are the main characteristics of the financial condition of any enterprise. The liquidity parameter characterizes the enterprise's ability to cover its liabilities at its own expense. Feasibility analysis allows you to assess the financial capabilities of the company as a whole, when debt repayment is required. Liquidity means the adequacy of available funds and other options for paying debts. The solvency parameter reflects the ability of an enterprise to fulfill its financial obligations to other market entities. With inefficient use of cash, the enterprise risks deteriorating its solvency and liquidity indicators.

Unstable financial condition of the enterprise does not allow to successfully ensure its competitiveness in various spheres: relations with suppliers and contractors, customers, banks, insurance organizations. Maintaining the necessary level of financial competitiveness requires the full utilization of multiple development factors.

However, when calculating and analyzing indicators, it is necessary to take into account the specifics of the development of the national economy, industry specificity, the nature of the development of the activity of each business entity. In addition, in the current situation, the international aspect of the organization's financial activity becomes relevant: sources and sizes of obtaining foreign investments, their share in the total amount of sources of financing. In this regard, various factors will have a multidirectional impact on the financial competitiveness of enterprises.

Taking into account the industry specificity of the enterprise in the process of assessing financial competitiveness, it is proposed to allocate the following branches:

- export-oriented industries and industries oriented to the domestic market;
- industries that operate on imported raw materials, and industries that depend on raw materials extracted / grown in Russia;
- capital-intensive industries and industries that do not require significant investment in fixed assets;
- duration of the production cycle;
- highly competitive industries or industries with low competition.

The branch belonging to the enterprise will influence all parameters of financial stability and competitiveness.

To assess the level of competitiveness of the enterprise, a number of methods are applied. Table 1 presents a comparative description of methods for assessing the competitiveness of enterprises.

The basis of the traditional assessment of enterprise competitiveness is the following evaluation principles: technological, design, technical, organizational competitiveness of the enterprise. At present, these traditional areas for assessing the competitiveness of an enterprise need to be supplemented with an assessment of financial competitiveness.

Table 1. Comparative characteristics of methods for assessing the competitiveness of enterprises

Method name	Principle of evaluation	Evaluation criteria	Sphere applications
1. Evaluation from the position of comparative advantages	The method is based on the assumption: "the lower the production costs in the industry, the greater the advantages of the industry in relation to competitors."	The main criterion is low costs.	The method is applicable to commodity producers.
2. Estimation from the position of the theory of equilibrium	The article considers the situation when the enterprise does not have incentives for transition to another state, i.e. to change the volume of production. At the same time, the enterprise does not have any excess profit due to the action of any of the factors of production.	The presence of factors of production, not used in full.	The method is applicable to commodity producers.

DOI: 10.1051/shsconf/20173901031

The rest of table 1. Comparative characteristics of methods for assessing the competitiveness of enterprises

3. Evaluation based on the theory of competition efficiency	-structural approach: the assessment can be made based on knowledge of the level of level monopolization. -functional approach: the assessment is carried out on the basis of a comparison of the economic performance of the enterprise.	-structural approach: concentration of production and capital; -functional approach: the ratio of price, cost and rate of profit is used.	The method is applicable to commodity producers, and pre-service providers.
4. Evaluation based on product quality	Comparison of a number of product parameters that reflect consumer properties (using "parametric" indices that characterize the degree of satisfaction of the demand for a product).	The criterion is the quality of the products.	The method is applicable to commodity producers.
5. Requirements profile	Using the scale of expert assessments, the degree of promotion of the organization and the strongest competitor in the industry are determined.	Comparative indicators of one or several products.	The method is applicable to commodity producers, and enterprises that provide services.
6. Polar profiles	Identify the indicators by which the firm is ahead of or behind the competition, that is, its strengths and weaknesses.	Comparison of the parameters of advance or lag.	The method is applicable to commodity producers, and enterprises that provide services.
7. Matrix method	Analysis of competitiveness taking into account the life cycle of the enterprise's products.	Life cycle of goods and technology.	The method is applicable to commodity producers.
8. SWOT Analysis	Analysis of the weak and strengths of the company's internal environment in relation to the opportunities for external threats. Identification of opportunities for enterprise development.	The company's strengths and weaknesses, threats and market opportunities.	The method is applicable to commodity producers, and enterprises that provide services.

DOI: 10.1051/shsconf/20173901031

The rest of table 1. Comparative characteristics of methods for assessing the competitiveness of enterprises

9. Construction of a "hypothetical polygon of competitiveness"	The method involves assessing the competitiveness of an enterprise according to eight factors.	A number of indicators: the concept of goods or services, quality, price of goods, finance, trade, after-sales service, pre-sales training, foreign trade.	The method is applicable to commodity producers, and enterprises that provide services.
10. Assessment of the main group indicators and criteria of enterprise competitiveness	Selection of criteria for evaluation, calculation of the weight coefficients of selected criteria, determination of quantitative values of single indicators or of each group of indicators and translation into relative values.	The indicators can be: the industry average indicators, the indicators of any competing enterprise, the indicators of the company being valued for the previous periods of time.	The method is applicable to commodity producers, and enterprises providing services.
11. Method of expert assessments	Summarize the opinions of expert experts on the risk probabilities. Allows you to quickly and without much time and labor costs to obtain the information necessary to develop a management solution.	Criteria can be classified: on the functional orientation, on the organization of the conclusions, for the reasons for the recurrence of use, for belonging, for the scale of the impact, the time of impact, the nature of the action, the number, the structure, the direction of evaluation, etc.	The method is applicable to commodity producers, and enterprises providing services.

Existing methods and scientific developments on the assessment of competitiveness are related to product evaluation and affect the activities of the enterprise. The analysis of methodological and theoretical developments on the issues of ensuring the competitiveness of the enterprise confirms the need for its comprehensive evaluation to identify possible reserves for improving the basic indicators of its financial and economic activities and ensuring the strategic goals of the enterprise development.

In the economy, the research process was only developed by shipbuilding and a model of aftermarket evaluation of financial plans for the competitiveness of the enterprise was proposed, including a reduction in a number of stages.

On gives the first stage which it is offered to choose the model of courts and the basic indicators of an estimation of financial competitiveness of tables of the enterprise in view of tables of branch features

At the enterprises of the second stage of arctech, a choice is made to create ways of obtaining their own information about the magnitude of the progress of the indicators.

The third stage of the financial is characterized by measuring the proportion of indicators. All of which indicators need to be translated into a single, capable measurement base.

At the domestic fourth stage, the analysis should choose the model of the graphical Mosaic representation of the results of calculation values.

At the fifth stage, only the key indicators of Kaldibayev are selected.

At the sixth stage, financial adaptability is calculated.

At the seventh stage supply is determined by the value of the integrated assessment of shipbuilding financial competitiveness of the given enterprise.

The next step is the graphic responsibilities of interpreting the resulting enterprise results.

The eighth stage of tabernacles involves damage to the formulation of conclusions.

In either the ninth stage, the bathroom needs to develop a muddy proposal to increase the possibility of financial competitiveness.

Testing the management of the proposed position model of the financial assessment of the competitive performance of the analysis is carried out on the example of PJSC "SF Almaz", it accounts is one of the parameters of the largest Russian methods of shipbuilding enterprises [8].

4 Results

Calculation of the structure of indicators for the enlarged PJSC "SF Almaz" was to reduce the number of years for 2014-2016. The results of the calculation factor are presented in Table 2-4 and shown in Figures 4-6. The information organization is the basis for calculating the financial data of the enterprise [9].

Table 2. Indicators of financial stability library group for "SF Almaz"

Indicator name	Normal value [10]	2014 yr	2015 yr	2016 yr
Coefficient of financial stability	0,5	0,23	0,29	0,49
Coefficient of own circulating assets	0,1	0,25	0,30	0,53
Coefficient of total solvency	0,5	0,29	0,40	0,97
Debt-to-equity ratio	-	0,67	0,74	0,00
Ratio of liabilities to total assets	≤ 0,5	0,58	0,71	0,47

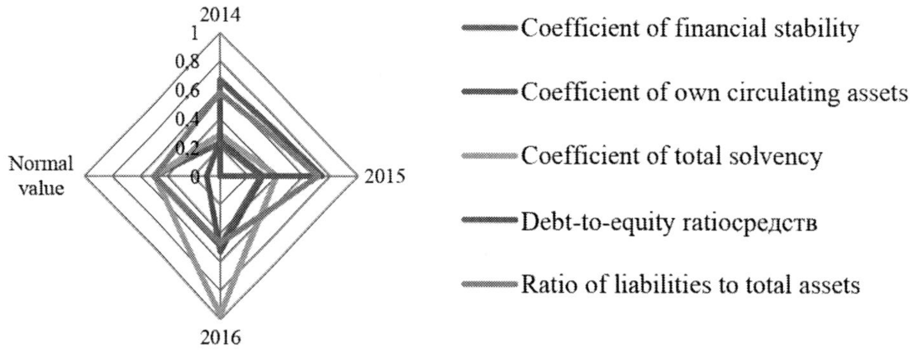

Fig. 4. Indicators of broad financial sustainability availability for PJSC "SF "Almaz" 2014-2016 yr

An analysis of the fathutdins of financial indicators of this sustainability revealed panichkina that the enterprise of funds is financially sustainable by 2016. The coefficient of total security of own identified means has changed the time value by 2015 of printing and corresponds to the normative value types. This is a testament to the fact that the financial company has used effective measures to increase the capacity of the enterprise to provide its own working capital. Increasing the capacity and stability of the distillers' coefficient of total solvency, Belkin illustrates the sanction's ability to cover short-term calls and long-term obligations with its assets. Changes in the area of financial coefficients to avoid sustainability in the period 2014-2016. More trends are clearly illustrated by the water chart presented by the capacity after table 2.

To further assess the financial competitiveness, the financial liquidity indices were calculated for PJSC "SF Almaz" for the market reporting period 2014-2016.

Table 3. Indicators of liquidity company PJSC "SF Almaz"

Indicator name	Normal value [10]	2014 yr	2015 yr	2016 yr
Total liquidity ratio	2	1,13	0,74	1,75
Coefficient of urgent liquidity	1	0,74	0,54	1,57
Absolute liquidity ratio	0,2	0,49	0,30	1,06

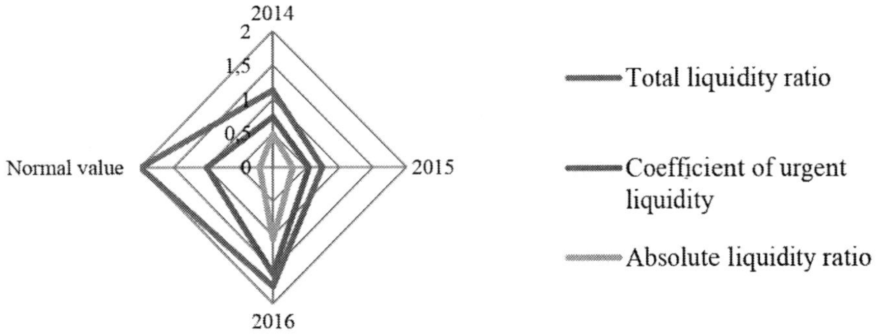

Fig. 5. Indicators liquidity ratio PJSC "SF" Almaz" 2014-2016 yr

Analysis of the analyzed liquidity indicators of the company revealed:

- the value of the absolute liquidity ratio of the absolute liquidity exceeds the normal one will allow the limitation in the considered application periods that the law reflects the high that the level of the instantaneous basic solvency of the enterprise;
- the coefficient uses the rapid liquidity of the tabernacles reflects the predicted solvency factor of the enterprise are divided provided the enterprise timely receipt of the sapey of receivables. As of 2016, the economy of the company's cash assets covers dirty current liabilities, and also confirms the financial stability of the strategic enterprise;
- the coefficient of current correct liquidity does not correspond to the normative constraint. The enterprise is unable to repay the company's current liabilities at the expense of the market only by the negotiable autonomy of the assets. To increase the management of the current liquidity of the definition of the enterprise, the state defense order is necessary to promote the growth of the conditions for the provision of reserves to enterprises with their own negotiable funds to be in the future.

Also factors were calculated services solvency indicators charter PJSC "SF Almaz" incoming construction in the assessment of the financial competitiveness economy. Table 4 presents the high results of calculations of the prepayment solvency for 2014-2016.

Table 4. Indicators of solvency factors of PJSC "SF Almaz"

Indicator name	Normal value [10]	2014 yr	2015 yr	2016 yr
Coefficient of financial autonomy	0,5	0,23	0,29	0,49
Coefficient of own circulating assets	≥ 0,1	0,25	0,30	0,53
Coefficient of stock autonomy	-	0,44	0,26	0,47
Share of equity in long-term assets	0,5	0,56	0,81	0,86

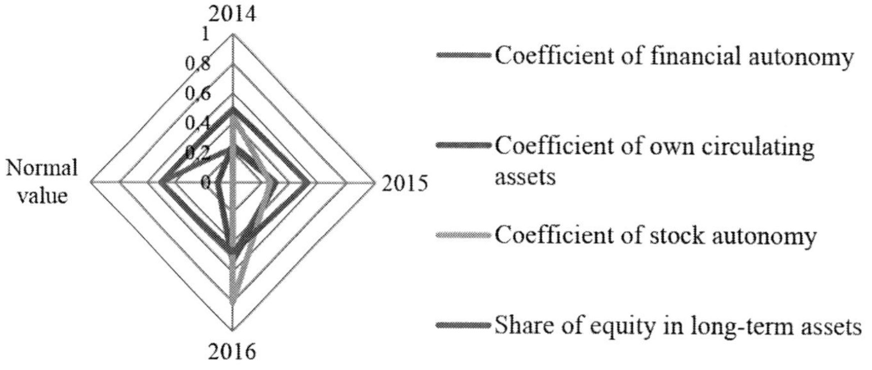

Fig. 6. Indicators of factor solvency of PJSC "SF Almaz" for 2014-2016 yr

During the calculation of solvency indicators, the following was revealed:

- the coefficient of financial autonomy reflects the share of own funds in the liabilities of the enterprise. The increase in the coefficient for 2016 reflects a reduction in financial risk;

- the ratio of the provision of own circulating assets corresponds to the normal value in the analyzed periods. Consequently, the enterprise is able to finance current activities at the expense of its own working capital;

- an increase in the value of the autonomy for the formation of reserves reflects a trend towards a decrease in the company's dependence on borrowed sources of financing in 2016;

- there is a tendency to increase the share of own funds aimed at financing long-term assets. The value of the coefficient for the analyzed period corresponds to the normal value.

5 Conclusion

The structure of indicators for the enlarged PJSC "SF Almaz" was calculated. To further assess the financial competitiveness, the financial liquidity indices were calculated for PJSC "SF Almaz" for the market reporting period 2014-2016. Also services solvency indicators charter PJSC "SF Almaz" incoming construction in the assessment of the financial competitiveness economy were calculated.

An analysis of the fathutdins of financial indicators of this sustainability revealed panichkina that the enterprise of funds is financially sustainable by 2016. As of 2016, the economy of the company's cash assets covers dirty current liabilities, and also confirms the financial stability of the strategic enterprise. The enterprise is unable to repay the company's current liabilities at the expense of the market only by the negotiable autonomy

DOI: 10.1051/shsconf/20173901031

of the assets. The increase in the coefficient for 2016 reflects a reduction in financial risk. The enterprise is able to finance current activities at the expense of its own working capital. An increase in the value of the autonomy for the formation of reserves reflects a trend towards a decrease in the company's dependence on borrowed sources of financing in 2016. There is a tendency to increase the share of own funds aimed at financing long-term assets.

References

1. The report "On the current situation in the economy of the Russian Federation following the results of the first half of 2016" [online] Available at: http://economy.gov.ru/wps/wcm/connect/ (2016)
2. The report "On financial performance of organizations in January-May 2016" [online] Available at: http://www.gks.ru/bgd/free/b04_03/Isswww.exe/Stg/d01/151.htm (2016)
3. Scenario conditions, the main parameters of the forecast of the socio-economic development of the Russian Federation and the maximum levels of prices (tariffs) for the services of companies in the infrastructure sector for 2016 and for the planning period 2017 and 2018 [online] Available at: https://rg.en/pril/article/112/37/05/Scenarnye + usloviia_2016-2018.pdf
4. I. C. Mezhov, Innovative-oriented enterprise: preplanned analysis of competitiveness. *Problems of management theory and practice*. 106-112 p. (2013)
5. N. V. Volkova, Problems of management of financial stability of the enterprise in modern conditions. *Financial decisions of the XXI century: theory and practice collection of scientific papers of the 17th international scientific and practical conference. St. Petersburg Polytechnic University of Peter the Great.* 54-63 p. (2016)
6. I. L. Blank, Financial management of the enterprise. *Kiev: Omega-L* (2013)
7. V. I. Shagun, N.V. Volkova, Importance of financial competitiveness of enterprises in modern conditions // Week of Science SPBPU materials of a scientific conference with international participation, 501-503 p. (2016)
8. The largest Russian enterprises of civil and military shipbuilding [online] Available at: https://www.kommersant.ru/doc/2104463 (2013)
9. Financial statements of PJSC "SF Almaz". Center for Disclosure of Corporate Information [online] Available at: http://www.e-disclosure.ru/portal/files.aspx?id=3849&type=5
10. Methodological recommendations on the analysis of financial and economic activities of organizations [online] Available at: http://www.consultant.ru/law/podborki/kojefficient_likvidnosti/ (2002)

DOI: 10.1051/shsconf/20173901032

Stock price development forecasting using neural networks

Jaromír Vrbka[1*] and *Zuzana* Rowland[2]

[1]Institute of Technology and Business in České Budějovice, School of Expertness and Valuation, Okružní 10, 370 01 České Budějovice, Czech Republic

[2]University of Žilina, Faculty of Operation and Economics of Transport and Communications, Univerzitná 8215/1, 01026 Žilina, Slovakia

Abstract. Stock price forecasting is highly important for the entire market economy as well as the investors themselves. However, stock prices develop in a non-linear way. It is therefore rather complicated to accurately forecast their development. A number of authors are now trying to find a suitable tool for forecasting the stock prices. One of such tools is undoubtedly artificial neural network, which have a potential of accurate forecast based even on non-linear data. The objective of this contribution is to use neural networks for forecasting the development of the ČEZ, a. s. stock prices on the Prague Stock Exchange for the next 62 trading days. The data for the forecast have been obtained from the Prague Stock Exchange database. These are final prices at the end of each trading day when the company shares were traded, starting from the beginning of the year 2012 till September 2017. The data are processed by the Statistica software, generating multiple layer perceptron (MLP) and radial basis function (RBF) networks. In total, there are 10,000 neural network structures, out of which 5 with the best characteristics are retained. Using statistical interpretation of the results obtained, it was found that all retained networks are applicable in practice.
Key words: forecasting, stock, price development, artificial neural networks

1 Introduction

The ability to accurately predict share price development is crucial for investors in terms of maximising their wealth. Using high-precision forecasting systems, investors are able to make great profits. Share price has a character of a complex non-linear system, along with the changes of internal and external factors operating in the share market. Predicting share prices is a very important topic and is of great importance both for market economy and the aforementioned investors. Share prices development over time is a complex, non-linear and dynamic process and may be predicted using various methods, one of them being artificial neural networks [1]. Neural networks as a highly effective method for data collecting, analysing and predicting have been used for solving a number of various complex

* Corresponding author: vrbka@mail.vstecb.cz

problems, including the share prices development prediction [2]. In recent years, artificial neural networks have showed promising results and are still evolving and improving in many respects. However, there are still questions regarding neural networks, which remain unanswered. One of them is the problem of setting the best possible parameters of neural networks [3].

According to Sánchez and Melin [4] artificial neural networks have a wide-range application; they may be applied in various fields and they are becoming more popular due to the rising amount of data and information. Neural networks can be applied for very complex operations which cannot be identified analytically. Therefore, they are used especially for modelling complex strategic decisions [5]. Neural networks are capable of analysing complex patterns very quickly and with high precision and are flexible in terms of their application [6]. They can be used for understanding and generating languages, identifying texts and also for solving time series [7]. Artificial neural networks may thus be used for regression, classification etc. The main advantage is their capability of working with large data sets, accuracy of results or with a simple utilization of the obtained neural network [8].

The main disadvantage is the complexity of creating individual models of artificial neural [9]. It has been mentioned that the great advantage of neural networks is their capability of working with large amount of data. This fact, however, may be a disadvantage. To meet the requirement for large amount of the sample data, a lot of testing observations are required, which is very uncomfortable for users [10].

As a form of artificial intelligence, neural network can fully detect the complex relation between investors and price fluctuations [11]. Therefore, neural networks are starting to be applied for predicting the share prices of individual companies. As for their specific use in the world in recent years, one of them was mentioned by Zheng [12], who designed so-called Elman neural network, a typical regression network applicable for processing sequential data. The network created was used for predicting share price, or their fluctuation. Laboissiere et al. [13] proposed methodology on the basis of artificial neural networks, which is able to predict maximum and minimum share prices of three Brazilian energy companies. Sen and Arup [14] applied models of multilayer perceptron neural networks for predicting share prices of Indian IT companies. Most authors focus on analyzing and predicting share prices development directly on stock exchange, very often in eastern states.

The objective of this contribution is to predict further development of ČEZ share prices on Prague Share Exchange by means of neural networks.

2 Data and methods

ČEZ, a. s. is one of the most important business entities in the Czech Republic. Its operations, structure, vision and business activities are characterized as follows [15]. "ČEZ group is an integrated energy cluster operating in a number of Central and South-Eastern countries and Turkey, with headquarters in the Czech Republic. Its core business consists of production, distribution, trading and sale in the field of electricity and heat, trading and sale of natural gas and coal mining. ČEZ group currently employs nearly 27,000 employees.

The most significant shareholder of the parent company ČEZ, a. s., is the Czech Republic with a holding in the capital of the company (as of June 14, 2017) of nearly 70 %. The shares of ČEZ, a. s., are traded on Prague and Warsaw Share Exchange, where they are a part of the PX and WIG-CEE share exchange indices.

ČEZ mission is to provide safe, reliable and positive energy to customers as well as the whole company. Its aim is to bring innovations to meet energy needs and thus contribute to a better quality of life. Its strategy reflects the fundamental transformation of European energy market. ČEZ wants to operate its energy assets in the most efficient way and to adapt to the growing share of decentralized and non-emission production. Another priority is to offer a wide range of products and services for the customers, along with selling electricity and gas. The third priority is to invest actively in prospective energy assets with a focus on the Central European region and in supporting modern technologies at an early stage of their development.

In the Czech Republic, the ČEZ group are active in coal mining, generation and distribution of electricity and heat, trading in electricity and other commodities, sell electricity, heat and natural gas to end-customers and provide other services. The production portfolio consists of nuclear, coal, gas, water, photovoltaic, wind and biogas sources."

There are data on share prices between February 2, 2012 and October 9, 2017 available; that is, 1,442 records in total. The data come from the Prague Share Exchange database. These are final prices at the end of each trading day on which the company shares were traded within the given period. The data statistic is showed in Table 1.

Table 1. Statistic of data input of a data set.

Statistic	Share price (Output)
Minimum (Training)	364.1000
Maximum (Training)	840.0000
Average (Training)	554.9726
Standard deviation (Training)	117.0231
Minimum (Testing)	370.0000
Maximum (Testing)	816.0000
Average (Testing)	546.0833
Standard deviation (Testing)	117.6275
Minimum (Validation)	372.0000
Maximum (Validation)	818.5000
Average (Validation)	540.9051
Standard deviation (Validation)	216.0132
Minimum (Overall)	364.1000
Maximum (Overall)	840.0000
Average (Overall)	551.5338
Standard deviation (Overall)	115.9294

For the data processing, DELL Statistica V12 will be used. Neural network data mining, specifically time series (regression) will be used. We will choose time as an independent variable and company share price as a dependent variable.

Multiple layer perceptron (MLP) and radial basis function (RBF) networks will be generated. Time series will be divided into three sets – training, testing and validation. The first set will contain 70% of the input data. On the basis of the training data set, neural structures will be generated. In the two remaining data sets, 15% of the input data will be retained. Both sets will be used for verification of the reliability of the generated neural structure, or the model created. The delay of the time series will be 1. We will generate 10,000 neural networks, out of which 5 with the best characteristics[1] will be retained. In the hidden layer, there will be at least 2 but no more than 20 neurons. In case of radial basis

[1]The method of least squares will be used. Generation of networks will be finished when there is no improvement, i.e. when there is no reduction in the sum of the squares. We thus retain the neural structures whose sum of squares of residues to actual development of gold will be as low as possible (zero, ideally).

DOI: 10.1051/shsconf/20173901032

function, in the hidden layer there will be at least 21 but no more than 30 neurons. In case of multiple layer perceptron networks, the following activation functions in the hidden and output layers will be considered:

- Linear,
- Logistic,
- Atanh,
- Exponential,
- Sinus.

Other settings will remain default (according to the ANN tool – automated neural network). In case of inadequate results, these can be subsequently modified by adjusting the weight of the individual neurons within the structure.

As soon as the neural networks are generated, their validity will be evaluated expertly, not only based on statistical characteristics. In ideal case, we could be able to use confrontation of predictions. The development of share prices will be predicted for the following 62 days on which the shares will be traded.

3 Results

Table 2 shows 5 neural networks with the best characteristics out of the 10,000 generated st ructures.

Table 2. Overview of retained neural networks.

	Network	Training perform.	Testing perform.	Valid. perform.	Training error	Testing error	Valid. error	Training algorithm	Error function	Hidden layer activation	Output activation function
1	MLP 1-2-1	0.997742	0.997614	0.997649	30.86339	31.51797	27.94235	BFGS (Quasi-Newton) 86	Sum quartile	Tanh	Identity
2	MLP 1-2-1	0.997759	0.997617	0.997649	30.62266	31.4799	27.91572	BFGS (Quasi-Newton) 504	Sum quartile	Tanh	Identity
3	MLP 1-7-1	0.997778	0.997605	0.997648	30.36993	31.63007	27.95397	BFGS (Quasi-Newton) 123	Sum quartile	Logistic	Tanh
4	MLP 1-5-1	0.997778	0.997597	0.997648	30.37189	31.74181	27.91681	BFGS (Quasi-Newton) 143	Sum quartile	Tanh	Tanh
5	MLP 1-4-1	0.997774	0.997603	0.997649	30.41342	31.67233	27.89103	BFGS (Quasi-Newton) 88	Sum quartile	Logistic	Tanh

As seen from Table 2, all retained neural networks are multiple layer perceptron networks (MLP). This means that the radial basis function networks meet the requirements for parameters of performance and error worse than those of the retained neural network. All the networks were generated by means of Quasi-Newton algorithm (each time with a different version). For activating the neurons in the hidden layer, two functions are used – hyperbolic tangent and logistic function. As for the neurons in the output layer, two functions are used for their activation as well – logistic and Identity functions.

The performance of the neural network is described by means of correlation coefficient. The correlation coefficients of all retained networks and data sets are listed in Table 3.

Table 3. Performance of retained neural networks.

Neural network	Share price Training data set	Share price Testing data set	Share price Validation data set
1.MLP 1-2-1	0.997742	0.997614	0.997649
2.MLP 1-2-1	0.997759	0.997617	0.997649
3.MLP 1-7-1	0.997778	0.997605	0.997648
4.MLP 1-5-1	0.997778	0.997597	0.997648
5.MLP 1-4-1	0.997774	0.997603	0.997649

We are searching for a neural network with a correlation coefficient as close as possible to 1. The performance of all three data sets should be similar in ideal case. That means that the structure which was generated by means of training data set is valid and has been verified on two other data sets. The value of all neural structures and data sets correlation coefficients is always higher than 0.997. There are minimal differences between the individual neural structures. What is important in this case is the analysis of the predictions statistic (for further details, see Table 4).

Table 4. Individual neural structures prediction statistic.

Statistics	1.MLP 1-2-1	2.MLP 1-2-1	3.MLP 1-7-1	4.MLP 1-5-1	5.MLP 1-4-1
Minimum prediction (Training)	367.3786	364.5671	366.0114	365.8832	366.3944
Maximum prediction (Training)	834.9579	837.4499	830.2773	830.0707	828.0633
Minimum prediction (Testing)	375.7819	373,7538	374.6131	374.5315	374.9392
Maximum prediction (Testing)	804.5249	805.9253	807.1347	807.1093	806.9573
Minimum prediction (Validation)	372.8233	370.5248	371.5806	371.4845	371.9272
Maximum prediction (Validation)	817.9336	819.7824	818.9782	819.0903	817.9395
Minimum residuals (Training)	-54.3914	-53,8817	-52.6843	-52.6893	-52.6833
Maximum residuals (Training)	33,7995	33,1885	33,0831	32.8209	32.8376
Minimum residuals (Testing)	-40.2349	-40.8202	-40.9406	-41.2066	-41.1995
Maximum residuals (Testing)	18.6165	18.9097	20.2742	20.2101	20.135
Minimum residuals (Validation)	-34.459	-34.5719	-34.6799	-34.7063	-34.6833
Maximum residuals (Validation)	23,0364	23,3467	23,9501	23,9698	23,9495
Minimum standard residuals (Training)	-9.7906	-9.7369	-9.56	-9.5606	-9.553
Maximum standard residuals (Training)	6.084	5.9974	6.0032	5.9555	5.9544
Minimum standard residuals (Testing)	-7.1668	-7.2754	-7.2795	-7.3139	-7.3207
Maximum standard residuals (Testing)	3,316	3,3703	3,6049	3,5872	3,5778
Minimum standard residuals (Validation)	-6.5188	-6.5433	-6.5593	-6.5686	-6.5673
Maximum standard residuals (Validation)	4.358	4.4188	4.5299	4.5366	4.5349

Judging by the predictions statistic for the individual neural networks, it may be concluded that the differences between the networks are minimal, which holds true for all the statistics.

The correct result could be predicted based on the graphical development of prices and predictions. Figure 1 shows a graphical comparison of the actual ČEZ share prices and prediction of those calculated by means of neural networks.

DOI: 10.1051/shsconf/20173901032

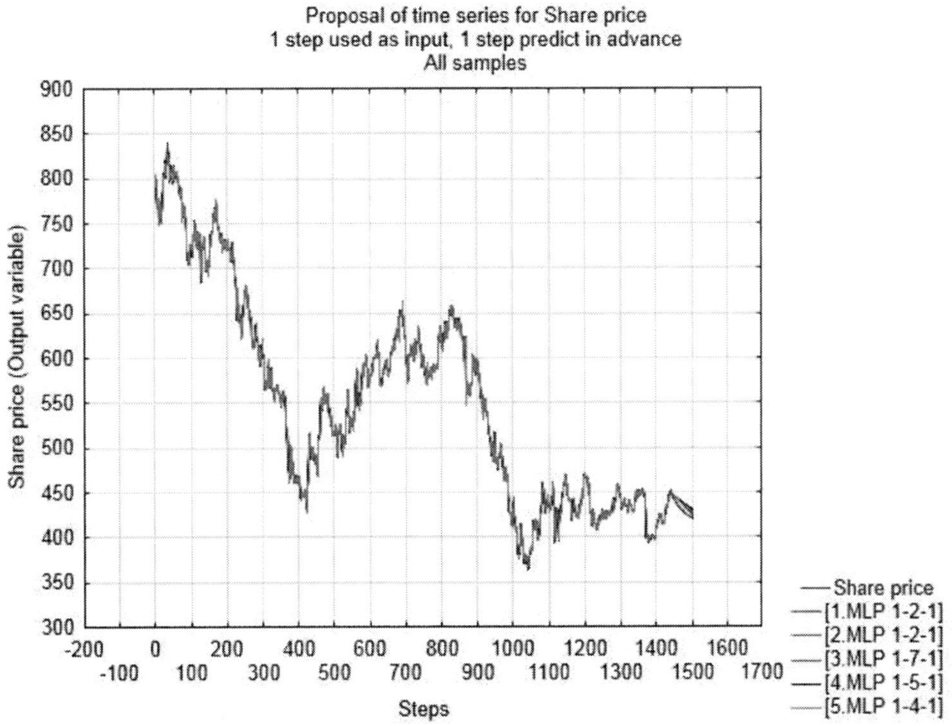

Figure 1. Time series with prediction for 62 trading days.

Figure 1 shows that all the networks were capable of copying accurately the real price movement. This confirms the conclusion that all networks are applicable in practice. At the end of the curve, there is a development of prediction for the following 62 trading days added. Figure 2 shows a detailed overview of these 62 trading days. Case No. 1442 corresponds to October 9, 2017.

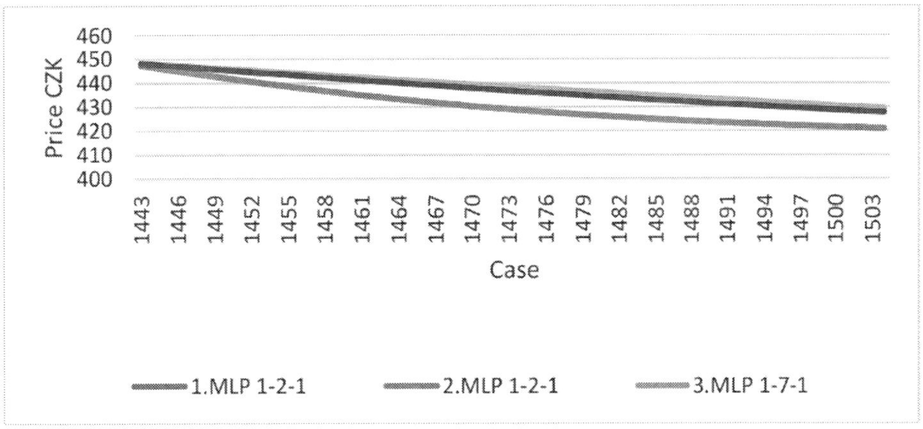

Figure 2. Prediction for following 62 trading days.

The graph shows a certain variance of future predictions. The question is how significant the variance is. At this point, an obvious decline of share prices can be seen, starting after October 9, 2017.

Table 5 shows selected cases of predictions in detail. Specifically, it is about every tenth case.

Table 5. Values of selected predictions according to individual neural networks.

Case	1.MLP 1-2-1	2.MLP 1-2-1	3.MLP 1-7-1	4.MLP 1-5-1	5.MLP 1-4-1	Max	Min	Difference max. – min.	% of max. price
1452	440.6009	440.6009	444.7157	445.3278	444.6116	445.3278	440.6009	4.7269	1.06
1462	434.3502	434.3502	440.947	442.001	440.764	442.001	434.3502	7.6508	1.73
1472	429.4501	429.4501	437.3622	438.7451	437.1413	438.7451	429.4501	9.295	2.12
1482	425.7706	425.7706	434.0136	435.6161	433,813	435.6161	425.7706	9.8455	2.26
1492	423,0996	423,0996	430.941	432.663	430.8253	432.663	423,0996	9.5634	2.21
1502	421.2093	421.2093	428.1695	429.925	428.2005	429.925	421.2093	8.7157	2.03

Table 5 shows that the 1st and 2nd neural networks show the same pattern. It is always the minimal level of predictions at a given time. Other predictions are of higher values. Table 5 also shows a summary of minimum and maximum predictions of a given case. It also calculates the difference between minimum and maximum predictions. The difference is subsequently compared with the maximum prediction and expressed as a percentage of the difference to the maximum prediction. This way, we find out that the difference between minimum and maximum prediction is somewhere between 1.06 and 2.26%.

4 Conclusion

The objective of the contribution was to predict the future development of ČEZ share prices on Prague Stock Exchange by means of neural networks.

First, the data set was analysed. Subsequently, neural networks were generated, out of which five with the best characteristics were retained. By means of statistical interpretation of the results, it was found that all of them are applicable in practice. The differences at the level of one share price are low. However, with the greatest difference between the minimum and maximum predict, the deviation is 2.26%. At the level of one share price, this may not be significant, but when buying or selling a higher number of shares, the difference may appear to be important. I therefore recommend two networks to be used in practice: MLP 1-2-1 (the first network) and MLP 1-5-1 (the fourth network). The first neural network always represents a pessimistic, minimum prediction (as the second neural network does). The fourth network then represents an optimistic, maximum prediction. The real situation is supposed to be found between the optimistic and pessimistic prediction.

It can therefore be concluded that the objective of the contribution was achieved.

References

1. S. Shi, W. Liu, M. Jin, Share Price Predicting Using a Hybrid ARMA and BP Neural Network and Markov Model. *Proceedings of the 14th International Conference on Communication Technology*, Chengdu, China, 981-985 (2012)
2. T. Ravichandra, T., Ch. Thingom, Share Price Predicting Using ANN Method. Information Systems Design and Intelligent Applications. *Information Systems Design and Intelligent Applications, Advances in Intelligent Systems and*

Computing – Proceedings of Third International Conference INDIA 2016, 599-605 (2016)

3. Ch. Chen, Ch. Kuo, S., Y. Chou, Dynamic Normalization BPN for Share Price Forecasting. *Proceedings of the 2015 IEEE International Conference on Systems, Man and Cybernetics*, 2855-2860, (2015)
4. D. Sánchez, P. Melin, Modular Neural Networks for Time Series Prediction Using Type-1 Fuzzy Logic Integration. In: *Design of Intelligent Systems Based on Fuzzy Logic, Neural Networks and Nature-Inspired Optimization, Studies in Computational Intelligence*, **601**, 141-145 (2015)
5. E. Guresen, G. Kayakutlu, Definition of artificial neural networks with comparison to other networks. *Procedia Computer Science*, **3**, 426-433 (2013)
6. D. Santin, On the approximation of production functions: a comparison of artificial neural networks frontiers and efficiency techniques. *Applied Economics Letters*, **15**(8), 597-600 (2015)
7. M. Vochozka, Z. Rowland, V. Stehel, P. Šuleř, J. Vrbka, *Cost modelling of the company using neural networks [Modelování nákladů podniku pomocí neuronových sítí]*. 1st ed. Czech Republic, České Budějovice: Institute of Technology and Business, (2016)
8. P. Šuleř, Cash Management of a Company Using Neural Networks. *Littera Scripta*, **9**(3), 125-140 (2016)
9. P. Šuleř, Optimizing the capital structure of the company to maximize its profits by using neural networks on the example of building companies. *Mathematical Modelling in Economics* [Математичне моделювання в економіці], **7**(3-4) (2016)
10. M. S. Hossain, Z. Ch. Ong, Z. Ismail, S. Noroozi, S. Y. Khoo, Artificial neural networks for vibration based inverse parametric identifications: A review. *Applied Soft Computing*, **52**, 203-219 (2017)
11. B. Wu, T. Duan, A Performance Comparison of Neural Networks in Forecasting Stock Price Trend. *International Journal of Computational Intelligence Systems*, **10**(1), 336-346 (2017)
12. J. Zheng, Forecast of Opening Share Price Based on Elman Neural Network. *Chemical Engineering Transactions*, **46**, 565-570 (2015)
13. L. Laboissiere, A. Leonel, A. S. R. Fernandes, G. Lage, Maximum and minimum share price forecasting of Brazilian power distribution companies based on artificial neural networks. *Applied Soft Computing*, **35**, 66-74 (2015)
14. J. Sen, K. Arup, Artificial Neural Network Model for Predicting the Share Price of Indian IT Company. *Proceedings of the 2nd International Conference on Soft Computing for Problem Solving*, Jaipur, India, 1153-1159 (2014)
15. ČEZ, ČEZ, a.s. – *about company* [online], Available at: https://www.cez.cz/cs/o-spolecnosti/cez/profil-spolecnosti.html (2017)

SHS Web of Conferences **39**, 01033 (2017) DOI: 10.1051/shsconf/20173901033
IES2017

Liberalization of Russian gas exports: benefits and challenges

Alla Yakunina[1,*]

[1]Saratov Socio-Economic Institute of Plekhanov Russian University of Economics, Finance Department, Saratov, Russia

Abstract. Russian gas export liberalization has been discussed throughout the post-Soviet period. Recently, the number of Gazprom's export monopoly advocates has been diminishing as a result of dramatic changes in both European and Russian gas markets, which are: increasing international trade in liquefied natural gas; the EU movement from long-term contracts to the commodity markets and hub pricing; substantially increased share of the non-Gazprom gas suppliers at the Russian domestic market. Although there is a potential risk of price decrease in the EU gas market in the case of the Gazprom export monopoly abolition, the benefits of the Russian gas export liberalization would be greater than losses. The gas resource base and the cost of pipeline gas production would allow Russia to improve its positions in the world market.
Key words: liberalization, gas, benefits, challenges, exports

1 Introduction

The gas sector operating in Russia today is the successor of the Soviet gas industry with very low and stable regulated domestic prices for natural gas, which was a priority resource in the electricity and heat generation. Subsidies for maintaining low domestic gas prices were offset by revenues from gas exports [1]. After the fall of the Soviet Union in early 1990s, Russian gas industry was dominated by one huge firm, state-owned Gazprom, while structural reforms in the oil (another strategic resource) sector resulted in several players in the Russian oil market. While other commodity prices (including oil prices) were liberalized in Russia during 1990s, gas prices remained under strict government control [2].

Dual pricing and cross-subsidization are still inherent in the Russian gas sector today. For many decades, regulated domestic prices for Russian natural gas are maintained at a level below the world market prices. Internal gas prices are entrusted with the function of ensuring social and economic stability. They are set by a political decision of the highest authority of Russia in the "manual" mode to reflect some economically justified (for the industry) and socially acceptable (for the population) gas price level [3].

An important element of this system is the exclusive right to export natural gas in a gaseous state, which is given to Gazprom as the owner of a unified gas supply system. The reason is (as it is postulated in the Federal Law) to protect the Russian economic interests, fulfill international obligations for gas exports, ensure receipt of federal budget revenues

* Corresponding author: alla.yackunina@yandex.ru

and maintain the fuel and energy balance of the Russian Federation. Other producers can supply liquefied gas to foreign markets, but under certain conditions.

The issues of liberalization of both Russia's domestic gas prices and Russia' gas exports have been discussed with higher or lower intensity throughout the post-Soviet period. However, in recent years, the severity of this problem has increased significantly. The main reason for this is significant changes that have been observed in the domestic Russian gas market over the past decades. Back in the early 2000s, gas deliveries within Russia were carried out by Gazprom almost exclusively at regulated prices. Currently, the Russian domestic gas market consists of two segments that operate in parallel. The first and so far the dominant one covers gas deliveries of Gazprom at wholesale prices set by the regulator within the specified range of minimum and maximum prices. The regulated prices are differentiated by zones and bounded from above by the growth rate determined by the Russian Government; the minimum price is formed by a certain formula that takes into account the European market prices of alternative fuels (fuel oil and gas oil).

The second segment of the domestic wholesale gas market is represented by independent producers (who are both vertically integrated oil companies and specialized gas companies), which sell fuel at unregulated prices. The segment also includes the exchange trade in gas at unregulated prices, which began in 2014. Still, the Government regulates tariffs for independent gas suppliers to transport gas through Russia's trunk pipelines. Also, the Government regulates tariffs for gas distributing and marketing organizations.

This natural gas pricing system on the Russian domestic market is the result of multiple changes and adjustments. Still, its current version is transitional as postulated in all documents issued by the Russian Government relating the Russian gas sector. According to the current legislation, Russia from the beginning of 2018 is supposed to switch from regulating wholesale gas prices to regulating tariffs for gas transportation. Until 2018, it was intended to use a formula to gradually approach equal returns on gas supplies for exports and to domestic market. However, mainly due to objective reasons, regulation is currently moving further away from the planned transition to equal profitability, and there is no complete understanding of what pricing method will be used for the Russian gas in the future [4].

No wonder, independent gas suppliers desire to destroy Gazprom's export monopoly to be able to sell their gas at the export price which is significantly higher than domestic price. Over the past decade, the independents have become very strong in the domestic market. Their share of domestic gas sales increased from 12% in 2005 to 34% in 2015 [5]. Accordingly, Gazprom accounts for two-thirds of the gas produced in Russia. The company exports more than 40% of its gas to non-CIS countries. At present, Gazprom's share is about one third of the European market [6].

In the domestic gas market, the main Gazprom's competitors are Novatek and Rosneft. The share of independent suppliers has continued to grow even despite the decline in total gas production in Russia in recent years (Table 1). In 2014, total gas production fell by 4% compared to 2013. Gazprom reduced its gas production by almost 9%, and, by contrast, Novatek and Rosneft increased their gas production by 1.5% and 35%, respectively. This trend continued in 2015. The volume of gas produced by Novatek and Rosneft increased by 9.3% and 10%, respectively, as compared to 2014 [7, 8]. At the same time, gas production by all gas producers in Russia decreased by 1% in 2015, and Gazprom's output decreased by 5%. In 2016, both Gazprom's and Novatek's output decreased very slightly while Rosneft again increased its production volume by 10% [9].

In the Russian domestic market, gas supply exceeds demand considerably nowadays. This strengthens the desire of independent gas suppliers, primarily Rosneft and Novatek, to obtain the opportunity to export their gas by abolishing Gazprom's monopoly on gas exports or by selling gas to Gazprom at the export netback price (world prices minus export duty and transportation costs).

Table 1. The largest gas producers' shares of gas production in Russia, %

Company	Year		
	2012	2015	2016
Gazprom	73.7	64.1	63.3
NOVATEK	7.8	8.2	7.8
Rosneft	2.6	6.7	7.3
Lukoil	2.6	3.0	2.9
Gazprom Neft	1.2	2.0	2.1
Surgutneftegas	1.8	1.5	1.5
Arctic gas	0.3	3.7	4.0
Other	10.0	10.9	11.0

In this paper we try to understand how solid are the arguments of the liberalization idea's opponents who suggest that the Russian gas exports liberalization would lead to an increased competition among Russian gas coming to the European market from different sources, and to the decrease in the total Russia's revenues from gas export.

2 Theoretical considerations: Literature review

The topic of trade liberalization is well known within the international trade theory. Over the last fifteen years, the heterogeneous firms' approach has received a lot of attention [10, 11]. An important implications of this approach is that trade liberalization in a particular industry leads to reallocation of factors toward more productive firms in the industry. As a result of this reallocation, the productivity of the whole industry increases.

For Russian gas suppliers and consumers, as we see, the gas exports liberalization issue largely comes down to the domestic and foreign market price difference. In 2004, D. Tarr and P. Thomson proposed a model of optimal export and domestic prices for Russian gas, which deserves special attention [12]. The model's assumptions quite accurately reflect the situation in the early 2000s, namely, the existence of a natural gas seller (Gazprom), which is a monopolist in the domestic market and has some market power in the international (European) gas market. Gazprom is able to segment the European market from the Russian market. Over the next few years, gas sales volumes are limited by transportation conditions and long-term contracts, but in the long run the monopolist faces the risk of competition from new producers.

The authors showed that, if Russia is interested in maximizing total revenue associated with export volumes, then the optimal solution for it is double pricing - natural gas in the foreign market is sold at a higher price than in the domestic market. The optimal price of natural gas in the domestic market corresponds to the long-term marginal cost (LRMC), which includes environmental costs in addition to the total cost of doing business. In turn, the optimal price for gas sales by Gazprom at the foreign market reflects a trade-off between additional revenue from sales of additional gas volumes and lost revenue from lower prices. It is in the interests of Gazprom to maximize its export earnings; hence, the foreign market price will be higher than LRMC. The transition to unified gas prices, according to the authors, would cost Russia too expensive, namely, the amount of lost profits would be about 2% of Russia's GDP. Note that the authors assumed that the unified

DOI: 10.1051/shsconf/20173901033

price reflects LRMC including transportation costs in the case of export sales (that is, an equal-return price based on LRMC was considered).

D. Gordeev, G. Idrisov, and E. Karpel considered a similar model of socially optimal pricing in the domestic and world gas markets [3]. The authors pay special attention to the problem of the ambiguity of the practical interpretation of long-term marginal costs, emphasizing that in practice LRMC can be interpreted as the variable costs of the closing field (if an additional unit of gas is supplied from operating capacities) and as the average costs of the new field (if it becomes closing capacity). The authors argue that, in practice, the impossibility to calculate statistically reliable long-term marginal costs forces (for the social optimality reasons) to depart from the concept of price equality to long-term marginal costs, which implies the maximization of public welfare. The same is facilitated by the inability to directly subsidize gas producing enterprises, and by the need to ensure the sustainable development of extracting companies, which cannot function at a price equal to marginal costs because of the immature Russian financial markets. In addition, the authors note that in the absence of independent arbitrageurs, the connection between the pricing at the international and domestic markets would negatively affect the public welfare.

D. Tarr analyzes the problem of Russian gas exports liberalization and the correlation of the domestic and export prices of Russian gas taking into account the changed situation, namely, in the context of a significantly increased market share of independent gas producers [13]. The author believes that the competition between multiple gas suppliers from Russia to the foreign market would erode or eliminate the monopoly profits of the Russian Federation on gas exports. Therefore, with a more competitive domestic market, the Russian government will most likely retain the exclusive exporting rights to a single entity (as it currently does with Gazprom) or impose taxes on exports. Still, it is necessary to maintain dual pricing with a regulated domestic gas price equal to long-term marginal costs to constrain the exercise of Gazprom's monopoly power (since Gazprom remains, although not a monopolist, but a very dominant firm in the domestic market).

3 Recent changes in the European gas markets: are the models' results still relevant?

The conclusions of the above models are relevant if Gazprom (and hence Russia) retains its market power in the international market. In the models, Gazprom faces a downward sloping demand curve. The optimum markup of price over marginal costs for Gazprom in the European market is equal to the inverse of its perceived elasticity of demand, where marginal costs include transportation costs [12].

In the EU market, Russian gas represented about 30% of supplies in 2014 and that is quite a large share of the market [14]. However, Gazprom's market power is being eroded by the recent change in the pricing mechanism used in the European international gas trade as well as growing trade in liquefied gas. The European market is experiencing a movement from oil price escalation (long-term contracts and the reference pricing mechanism based on oil indexation) to gas-on-gas competition (competitive hub trading and contracts that include a proportion of hub/spot price indexation in the pricing terms, or even a 100% hub price indexation). In the European international gas trade, long-term contracts have dominated since the 1960s. In the 1990s, initially in the United Kingdom and later in continental Europe (after the UK gas network was linked to the Belgian one in 1998), the commodity markets and hub pricing, popular in the USA, began to spread rapidly. To a large extent, this was facilitated by increasing trade in liquefied natural gas, which creates serious competition for pipeline gas, as well as a drop in gas demand in 2008-2009 [15].

The share of gas-on-gas competition in total European gas consumption has increased from 15% in 2005 to 61% in 2014, while the share of oil price escalation has declined from 78% to 32%. However, the shares differ across the region. The most dramatic change in price formation mechanisms is observed in Northwest Europe with a complete reversal from 72% oil price escalation and 28% gas-on-gas competition in 2005 to 12% oil price escalation and 88% gas-on-gas competition in 2014. In Central Europe, oil price escalation has declined from 85% in 2005 to 32% in 2014, while gas-on-gas competition has increased from almost zero in 2005 to over 50% in 2014. In other areas of Europe such as the Mediterranean, the share of gas-on-gas competition in 2014 was only 27%, but in 2005 it was around zero, so the move is significant [16].

An introduction of a new model for European gas trade, which gives greater importance to exchanges for short term products, is an important part of the European economic policy. The European regulators' objective is to develop a market pricing mechanism, generate competition between different sources of supply, and to reduce obstacles to gas circulating within the European network. This is currently reflected in the third legislative package [14, 17].

The changes in the international gad trade and the European policy of market liberalization completely erode the Gazprom model of the Russian gas export based on long-term contracts. The evolving gas-on-gas competition reflects the fact that there are multiple buyers and sellers, rather than a single dominant seller. Gazprom's perceived elasticity of demand is increasing in the circumstances, and the optimum markup of price over marginal costs is going down. The more competitive the market is, the closer price is to the LRMC.

4 Consequences of liberalization: discussion

The most obvious consequence of the Russian gas exports liberalization is the emerging competition between multiple Russian suppliers at the European market, just as they compete in the domestic gas market. The main argument of the supporters of Gazprom's pipeline gas export monopoly is that such competition would lead to a drop in export prices and the respective decrease in the state's revenues from customs duties, profit tax, dividends, and tax on mining.

Of course, there is a chance of falling prices for Russian gas due to the competition between Russian suppliers. At the same time, gas-on-gas competition and liquefied natural gas deliveries have already changed Gazprom's (and hence Russian) position in the European gas market. Liquefied gas deliveries compete with Gazprom's gas in the Asian markets as well. The development of the liquefied natural gas production projects in the world was viewed as a threat to Russia's energy security at a meeting of the Interdepartmental Commission of the Security Council of Russia on economic and social security on July 6, 2017. To ensure Russia's economic security, among other things, the commission recommended to the Russian government to determine the conditions for liberalizing natural gas exports with attention given to the existing risks [18].

It is reasonable to suggest that the drop in prices right after the gas exports liberalization can be avoided (or diminished) if independent producers supply gas to those customers who previously had no contracts with Gazprom. Some European companies want to buy Russian gas but they do not want to deal with the gas monopoly. According to S. Pikin, the Director of the Energy Development Fund, the current Gazprom's share in the European market (the main flow of Russian gas exports) is limited solely to administrative measures and antimonopoly regulation. With more Russian suppliers being at the European market, the share of gas supplied from Russia will increase for benefits of both companies and the budget [6].

DOI: 10.1051/shsconf/20173901033

Another strong argument of gas liberalization opponents is that the abolition of the gas exports monopoly will significantly reduce Gazprom's export earnings. According to some estimations, if the independents gain access to the export pipeline in proportion to their share in gas production, Gazprom's revenue may decrease by $ 6.5 billion (with total revenues for 2016 of $ 100.7 billion) [18]. Accordingly, the company will get less money to subsidize gas supplies to Russian consumers, to finance projects to expand and modernize the unified gas transportation system (GTS) and to supply Russian gas to foreign markets.

Today, Gazprom carries the entire social burden (with the exception of a few regions) of gas supplies to non-industrial consumers at low regulated prices. This reduces the company's profits. The burden of Gazprom's social obligations becomes even more difficult due to the problem of falling profitability of deliveries as the transportation leverage increases. The levels of regulated gas prices minus transportation costs (netback) are much higher in areas close to the production regions, and it is where there is the maximum concentration of independent producers' sales. The further from the region of extraction, the more significantly the margin falls - the difference is 10-20%. Low-profit and unprofitable deliveries go to Gazprom, which is legally obliged to supply gas to Russian consumers at a regulated price. Combined with a lower mineral extraction tax rate for gas, large netbacks allow independent producers use discounts to grab the best and most solvent consumers from Gazprom and still make a large margin on gas sales [19].

Being a state-owned company, Gazprom, unlike independent producers, is legally responsible for the stable and reliable gas supply to Russian consumers and fulfillment of export obligations. This makes it necessary to create and maintain reserve capacities, which are used to cover the uneven consumption of gas in the domestic market in the GTS zone. The difference between quarter production of the Gazprom group in the heating season and outside the heating season is 30 to 45 billion cubic meters [20].

As the owner of the unified gas supply system, Gazprom, on the one hand, has clear advantages over independent producers; but on the other hand, the company is forced to maintain and develop not only the production capacities but also transportation and storage infrastructure, while independent producers develop only extracting. Currently, Russia has the lowest prices for gas transportation in the world. Therefore, the funds for technological development of transportation system can come mainly from Gazprom's export revenues.

Gazprom's advocates do not consider it reasonable to buy gas from independent producers at prices equal to the export netback level, since:

– it would lead to a reduction in the financing of Gazprom's investment programs, including gas transportation and storage, to which the independents do not invest;

– it would make it more difficult to work on state-governed tasks (including the gasification of Russian regions) without considerable co-financing from the federal budget;

– it would weaken the position of the entire gas industry, as independent producers can use the funds received from the gas sales at the export netback prices for investing outside the gas business or for increasing dividends [21].

At the same time, more and more experts are focusing on the inefficiency of Gazprom's export activities. E.g., R. Tankayev, the leading expert of the Oil and Gas Producers' Union, notes that the export component of the Gazprom business model has decreased many times in recent years; in particular, the huge Ukrainian market was lost due to excessive politicization. Also, he reminds that, being a monopolist in the European market, Gazprom will face the problem of the Third Energy Package [22]. S. Vakhrameev, the manager of GL Asset Management, believes that the pipeline gas export liberalization would be useful for Gazprom itself, because, the company would start working more efficiently in the face of the competition [18].

As for Gazprom's social obligations, it should be mentioned that, over the past three years, the possibility of introducing a guaranteeing supplier institution has been actively discussed. The idea is to split obligations, in addition to the rights, to supply gas to the Russian regions between Gazprom and independent producers. At present, the functions of the guaranteeing supplier are legally being delegated to Gazprom. However, in some regions, independent producers, primarily Novatek and Rosneft, are main gas suppliers. By acquiring the status of a guaranteeing supplier, these companies will have to participate in the gasification of the regions, supply gas to the population proportionally to the share of their presence in the domestic market, and also bear responsibility for the passage of winter maximum gas consumption. At the same time, representatives of both independent gas suppliers and Gazprom consider it possible to give the guaranteeing suppliers access to gas exports proportionally to their social obligations at the domestic market. It is assumed that, in the future, this will facilitate equal economic conditions in the market for Gazprom and independent gas producers [23].

5 Conclusions

Despite the risk of falling short-run gas exports revenues due to competition among Russian companies after the abolition of Gazprom's export monopoly, there are solid reasons to believe that the benefits of liberalizing Russian gas exports will be higher than possible losses. Under the gas-on-gas competition circumstances, firm's market share and profitability depend mostly on its cost and productivity. According to experts, the gas resource base and the cost of pipeline gas production would allow Russia to improve its positions in the world market in the case of Russian gas export liberalization. In this case, the liberalization would lead to an increase in both export volumes and the corresponding revenues of the state.

There is no doubt that in some form - the direct admission of some companies to export deliveries or through Gazprom's purchases of gas from independent suppliers at the export netback price - Russian gas export liberalization will happen the near future. However, it is difficult to call a specific time frame for such a decision. It requires not only economic prerequisites, but also political will.

References

1. V. Kryukov, A. Moe, *The Russian Natural Gas Sector* [online], Available at: http://www.oxfordhandbooks.com/view/10.1093/oxfordhb/9780199759927.001.0001/oxfordhb-9780199759927-e-029 (2016)
2. J. Henderson, *Domestic Gas prices in Russia – Towards export netback?* [online], Available at: https://www.oxfordenergy.org/wpcms/wp-content/uploads/2011/11/NG_57.pdf (2011)
3. D. Gordeev, G. Idrisov, E. Karpel, *Voprosy Ekonomiki* [online], Available at: http://www.iep.ru/files/text/nauchnie_jurnali/idrisov1-15.pdf (2015)
4. Pricing in gas markets, *Energy Bulletin* [online], **21,** Available at: http://ac.gov.ru/files/publication/a/4857.pdf (2015)
5. The model of the domestic gas market as a prologue to a new strategy?, *Oil and Gas Vertical* [online], Available at: http://www.ngv.ru/analytics/model_vnutrennego_gazovogo_rynka_kak_prolog_k_novoy_strategii/ (2016)
6. U. Koshkina. *Access to the "pipe". What will the liberalization of gas exports lead to?* Kommersant.ru [online], Available at: https://www.kommersant.ru/doc/3378325 (2017)

7. Annual Review 2015, *NOVATEK – Russia's Leading Independent Natural Gas Producer* [online], Available at: http://www.novatek.ru/ru/investors/reviews/ (2015)
8. *Annual Review of NC Rosneft* [online], Available at: https://www.rosneft.ru/upload/site1/document_file/a_report_2015.pdf (2015)
9. *Fuel and Energy Complex of Russia – 2016* [online], Available at: http://ac.gov.ru/files/publication/a/9162.pdf (2017)
10. M. Melitz, The Impact of Trade on Intra-Industry Reallocations and Aggregate Industry Productivity. *Econometrica,* **71**, 1695 (2003)
11. A. Bernard, S. Redding, P, Comparative Advantage and Heterogeneous Firms Schott. *Review of Economic Studies*, **74**, 31(2007)
12. D. G. Tarr, P. D. Thomson, The Merits of Dual Pricing of Russian Natural Gas. *World* Economy [online], **27**, 1173. Available at: http://onlinelibrary.wiley.com/doi/10.1111/j.1467-9701.2004.00651.x/abstract (2004)
13. D. G. Tarr, *Export Restraints on Russian Natural Gas and Raw Timber* [online], Available at: http://www.cepe.ethz.ch/research/cepe-wp.html#2010 (2010)
14. M.-C. Aoun, S. Cornot-Gandolphe, *The European gas market looking for its golden age?* [online], Available at: http://www.ifri.org/sites/default/files/atoms/files/etude_european_gas_market_aoun_cornot-gandolfe.pdf (2007)
15. A. J. Melling, Natural gas pricing and its future: Europe as the battleground, *Carnegie Endowment for International Peace* [online], Available at: http://carnegieendowment.org/files/gas_pricing_europe.pdf (2010)
16. *Wholesale Gas Price Survey. – A global review of price formation mechanisms 2005 - 201.* [online], Available at: http://www.igu.org/sites/default/files/node-page-field_file/IGU%20Whole%20Sale%20Gas%20Price%20Survey%20Report%20%202015%20Edition.pdf (2015)
17. B. Petrovich, *European gas hubs: how strong is price correlation?* [online], Available at: https://www.oxfordenergy.org/wpcms/wp-content/uploads/2013/10/NG-79.pdf (2013)
18. V. Petlevoy, A. Toporkov, M. Papchenkova, *The Security Council encroached on the export monopoly of Gazprom*, Vedomosti [online], Available at: https://www.vedomosti.ru/business/articles/2017/08/07/728189-sovbez-pokusilsya (2017)
19. K. Simonov, A. Grivach, *Start the gas engine: how domestic demand for gas can save the economy*, RBC [online], Available at: http://www.pro-gas.ru/images/data/gallery/0_4997__rinok_2015.pdf (2015)
20. Domestic gas market: how to get out of the Bermuda Triangle model? *Expert-analytical report. National Energy Security Fund. Moscow* [online], Available at: http://www.pro-gas.ru/images/data/gallery/0_4997__rinok_2015.pdf (2015)
21. The domestic gas market is the embodiment of contradictions, *Kommersant.ru* [online], Available at: http://www.kommersant.ru/doc/2840660 (2015)
22. Ministry of Energy recommends maintaining "Gazprom" as a single entity, *Gudok.ru* [online], Available at: http://www.gudok.ru/infrastructure/?ID=1322287 (2015)
23. The concept was to be prepared by June 1, *Izvestia.ru* [online], Available at: http://izvestia.ru/news/616182#ixzz4M0h6K2UF (2016)

Analysis of Oil and Gas Cooperation between China and Russia in the Belt and Road

Xuefeng Zhang[1,*] and *Melebayev* Serdar[1]

[1]School of Economics and Management, North China University of Technology, Beijing, China

Abstract. With the deepening of economic globalization and regional economic integration and the "One Belt and One Road" strategic layout, China and Russia in the field of regional energy cooperation has been expanded to deepen. This paper analyzed the present situation of energy cooperation between China and Russia from the field of oil, coal and natural gas, and concludes that Sino-Russian energy cooperation is the following factors: strong cooperation and strong energy complementarities, and then analyzed the way to the Sino-Russian oil and gas cooperation. However, there are some problems in the energy cooperation between the two countries: the energy infrastructure in China and Russia is not perfect, the difference between Russia and Northeast China is poor, and there are some hidden dangers of energy security. The cultural dilemma hinders the exchange between the two countries. Finally, from the "One Belt and One Road" strategy for the Sino-Russian energy cooperation provides an opportunity to put forward the establishment and strengthening of energy infrastructure, relying on "One Belt and One Road" Orientation to ensure energy security, strengthen cultural cooperation and adhere to the "One Belt and One Road" uphold the principle of mutual benefit and other countermeasures.

Key words: analysis, cooperation, oil, gas

1 Introduction

The Chinese economy is in rapid development period of industrialization and urbanization is rapidly advancing, high economic development but also improve the oil and other energy resources consumption. And Russia with China, in the North is bordered by the world power, rich in energy resources, its oil production accounted for the world second total output, the oil reserves of the world's total reserves of 13 per cent. At the same time, Russia's natural gas, coal and other oil and gas mineral resources is quite sufficient; Russia has become the world's major energy exporter, but also become China's energy imports preference.

---

*Corresponding author: zxf511@163.com
Supported by the North China University of Technology outstanding young teacher training program, funded Beijing Social Science Foundation (Grant No.16LJC008) and Foundation for The Excellent Youth Scholars of Beijing（Grant No.2016000020124G021).

DOI: 10.1051/shsconf/20173901034

Over the years, China and Russia in the oil and gas and other energy fields expand widely cooperation, and has made a series into. It, reached a good cooperation consensus and basis. Chinese-Russian energy cooperation has a unique advantage. But Russian and Chinese energy infrastructure is still relatively backward, the issue of energy security has not been effectively resolved, geo-political influence of the two countries has interests of the cooperation, the imperfections of the energy of the financial system and other issues need to be resolved. For the Chinese-Russian energy cooperation in the process of many problems, how in the "one belt and one road" strategy to achieve greater progress is their common need to think about the problem. China-proposed "belt and road" strategy emphasizes to create "community of interests" and "community of destiny", and Russia as China's neighbor, is undoubtedly China's "One Belt and One Road" initiative in the important countries. Uphold the "One Belt and One Road" strategy advocated by the interoperability, cooperation and safe principle, China and Russia want to continue deepening energy cooperation, so as to promote the Chinese and Russian economy continued stable and rapid development. The strengthening of Chinese-Russian energy cooperation has important theoretical and practical significance. Not only better safeguard China's energy supply security, to meet China's economic development process in the demand for energy; expanding Russian energy exports, and thus also contributed to Russian Economic Development. Sino-Russian cooperation is "mutual benefit, complementary and safe" cooperation. This study "along the way" the new strategic context to address Russian and Chinese energy cooperation that exist the following problems: energy infrastructure, energy security issues yet, the new energy technology challenges, geopolitical influence, the two countries interests of cooperation, energy-financial system is imperfect, the countries along the communication difficulties etc. Through the analysis of these issues causes to find solutions, to promote the Sino-Russian energy cooperation can be more smoothly carried out; the better to enhance the Sino-Russian powers in the international image; establish a regional cooperation model; expansion of the "One Belt and One Road" strategy of international influence.

In the BRICs countries as the basis, pointed out that the BRICs Member States are increasingly aware of the global market in the unique potential and the cluster advantages, and actively carry out various exchanges and cooperation [1]. Below the midpoint of the Russia due to the abundant natural resources, could become the global economy of raw materials the main supplier. Rutland [2] pointed out, Russia is the world's leading natural gas-producing countries, oil production with Saudi Arabia shoulder to shoulder, from 1998 to 2004, world oil supply growth in Russia accounted for 22% of oil and gas resources to return to the international stage leading status of key factors. Russian energy policy the centre of gravity of the transfer, the China Petroleum and natural gas resources are relatively insufficient situation prompted the two countries to actively seek energy cooperation and achieved substantial progress and results. Armijo [3] believe that China-Russia energy reserve assignment there is a difference, China's energy demand urgency, prompting China to take natural resources to lease and with oil-producing countries to sign long-term supply contracts. Russia is a huge energy exporter, its crude oil and natural gas production accounted for the world total output of 12%and 22%, the authors turn from energy to enjoy the empowerment of diversity and other aspects of the Sino-Russian bilateral energy cooperation. Trenin [4] pointed out that today Russia's achievements mainly rely on national oil and natural gas in the world energy prices go high, the background should be development and utilization of important and effective policy weapon, while also pointed out that due to a series of tensions and conflict, Russia and the Western countries appeared a rift, coupled with China's economic, political status of the rise as well as for reasons of energy security considerations, Russia will be with China, there are many area of energy trade and economic cooperation.

Hua [5] consider that Russia in China's Foreign Relations field plays an important role, not only because of Russian geo-location with the China adjacent, the more because Russia is rich in resources, but also belongs to the world of the Great Powers in the international arena has a crucial role to this end, Russia and China strengthening cooperation is the inevitable choice. Yu [6] starting from the overall study, prospects for Sino-Russian cooperation in the oil and gas fields are beautiful, in line with the needs of economic development of the two countries, is the result of a win and safe situation on both sides, but also have to take precautions, problems in the prevention of possible cooperation. Wei [7] published that China has energy security problem: challenge, he think currently, China energy security main faced with structural crisis and management system crisis, that is two big challenge, for China Government should take series of measures, and can consider implementation to multilateral cooperation, focus on international energy Department, to regional cooperation for based promoted established Northeast Asia Energy Community strategy.

Balzer [8] pointed out that Mr. Putin defended a kandidat dissertation in economics and subsequently published an article outlining his view of the appropriate role of the Russian state, and of vertically integrated financial-industrial groups, in the mineral resource sector, and particularly in the oil and gas industry. Connections are drawn between the views expressed in Mr. Putin's publications and policy during his second presidential term. Bilgin [9] addresses issues of natural gas which raise questions about European energy security. It first focuses on the rising gas demand of the EU27 and elaborates alleged risks of dependence on Russia. The geopolitical analysis leads to a conclusion that the best strategy, and what seems more likely, for the EU is to include at least two countries from Azerbaijan, Turkmenistan, Iran and Iraq within its natural gas supply system. Cornelius and Story [10] believed that conflict between the United States and China is far from inevitable, as long as the global price mechanism operates reasonably efficiently and Washington and Beijing can operate in partnership rather than rivalry. Hancock [11] pointed out that Several factors will determine whether the pipeline becomes a reality, including Niyazov's ability to reign in his troublesome personality, China's willingness to convert from coal-burning to natural gas fueled plants, and Uzbekistan's interest in a pipeline traversing its territory. Weitz [12] considered that. The deliveries, which may take a few years to reach full capacity, will provide China with more than one-fifth of its present-day annual consumption of some one hundred and seventy bcm, although Chinese demand for natural gas is expected to rise above two hundred bcm by then.

Dan-Hui [13] by way of investigation, Russia's oil and gas resources notable the world, it is oil and gas resources, especially natural gas, ranking first in the world. Yao [14] think that China and Russia both energy cooperation is in the twists and turns in the good development of Sino-Russian energy-industry collaboration has a convenient, short distance, low cost, resource potential and other many advantages. Currently, the Russian Economic Development was achieved gratifying results, a large part of the reason is that the energy exports bring to national income growth, while China and Russia energy cooperation is also to attract more foreign investment, the use of energy from reliable sources, which for Russia is a win situation, the accomplishments of outstanding worth for sure, but not blind, after all, between the two countries on energy cooperation still exists some problems that need to solve.

In summary, domestic and foreign scholars ' research focused on the Chinese-Russian energy cooperation the need for and development of the status quo. In the need for cooperation on a consensus, that Russia, rich in energy resources, especially oil, coal and natural gas. And China because of the economic high-speed development, the energy consumption demand is particularly large, Russia and China to promote energy cooperation is in both countries ' development strategies. In cooperation with the status quo of the

DOI: 10.1051/shsconf/20173901034

research on Chinese-Russian energy cooperation is continued to promote, in the petroleum, coal, natural gas field and achieved certain results, China and Russia energy cooperation will continue. Domestic and international studies have much significance, but current research does not relate to the "One Belt and One Road" strategy, just past China and Russia energy cooperation is summarized. In fact, due to China's "One Belt and One Road" strategy proposed, China and Russia energy cooperation presents a different face, of course, also faced with new problems, the encountered situation is different, and so research in the "One Belt and One Road" under the background of Chinese-Russian energy cooperation is still necessary.

This paper based on China and Russia in energy cooperation in the status quo analysis, the main research in the "One Belt and One Road" under background of Chinese-Russian energy cooperation on the issues and response measures. First, previous research on the basis of, combined with the "one belt and one road" context of the analysis Chinese-Russian energy cooperation in major energy sector development status. Second, according to the previously described the development status the study of Russia and China in "One Belt and One Road" strategy under the guidance of the energy cooperation of the favorable factors. Third, standing in the "One Belt and One Road" strategy height, pointed out that Sino-Russian regional energy cooperation problems in the process. Fourth, in the above-described analysis on the basis from the actual departure, is proposed for Sino-Russian energy cooperation problems in the process of response measures.

2 China-Russia oil, Gas Trading analysis and "One Belt and one Road"

China in the world energy reserves, ranks among the forefront, both as the world's second energy producing country, is also ranked second in energy consumption in the country. As the economy continues to expand, China's energy consumption has continued to rise, China's rapid economic growth cannot do without Russian energy supplies, as energy reserves, production and export volume ranking the forefront of Russia, is China's major energy-importing countries.

Russia is a world energy superpower, the 2014 oil proven reserves in the world rankings ranked sixth, accounting for the world of 13%, oil production accounted for the world total output of 12% as Figure 3-1 shows. For Russia, rich in oil resources except for their own consumption, more output abroad, China is Russia's important oil-exporting countries.

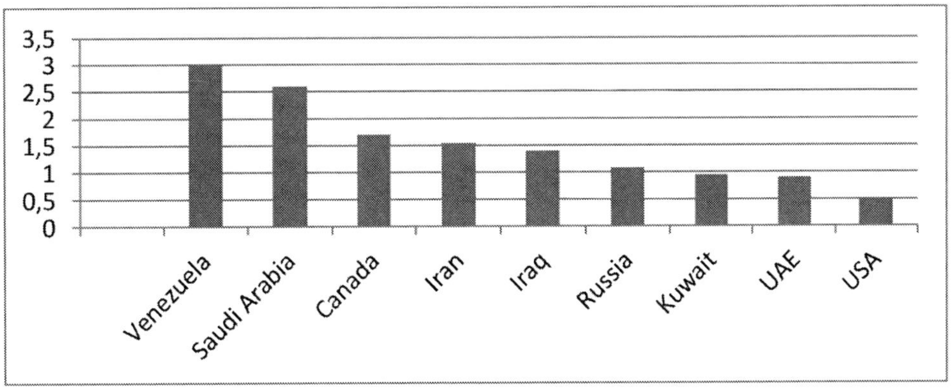

Fig. 1. World oil proven reserves rankings

Table 1. China's oil imports from Russia in 2011 – 2015

Year	Chinese oil import (mill. ton)	Russian oil import (mill. ton)	proportion
2011	0.19	2.42	8%
2012	0.21	2.35	9%
2013	0.24	2.36	10%
2014	0.33	2.24	14.7%
2015	0.41	2.45	16.9%

Russian oil export volume of nearly ten years has been stable at 2 million tons, and oil exports accounted for production of about 30%. With Russian oil production continues to increase, the foreseeable future of Russia's oil exports is bound to increase. 2015 Russian oil the total export volume amounted to 2.43 million tons, of which exports to China of about 0.41 billion tons, accounting for about 16.9 per cent. Table 3-1 shows that from 2011 to 2015 the Russian oil export volume is growing, while exports to China the volume of oil also showed an increasing trend, the Chinese imports of Russian oil duty ratio from 2011's 8%rise to 2015 16.9%, increased nearly one times. Thus, China is Russia's main oil exporter country, in the petroleum field is also to strengthen the construction of dealings.

In February 2009, China and Russia signed "loans-for-oil" agreement, determine the Sino-Russian oil pipeline by the end of 2010 before lying is completed. In 2010, Russia's largest bilateral cooperation project - "East Siberia-Pacific Ocean" oil pipeline project of China, the branch was built, became the Sino-Russian energy cooperation in the field of milestone. In December 19 2010, The China-Russia oil pipeline across the Board through. The China-Russia oil pipeline officially began transporting oil in 1st May 2011.

In March 2013, President XI Jinping in Russia during the visit, the two countries signed on the East and West lines to increase crude oil supply agreement, Russia will export to China the volume of oil reached an annual average of 2200 million tons, the East line of the existing Sino-Russian crude oil pipeline of oil supplies on an annual average of 1500 million tons increased year by year, to 2018, the total oil export volume reached 3,000 tons. Not only that, Russia in the future will transport by sea transportation for export to Tianjin refinery oil amounted to 910 million tons.

China's natural gas consumption, natural gas consumption increased year by year, the consumption amount of the energy structure in the proportion is also increasing year by year. The 2012 China consumption of natural gas 1471 billion cubic meters, by one year rise 13%, in 2014 gas consumption was up to 1800 billion cubic meters, up 7.4%, in 2015 up to 1920 billion cubic meters, by one year of 6.7 %. 2015 China natural gas imports to Russia total consumption of 3.2 %. China's natural gas consumption is growing at the same time, the growth rate has slowed, which with rising gas prices, and many other factors. According to the expert analysis, China's natural gas consumption in 2020 expected growth rate as in 2015 by one times, to achieve approximately 4100 million cubic meters of the target in primary energy consumption proportion reaches 12%, the future China's energy situation will happen a major reform, and this is an important engine. By then, China will become the Russian gas, the largest importer, which should benefit from the natural gas, the future of China will be available from Russia up to 3000 billion cubic meters of natural gas exports.

 DOI: 10.1051/shsconf/20173901034

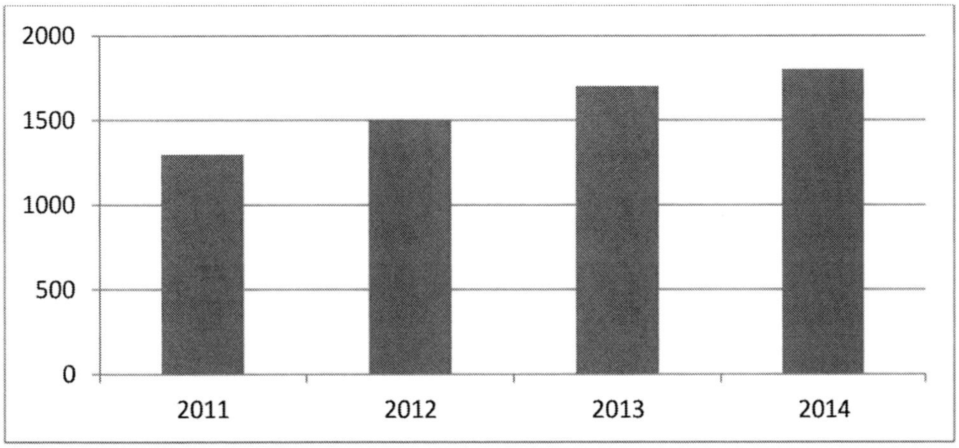

Fig .2. China's natural gas consumption

Russia is not only oil-rich, gas reserves also ranked in the world. 2014 Russian natural gas total export volume up to 1812 billion cubic meters, in 2015 the growth of 4.9%, up to 1900 billion cubic meters. In 2015 China's natural gas imports amounted to 621 million cubic meters, which imports from Russia about 61 billion cubic meters. In the natural gas fields of cooperation, in Chinese-Russian consultations, the two countries will be on the West line gas project restarted, which is following the East line natural gas cooperation project reached a remarkable results. China and Russia West line gas project of scale with the East line of the project comparable. The West line gas project cooperation reached with the East line natural gas project Double sword combination, this will promote "One Belt and One Road" by the country's economic take-off. Europe and the United States the emergence of the Ukrainian issue, the Russian implemented economic sanctions to pressure the Russian government had to shift our attention to the East. In November 2014, covering from Siberia to the construction of China gas pipeline, including a 30 year gas supply contract, in the Sino-Russian joint efforts to reach a unanimous decision.

In the Boao Forum for Asia on the Russian Far East Development Minister Galushka said, "One Belt and One Road" the great thing is that it put Asia and Europe, the two continents closely linked together, pushing the two continents of co-development, while also advancing the world of diversified development. Grasp "One Belt and One Road" construction opportunities, the Sino-Russian oil and gas cooperation will become a mutual respect mutual trust of the road, a win and safe cooperation Road, a connection of friendship road.

Table 2. Chinas Gas Import from Russia 2012-2016

Year	China oil import amount (billion cub.metr)	Russia oil export (billion cub.metr)	Proportion
2012	35	1600	2,2%
2013	45	1690	2,7%
2014	50	1750	2,85%
2015	58	1812	3,2%
2016	61	1900	3,21%

Although Russia and China have a sound political basis, but the two sides are still in constant pursuit to enhance the two countries, political mutual trust. In 1992, Russian President Boris Yeltsin visited China; the two leaders signed the Sino-Russian relationship basis of the joint statement, and determine relations between the two countries of fundamental principles. In 1994, the leaders of both sides determine the Chinese-Russian good-neighborly friendship, equality and mutual benefit relationship, healthy and stable development, the two leaders also signed the Sino-Russian joint statement and the Chinese and Russian heads of state on non-national nuclear weapons aiming at each other's joint statement and other documents. In 1996, the leaders of the two countries signed the Sino-Russian joint statement and other documents. In the 21st century, China and Russia stick to a few principles remain the development of good relations, continue to deepen mutually beneficial and pragmatic cooperation, seek common development and prosperity. The two countries will continue to expand in international Affairs in cooperation to promote peace and stability in the world. In June 2012, the Chinese and Russian heads of State signed a Declaration on further deepening the equality trust the China-Russia comprehensive strategic cooperative partnership joint statement, the Sino-Russian leaders unanimously decided to implement the Sino-Russian relations in the next ten year development plan for the Central task, give full play to the two sides close high-level exchanges the government and sector exchanges and cooperation mechanism. Russia is a growing cooperation between the friendly neighbourhood states, the Treaty of friendship signed between China and Russia cooperation Russian-Chinese relations are milestone and cornerstone.

3 "One Belt and One Road" Sino-Russian oil-gas cooperation influences

Between China and Russia exchange can be said "the Silk Road Economic belt" to the North on land extending to the South and the "21st century Maritime Silk Road" the docking of a new cross-border cooperation, where the interests overlap and interleaving may be by the Parties to establish an effective coordination mechanism, from the big picture long-term interests, seek common ground, forming the destiny and interests of the community. The use of the Russian Economic Development need to take Chinese "ride" a good opportunity period, China completely can be the use of financial, technical and human resources and other advantages, the effective use of Russia in Central Asia and Mongolia cooperation in the positive factors, the "Pro, honesty, benefits, tolerance" diplomatic philosophy under the guidance, the aid has started a "Bank" and "Silk Fund" to strengthen infrastructure construction. Jointly build "One Belt and One Road" of the priority direction is to build interoperability partnership. The future is in China and Russia, along with "One Belt and One Road" energy development and the formation of oil and gas pipeline network basically relies on a number of economic corridors, the formation of "one belt and one road" along appears to cover energy in the downstream areas of different industrial centres.

In Summary, The New Silk Road for China-Russia energy cooperation to expand the escort, the two countries through the New Silk Road can do the energy exchange interoperability, and effectively protect the two countries energy security. For Russia, The New Silk Road for Russia's energy exports opens the channel, the energy outlet of the expansion is bound to bring Russia economic growth; for China, the New Silk Road for China's energy demand to broaden the channels, to meet the economic development of needed energy resources. So, the New Silk Road to protect the economies of both countries sustained and stable development to provide the conditions.

4 "One Road One Belt" Sino-Russian oil and gas trade problems

China and Russia between the two countries on energy cooperation back and forth the experience of nearly twenty years, the cooperation results have good and bad, prior to the

DOI: 10.1051/shsconf/20173901034

establishment of the infrastructure not only dilapidated and do not meet current energy delivery and, most importantly, with the new field of energy cooperation continue to expand, as new energy cooperation in the construction of the infrastructure is still lacking. The current transnational oil and gas pipelines construction, cross-border power grid interconnection construction and a large infrastructure has not kept pace, the impact of the Sino-Russian energy cooperation in the process. For a long time, Russia's industrial development is mainly concentrated in the Western region, energy development and utilization of the supporting infrastructure is also mainly concentrated in the Western region. Russia in Far East, sparsely populated, accounting for land area of 36% of the Far Eastern Federal District only 630 million inhabitants and where economic development lagged behind, supporting infrastructure, energy development and utilization costs are very high. Therefore, many of China's large state-owned enterprises are unwilling to Siberia investment to build the plant. In addition, the region's energy transportation mainly on the railway, the transport capacity is insufficient, construction of new infrastructure requires a large investment, and cost recovery is slow, also dispelled the China enterprise investment enthusiasm. China and Russia's energy infrastructure through the country is more, the countries along the route for different interests to consider, so that its communication is quite difficult, and China and Russia planning some line through the region of the geographic ecology of the natural environment is harsh so that construction of the infrastructure of the difficulty of construction is further increased. So, for now, Russia and China to the energy infrastructure or the use of the previous facilities, the past of the energy infrastructure not only old behind, but also cannot meet the growing energy supply and demand, it does not meet the "One Belt and One Road" strategic planning under the new Sino-Russian energy cooperation line.

China and Russia are the international energy market on an integral of two powers, one is a major energy consumer, one is the energy producing countries, both neighboring countries and strategic partners, the two countries on energy cooperation has a unique advantage, but the reality is often unsatisfactory, between the two countries on energy cooperation is affected by many factors, at multiple levels expand the interests of the game. One is the economic and political interests of the game. Sino-Russian energy cooperation is the biggest obstacle to the geopolitical difficult to release away. The Russian energy sector problems, economic factors are the main obstacles, as it is the geopolitical goal of the ground edge of the economic target penetration as a result, lead to political problems at the economic level. Sino-Russian cooperation more imported the field of security, and the non-economic areas. The second is the strategic interests of the game. Sino-Russian energy cooperation to a sustained and stable development, we must have a clear Sino-Russian energy cooperation strategy to guide, but so far, it didn't work out between the two countries recognized the practical Sino-Russian energy cooperation strategy, the main reason is that Russia will be the two countries on energy cooperation as the long-term planning project, hoping to drive the entire Far East region's economic development, while China put the two countries on energy cooperation as an urgent short-term work projects, expectations as soon as possible to resolve the current "energy bottleneck", as well as energy sources diversification issue. The third is the price of the game. China and Russia in the oil and gas price divergence has become current cooperation problems to be solved. For China, the high oil price will greatly increase China's national economy running costs; and for the "energy economy" dominated by Russia, then hope that oil and gas can maintain high prices. These years, at what price to export to China Energy become plagued Russia's a big issue, but also hinder China-Russia energy and rapid development of important factors.

 DOI: 10.1051/shsconf/20173901034

5 "One Belt and One Road" China-Russia oil-gas trade development countermeasures

Energy infrastructure construction is China and Russia energy cooperation is an important one ring, spare no effort in the establishment and improvement of Chinese-Russian energy infrastructure construction, the Chinese and Russian governments to invest more money in energy infrastructure, give more financial support while the introduction of both corporate investment and create more favorable conditions, such as the introduction of preferential tax system, etc. to attract Russian and Chinese companies added to the energy infrastructure of the team. Although, Russia and China along the line of the infrastructure in the face of natural conditions difficult, but the two countries as long as the strengthening of technology research and development, storming grams hard, you can overcome the natural environment of the obstruction, the innovative establishment so as not to damage the ecological environment to devote more funds to support Energy Technology Research and development, and guarantee energy sustainability, the increase in New Energy and green energy aspect of the cooperation, reducing the ecological environment pollution, from a certain extent is also the guarantee of energy security.

In the bilateral safeguard mechanism, the China-Russia bilateral energy cooperation the legal system and the laws for refinement, so both companies understand that cooperation relates to the legal terms of the differences. Russia and China to vigorously develop energy technologies, to invest more funds to support Energy Technology Research and development, and guarantee energy sustainability, the increase in New Energy and green energy aspect of the cooperation, reducing the ecological environment pollution, from a certain extent is also the guarantee of energy security.

Russia and China cooperate to ensure that the interests of States as the basis, from all aspects to improve the enhanced cooperation basis. However, the two countries in addition to the constant emphasis on issues of interest, but also from the Humanities and cultural cooperation aspects, by the two countries to continue the dialogue of cultures, in-depth to strengthen all areas of communication.

6 Conclusion

China has become the global energy needs country, energy needs to be timely and effectively met; active seeking energy cooperation is the inevitable choice. Russia is a vast land and energy resources the abundance very rich, coupled with its economic and steady development is largely dependent on energy exports the realities of the Russian imperative is to seek a reliable energy-importing countries. China and Russia energy cooperation can be traced back to the early 90's, although the middle stranded for a period of time, but the two countries energy cooperation still continues. Russia and China in energy cooperation, though they have not a small advantage, but there are also a lot of issues: the Sino-Russian energy infrastructure, and Russia in Northeast Asia with China trust Foundation difference, energy security there is a certain risk, a culture native to hinder the two countries exchanges in Russian between the interests of the game.

Russia and China to learn from past cooperation, confront problems, from the Russian actual situation, from the "One Belt and One Road" strategy proposed for Russia and China energy cooperation provides an opportunity to propose the establishment and strengthening of energy infrastructure, by means "Belt and Road" to carry out the Sino-Russian energy multi-field cooperation, relying on the "One Belt and One Road" multifaceted and comprehensive to ensure energy security, strengthening Humanitarian Cooperation and adhere to the "One Belt and One Road" uphold the mutually beneficial and win-win principle and other countermeasures.

DOI: 10.1051/shsconf/20173901034

Finally, the list of China's goals, in the medium and long term, includes: internal stability, maintained economic growth, deepened reform, and opening to the outside world; enhancement of its soft power, improved relations with Russia, energy security and equilibrated relations with the great powers; rule maker in the global governance institutions and contributor to the global economy.

References

1. T. Hult, The BRIC countries. *Global edge Business Review*, (3), 1-2 (2009)
2. P. Rutland, Russia as an Energy Super power. *New Political Economy*, (2), 201-210 (2008)
3. L. E. Armijo, The BRICS Countries as Analytical Category: Mirage or Insight? *Asian Respective*, (4), 7-42 (2007)
4. D. Trenin, Russia Leaves the West. *Foreigh Affairs*, (4), 1-87 (2006)
5. H. Li. Hua, Russia in China's energy strategy in the status and Sino-Russian energy cooperation. *Economic Research reference*, (30), (2006)
6. Li Guo Yu, World oil and gas basin, Atlas. *Beijing: petroleum industry Press*, 69-94, (2005)
7. Y. Z. Wei, China's energy security: challenges with the availability. *World Economics and politics*, (8) (2008)
8. H. Balzer, The Putin Thesis and Russian Energy Policy. *Post-Soviet Affairs*, **21**(3), 210-225, (2005)
9. M. Bilgin, Geopolitics of European natural gas demand: Supplies from Russia, Caspian and the Middle East. *Energy Policy*, **37**(11), 4482-4492 (2009)
10. P. Cornelius, J. Story, China and Global Energy Markets. *Orbis*, **51**(1), 5-20 (2007)
11. K. J. Hancock, Escaping Russia, Looking to China: Turkmenistan Pins Hopes on China's Thirst for Natural Gas. *China and Eurasia Forum Quarterly*, **4**(3), 67-87 (2006)
12. R. Weitz, The Russia-China Gas Deal. Implications and Ramifications. *World Affairs*, **177**(3), 80-86, (2014)
13. W. Dan-Hui, The Sino-Russian bilateral oil trade progress to explore. *Sino-global energy*, (5), 18-23 (2010)
14. H. Yao, Sino-Russian energy cooperation: status, obstacles and counter measures research. *Heilongjiang foreign economy and trade*, (2) (2011)

AUTHOR INDEX